A SECOND COURSE ON
REAL FUNCTIONS

A SECOND COURSE ON REAL FUNCTIONS

A.C.M. VAN ROOIJ AND
W.H. SCHIKHOF

CAMBRIDGE UNIVERSITY PRESS
Cambridge
London New York New Rochelle
Melbourne Sydney

CAMBRIDGE UNIVERSITY PRESS
Cambridge, New York, Melbourne, Madrid, Cape Town, Singapore, São Paulo, Delhi

Cambridge University Press
The Edinburgh Building, Cambridge CB2 8RU, UK

Published in the United States of America by Cambridge University Press, New York

www.cambridge.org
Information on this title: www.cambridge.org/9780521239448

First published 1982
Re-issued in this digitally printed version 2008

A catalogue record for this publication is available from the British Library

Library of Congress Catalogue Card Number: 81-9933

ISBN 978-0-521-23944-8 hardback
ISBN 978-0-521-28361-8 paperback

CONTENTS

PREFACE

This book is devoted to the study of real functions of a single variable and is meant to present a survey of the topic, touching the most important notions. It is intended for advanced undergraduate students or anyone on a higher level who wants to deepen his or her knowledge obtained from elementary analysis courses.

It has become standard to treat real analysis with an eye to the modern theories of general topology, measure theory and functional analysis. This is, of course, a perfectly valid view. Hewitt & Stromberg's *Real and Abstract Analysis* (1965) is an excellent example of a book originating from this philosophy.

But real analysis is also an enthralling subject in its own right. Until about 1930 the theory was developed by people who were interested in the structure of the real line *itself*. The resulting abstractions were considered, in the first instance, to be merely by-products, whereas now they have grown very important. In this light it is not surprising that parts of the classical theory that did not admit interesting generalizations have virtually disappeared from the scene (e.g. monotonicity, Darboux cón-tinuity, derivative functions). Nevertheless, anyone who pays some attention to them will see that they can be fascinating.

In our book we follow the founding fathers by choosing the direct approach. We have at our disposal the real number system with its rich structure. Let us use all of that and try to obtain a deeper insight into the properties of $\mathbb{R}$, its subsets and the functions defined on them, and let us not go further into abstractions and generalizations than is necessary for our immediate purpose. The books of Baire, Borel and Hobson and the papers that appeared in *Fundamenta Mathematicae* in the period 1920–40 are examples of classical works that have inspired us both mathematically and philosophically.

Let us consider a few consequences of our choice.

Firstly, it brings about interest in subjects that are given scanty attention in current literature. We have already mentioned some of these; we may add the Baire classification, one-sided continuity, Riemann integration, the Dini derivatives.

Our philosophy leads to a study of pathology. For instance, we view differentiability as a phenomenon and not merely as a tool (say, to compute extreme values). Thus, we shall discover how badly even differentiable functions can behave; we shall construct functions that are continuous but nowhere differentiable, etc. (For the reader's convenience we have inserted a list of examples of such functions that occur in the book, often as exercises. They are in the same vein but in general less elementary than the ones found in Gelbaum & Olmsted's book *Counterexamples in Analysis* (1964).)

We have to pay for our approach with a certain loss of perspective. For example, is it the Bolzano–Weierstrass property (every sequence has a convergent subsequence) or the Heine–Borel property (every open covering has a finite subcovering) of the closed interval $[a, b]$ that is ultimately responsible for the theorem saying that every continuous function on $[a, b]$ has a maximum? Such a question makes sense only in so far as a closed interval can be seen as an instance of a more general mathematical object to which the theorem can or cannot be generalized. It is the very interplay of all the algebraic and topological properties of $\mathbb{R}$ that restricts the kind of questions we can ask.

Experience with courses that we have taught in the above spirit has confirmed our conviction that both future teachers (at high schools and at universities) and professional mathematicians can profit from a set-up such as we have sketched. The reader does not have to master many new notions but can pay full attention to re-investigating an area in which he already feels comfortable ('nice to be back home'). The more surprising then may be the discovery of beautiful theorems on the one hand and counterexamples that destroy 'obvious' conjectures on the other.

A few words on the practical organization of this book.

We assume the reader to possess some basic knowledge of continuity, differentiability and Riemann integration. To see what this means in practice, have a look at the first pages of Sections 8, 12, 13, 15 and at Appendix A. Moreover, the reader must be conversant with mathematical rigour, so that just a calculus course will not be a good enough background. Further, countability is assumed to be a known concept (as is, to a lesser extent, the cardinality of the continuum). We have added Appendix B as a refresher course.

In the first part of the book (Sections 1–15) we reconsider the subject matter of most calculus courses. After that, we treat questions that have come up in the first part and can be solved satisfactorily by raising the level of abstraction. Here one encounters Borel and Lebesgue measurability, absolute continuity and the Perron integral.

The Introduction is really meant for the first half only, while Section 15 may serve as an introduction for the second.

One subject that seemed too abstract to be put in the main text (making use of transfinite induction) has been exiled to Appendix C.

The book contains many exercises of varying degrees of difficulty. We give them not only for training purposes but also to indicate interesting by-paths. We hope they may serve as a mine yielding a high output to those who are willing to invest work in them. Some of the exercises lead to results that are used in the main theory; these are marked *.

At the ends of some sections we have gathered stray comments and references for further reading. Because of the wealth and variety of material we cannot even attempt to be systematic; we simply allow ourselves to be led by personal preference. Only occasionally do we mention details about historical backgrounds. The story of real analysis is a very interesting one but it does not fit into the scope of a book like this. For good background accounts we refer the reader to the books by Hawkins and Saks. In this context we also recommend two papers dealing with the history of the journal *Fundamenta Mathematicae*, viz. the Editors' note in *Fund. Math.* **100** (1978), 1–8 and M. G. Kuzawa's paper in *Am. Math. Monthly* **77** (1970), 485–92 (and **78** (1971), 874).

Logical dependence of sections:

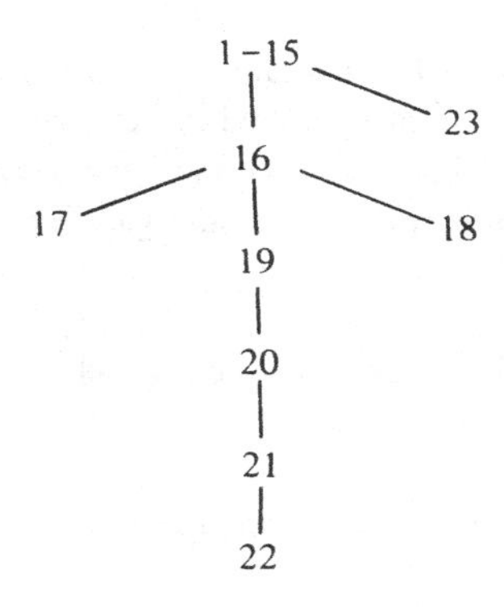

EXAMPLES AND COUNTEREXAMPLES

(The examples listed below that are not provided with a number can be found in the Introduction.)

A discontinuous Riemann integrable function with an antiderivative.

A Riemann integrable, Darboux continuous function that has no antiderivative.

A function that has an antiderivative but is not Riemann integrable.

A Darboux continuous function that has no antiderivative and is not Riemann integrable.

A Darboux continuous function that has no antiderivative on any interval and is not Riemann integrable over any interval.

A function on [0, 1] that maps any subinterval of [0, 1] onto [0, 1].

1.2. A monotone function whose discontinuity points form a prescribed countable set.

1.P. For a prescribed closed nowhere dense set A, a strictly increasing continuously differentiable function f such that A is the set of zeros of f'.

3.D. A differentiable function of bounded variation whose derivative is unbounded.

3.F. A continuous function g that is not of bounded variation, whereas g^2 is.
A differentiable function that is not of bounded variation.

4.B. The interval [0, 1] is not a null set.

4.C. A bounded null set that does not have zero content.

4.2, 4.K. (The Cantor set) An uncountable closed null set without isolated points.

4.3, 4.E. For a given null set E, a continuous strictly increasing function that is differentiable at no point of E (related to Lebesgue's theorem (4.10)).

5.5. A decomposition of $\mathbb{R}$ into a null set and a meagre set.

5.I. A function that maps every interval onto [0, 1], yet vanishes almost everywhere.

6.4. $\mathbb{Q}$ is not a G_δ.

6.5. A set A which is both an F_σ and a G_δ but which is not the intersection of an open and a closed set.

6.H. There is no monotone function f: $\mathbb{R} \to \mathbb{R}$ such that $f(\mathbb{R}) = \mathbb{R} \setminus \mathbb{Q}$. (For every monotone f: $\mathbb{R} \to \mathbb{R}$ we have that $f(\mathbb{R})$ is a G_δ.)

7.A. Given a **countable** set $S \subset \mathbb{R}$, a continuous function whose set of local extrema is just S. (For every $f: \mathbb{R} \to \mathbb{R}$ the set of local extrema of f is countable (7.2).)

7.B. A function f for which each value is a local extremum while f is constant on no interval.

7.D. A continuously differentiable function $f: \mathbb{R} \to \mathbb{R}$ having no local extrema but for which $f(\{x : f'(x) = 0\})$ is uncountable.

7.H. A function whose points of discontinuity form a prescribed F_σ-set.

7.M. A continuous function whose points of nondifferentiability are just the rationals.

7.16, 7.P, 8.H. A continuous function that is nowhere increasing, nowhere decreasing and nowhere differentiable.

8.D. A discontinuous function with a closed graph. (Compare Theorem 8.1.)

8.E. A discontinuous function with a connected graph. (Compare Theorem 8.2.)

8.G. For $0 < \alpha < \beta \leq 1$, a function in Lip_α that is not in Lip_β.
A function in $\mathrm{Lip}_{1/2}$ that is not of bounded variation.
A continuous monotone function that is in no Lip_α.

9.C. A Darboux continuous function with a disconnected graph. (Compare Theorem 9.4.)

Two functions whose graphs are homeomorphic, whereas one function is Darboux continuous and the other is not.

9.H. A right continuous Darboux continuous function that is not continuous.

9.K. A function that is not a uniform limit of Darboux continuous functions. (Compare Exercise 9.J.)

9.L. A uniform limit of Darboux continuous functions need not be Darboux continuous.

9.P. A continuous function f on $[0, 1]$ with $f(0) = f(1)$ whose graph has no horizontal chords of length $\alpha \notin \{1/n : n \in \mathbb{N}\}$.

Remark after 10.J. A semicontinuous bijection $[0, 1] \to [0, 1]$ whose inverse is not semicontinuous.

10.S. A function that has a local maximum at every point of its domain but has infinitely many values.

11.H. For a given meagre F_σ-set A, a semicontinuous function whose points of discontinuity are just the points of A. (Compare Theorem 11.4.)

11.L. A function whose discontinuity points form a meagre closed set, but which is not of the first class of Baire. (Compare Theorem 11.4.)

11.P. For a closed nowhere dense set A, a function whose graph is closed and for which the set of discontinuity points is just A.

11.Y. A function that is not of the first class of Baire although its graph is a G_δ. (Each function of the first class has a G_δ-graph.)

12.H. A function on $[a, b]$ that is equal almost everywhere to a continuous function but which itself is not Riemann integrable.

12.I. A semicontinuous function that is not Riemann integrable.
A Riemann integrable function that is not of the first class of Baire.

12.J. A continuous f and a Riemann integrable g such that $g \circ f$ is not Riemann integrable.

12.Q. For a given null F_σ-set X, a Riemann integrable function whose points of discontinuity form just X. (Compare Theorem 12.1.)

13.D. An injective function defined on the Cantor set, which is in Lip_α for all $\alpha \in (0, \infty)$.

13.E. A function f defined on a closed set $X \subset \mathbb{R}$ without isolated points, for which

$$\lim_{\substack{x \to y \\ x \in X}} \frac{f(x) - f(y)}{(x - y)^3} = 1 \qquad (y \in X)$$

13.2. A differentiable function $L: \mathbb{R} \to \mathbb{R}$ whose derivative is bounded while L is monotone on no interval.

13.3. A strictly increasing differentiable function whose derivative vanishes on a dense set.

14.B. A function f which is a derivative whereas $|f|$ is not.
A function g which is a derivative whereas g^2 is not.

14.C. A derivative function f with $1 \leqslant f \leqslant 2$ such that $1/f$ is not a derivative.

14.3. A bounded derivative that is not Riemann integrable, and, in addition, is lower semicontinuous (14.4) and discontinuous almost everywhere (14.5).

15.H. A continuous function f which is not decreasing although

$$\underline{\lim}_{y \to x} \frac{f(y) - f(x)}{y - x} \leqslant 0$$

for all x.

15.8. A differentiable nonconstant function whose derivative vanishes everywhere outside a meagre set.

15.9. (The Cantor function) An increasing nonconstant continuous surjection $\Phi: [0, 1] \to [0, 1]$, such that $\Phi' = 0$ almost everywhere.

15.10. A strictly increasing continuous function whose derivative vanishes almost everywhere.

15.P. Functions g, h such that (if we admit the values $\pm\infty$) $g'(x) = h'(x)$ for all x, while $g - h$ is not constant.

16.T. (A Peano curve) A continuous surjection $[0, 1] \to [0, 1] \times [0, 1]$.

18.16. An analytic subset of $\mathbb{R}$ that is not Borel.

19.E. A function that is improperly Riemann integrable but not Lebesgue integrable.

19.R. A bounded Lebesgue integrable function on $[0, 1]$ that is not equal almost everywhere to a Riemann integrable function.

20.C. A bounded Lebesgue integrable function that is not equal a.e. to a function of the first class of Baire.

20.10. A non-Lebesgue measurable set. (Vitali's example)

20.I. A Lebesgue measurable f and a continuous g such that $f \circ g$ is not Lebesgue measurable.

20.P. A positive Lebesgue measurable function on $\mathbb{R}$ that is integrable over no interval.

20.S, 20.T, 20.U. A continuous function $f : [0, 1] \to [0, 1]$ such that for each $y \in [0, 1]$ the set $\{x : f(x) = y\}$ is uncountable.

21.G. Two functions having property (N) (Definition 21.8) whose sum does not.

21.L. Two absolutely continuous functions from $[0, 1]$ to $[0, 1]$ whose composition is not absolutely continuous.

Remark after 21.29. A Lebesgue measurable set X such that $0 < \lambda(X \cap J)/\lambda(J) < 1$ for every bounded interval J.

22.G. A Perron integrable function whose absolute value is not Perron integrable.

23.M. There is no function g on $[0, 1]$ such that $f'(0) = \int_0^1 f\,dg$ for all polynomial functions f on $[0, 1]$.

INTRODUCTION

With this introduction (to Sections 1–15) we try to achieve several goals. Not only do we hope that it will give the reader an impression of the kind of problems that are treated in the sequel but, in addition, it includes the program that will be followed throughout Sections 1–15. With this program in mind the reader should find that at each stage the theory provides answers to questions that have come about in a natural way.

In the introduction we make an actual start with the theory, so in this sense it is an ordinary section, not to be disregarded by the reader.

The starting point

Throughout the introduction by a *function* we mean a real valued function defined on the closed unit interval $[0, 1]$.

Let us first collect a few results known from elementary courses on analysis and calculus. We have:

A monotone function is Riemann integrable.

A continuous function is Riemann integrable.

A differentiable function is continuous.

A continuous function has an antiderivative.

Less common is the notion of Darboux continuity (sometimes called the intermediate value property), so we explain it here.

DEFINITION. A function f is *Darboux* continuous if for all $p, q \in [0, 1]$ and for each c between $f(p)$ and $f(q)$ there is an x between p and q such that $f(x)=c$.

(A real number s is said to lie *between* the real numbers t and u if either $t \leqslant s \leqslant u$ or $u \leqslant s \leqslant t$.)

In many analysis courses one proves that continuous functions are Darboux continuous (possibly without explicitly mentioning the term 'Darboux continuous'). Here we prove a slightly stronger statement.

THEOREM *Every function that has an antiderivative is Darboux continuous.*

Proof. Let f be a differentiable function. We show that f' is Darboux continuous. Let $p, q \in [0, 1]$, $p<q$ and let $c \in \mathbb{R}$ be between $f'(p)$ and $f'(q)$. Without loss of generality, assume $f'(p)<c<f'(q)$. Now define $g : [p, q] \to \mathbb{R}$ by

$$g(x):=f(x)-cx \qquad (x \in [p, q])$$

Then g is differentiable. It attains a minimum value at, say, $r \in [p, q]$. Since $g'(p)<0$ and $g'(q)>0$ we have $r \neq p$ and $r \neq q$, so $r \in (p, q)$. But then $g'(r)$ must be 0, i.e. $f'(r)=c$ and we are done.

We shall see in a moment that Darboux continuity does not imply continuity. (The impatient reader may take the function $x \mapsto \sin x^{-1}$ with an appropriate interpretation at 0, but at the end of this section we shall present quite a dramatic example.) However, we have

Exercise. A monotone Darboux continuous function is continuous. (Hint. For a monotone function f we have the existence of $f(a+):=\lim_{x \downarrow a} f(x)$ for $0 \leqslant a < 1$ and $f(a-):=\lim_{x \uparrow a} f(x)$ for $0<a \leqslant 1$. Define $f(0-):=f(0)$ and $f(1+):=f(1)$. If a is a point of discontinuity of f, then $f(a-) \neq f(a+)$.)

In order to obtain a clear view on the above statements, let us first introduce symbols to indicate various sets of functions as follows.

$\mathcal{M}on$ the set of monotone functions
$\mathcal{D}$ the set of differentiable functions
$\mathcal{C}$ the set of continuous functions
$\mathcal{R}$ the set of Riemann integrable functions
$\mathcal{D}'$ the set of derivative functions (functions with antiderivatives)
$\mathcal{DC}$ the set of Darboux continuous functions

The results we have discussed so far can now be expressed concisely by: $\mathcal{M}on \subset \mathcal{R}$, $\mathcal{C} \subset \mathcal{R}$, $\mathcal{D} \subset \mathcal{C}$, $\mathcal{C} \subset \mathcal{D}'$, $\mathcal{D}' \subset \mathcal{DC}$, $\mathcal{M}on \cap \mathcal{DC} \subset \mathcal{C}$. We illustrate these implications by means of the diagram in Fig. 1.

Each of the sets $\mathcal{M}on$, $\mathcal{D}$, $\mathcal{C}$, $\mathcal{R}$, $\mathcal{D}'$, $\mathcal{DC}$ is represented by a rectangle; at the lower left corner of the rectangle we have placed the name of the corresponding set of functions. The reader may verify that the picture reflects our present knowledge faithfully.

Do there exist implications, hidden so far, that will simplify the picture? In other words, do some of the areas 1–11 represent the empty set? The answer, as the reader may have guessed already (if only for psychological reasons), is negative. So our next purpose will be to prove the following

STATEMENT. *Each of the areas 1–11 in Fig. 1 represents a nonempty set of functions.*

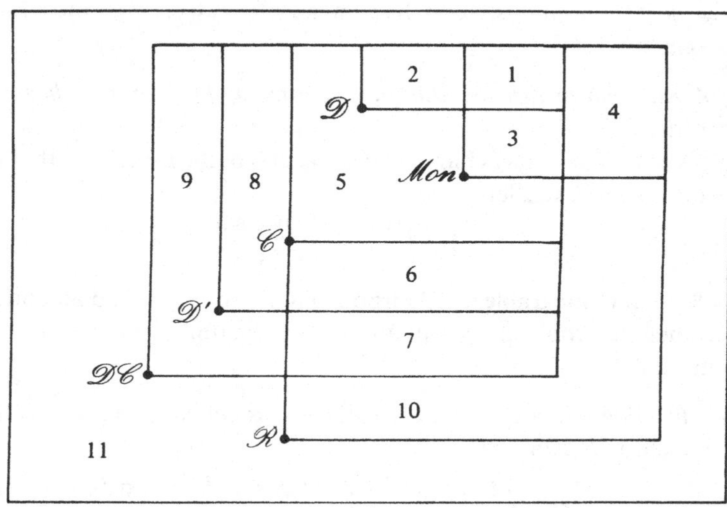

Fig. 1

For some of the areas we leave the inspection to the reader:

Exercise. For each of the areas 1, 2, 3, 4, 5, 10, 11 find an example of a function belonging to the corresponding set.

We proceed to prove that areas 6, 7, 8, 9 are not empty either.

(6) *A discontinuous Riemann integrable function with an antiderivative.*

Define f by

$$f(x) := \begin{cases} \sin x^{-1} & \text{if } x \in (0, 1] \\ 0 & \text{if } x = 0 \end{cases}$$

Then clearly $f \notin \mathscr{C}$ and one easily proves $f \in \mathscr{R}$. To show that f has an antiderivative we use a trick. Define

$$\phi(x) := \begin{cases} x^2 \cos x^{-1} & \text{if } x \in (0, 1] \\ 0 & \text{if } x = 0 \end{cases}$$

Then ϕ is differentiable and

$$\phi'(x) := \begin{cases} 2x \cos x^{-1} + \sin x^{-1} & \text{if } x \in (0, 1] \\ 0 & \text{if } x = 0 \end{cases}$$

We see that $\phi' = \psi + f$ where

$$\psi(x) := \begin{cases} 2x \cos x^{-1} & \text{if } x \in (0, 1] \\ 0 & \text{if } x = 0 \end{cases}$$

Now ψ is continuous, so it has an antiderivative. But then $f=\phi'-\psi$ has one too.

(7) *A Riemann integrable, Darboux continuous function that has no antiderivative.*
This is easy. We simply change the value at 0 of the function f that did the job for area 6. Thus, let

$$g(x):=\begin{cases}\sin x^{-1} & \text{if } x\in(0,1]\\ 1 & \text{if } x=0\end{cases}$$

g is Riemann integrable and Darboux continuous. If g had an antiderivative, then so would $g-f$. But the latter function is not even Darboux continuous!

(8) *A function that has an antiderivative but is not Riemann integrable.*
The function h defined by

$$h(x):=\begin{cases}2x\sin x^{-2}-2x^{-1}\cos x^{-2} & \text{if } x\in(0,1]\\ 0 & \text{if } x=0\end{cases}$$

is unbounded, hence certainly not Riemann integrable. But it is the derivative of H, where

$$H(x):=\begin{cases}x^2\sin x^{-2} & \text{if } x\in(0,1]\\ 0 & \text{if } x=0\end{cases}$$

(9) *A Darboux continuous function that has no antiderivative and is not Riemann integrable.*
As in (7), we slightly adjust the function of the previous example. Let

$$j(x):=\begin{cases}h(x) & \text{if } x\in(0,1]\\ 1 & \text{if } x=0\end{cases}$$

Reasoning similar to the one used in (7) tells us that j has no antiderivative. Being unbounded, j is not Riemann integrable. The Darboux continuity of j follows from the Darboux continuity of h and from the fact that for every $\delta>0$, $j([0,\delta])\supset h((0,\delta])=\mathbb{R}$.

We have now completed the proof of the Statement. Fig. 1 gives a fair representation of the set theoretic inclusions concerning the six classes of functions we have under consideration. We shall take this picture as a starting point for further explorations.

A program for Sections 1–15

A large part of the theory of the first fifteen sections will emerge from looking at the picture in Fig. 1 from several directions. We mention three points of view here that will recur frequently in the subsequent sections (although now always explicitly).

(1) We shall – more or less systematically – ask whether the sets $\mathcal{M}on$, $\mathcal{D}$, etc. are closed under addition, (scalar) multiplication, formation of suprema and infima, composition and limits. In cases where the answer is 'no' we are led to further investigations and the introduction of new sets of functions. For example, $\mathcal{M}on$ is not a vector space; can we describe its linear span? (Yes, see Section 3; the functions of bounded variation are born.)

(2) Classes of functions other than the ones mentioned in Fig. 1 arise naturally (not only out of the investigation of (1) above) such as bounded functions, semicontinuous functions, ... How are they related to the six classes we already have? As an illustration, let us go back to area 8 in Fig. 1. We have found a function h that has an antiderivative but is not Riemann integrable. In a sense this is a rather poor example since it is the *unboundedness* of h that prevents h from being integrable. We feel that this solution is too cheap. Can we find a *bounded* derivative that is not Riemann integrable? (Yes: Examples 13.2 and 14.3.)

(3) Apart from adding the new notions (as a consequence of the first two approaches mentioned above) we can also stick to our original picture and ask how extreme the examples of areas 1–11 can be made. To see what is meant by that, let us look at area 5, which represents the continuous nondifferentiable nonmonotone functions. In our second exercise we asked the reader to find an example. This is not hard to do. Indeed, the map $x \mapsto |x - \frac{1}{2}|$ ($x \in [0, 1]$) will do, but this function is still piecewise monotone and differentiable everywhere except at the point $\frac{1}{2}$. A much more satisfactory example would be a continuous function that is nowhere differentiable and 'nowhere' monotone. (We do not yet have a concept like 'monotonicity at a point', so we should read 'nowhere monotone' as 'monotone on no subinterval of $[0, 1]$'. See, however, Exercise 1.A.) To find such a function is a challenging task! We shall carry it out in Example 7.16.

The approach sketched here leads to interesting (counter)examples, but it does more than that. Proceeding with area 4 in the same spirit as we did above for area 5 we arrive at the following question. Does there exist a monotone function that is nowhere continuous? This time the answer will turn out to be 'no'. Then further inquiries about 'how discontinuous' a monotone function can possibly be will follow (Section 1).

Throughout Sections 1–15 the above will serve as a guide-line. Yet, the reader should not get the impression that from now on we are going to follow it too strictly. Doing so would make us narrow-minded and we would miss much of the beauty of the theory. Hence, while carrying out the program we shall also let the functions we are dealing with speak for themselves.

We wish to end our introduction with an 'extreme' example concerning area 9. Thus, we shall prove:

There exists a function $k : [0, 1] \to [0, 1]$ *having the following properties.*

(i) *k is Darboux continuous.*

(ii) *k is Riemann integrable over no closed subinterval of* $[0, 1]$.

(iii) *For no subinterval I of* $[0, 1]$ *does the restriction of f to I have an antiderivative.*

To construct such a k, define a map $t \mapsto t^*$ of $[0, 1]$ into $[0, 1]$ as follows. Let $D := \{m10^{-n} : m \in \mathbb{Z},\ n \in \mathbb{N}\}$. For $t \in [0, 1]$, let $0.t_1t_2t_3\ldots$ be its development as a decimal fraction. (If $t \in D$, $t \neq 1$, we let t_k be 0 for all sufficiently large k.) Then, set

$$t^* := 0.t_1t_1t_2t_1t_2t_3t_1t_2t_3t_4t_1t_2\ldots$$

Observe that, if t, $s \in [0, 1]$ are distinct, then $t^* - s^* \notin D$. Consequently, for every $x \in [0, 1]$ there exists at most one $t \in [0, 1]$ with $x - t^* \in D$.

Now define $k : [0, 1] \to [0, 1]$ by:

$$k(x) := \begin{cases} t & \text{if } t \in [0, 1] \text{ and } x - t^* \in D, \\ 0 & \text{if there does not exist a } t \in [0, 1] \text{ with } x - t^* \in D. \end{cases}$$

(i) Let $t \in [0, 1]$ and let J be any subinterval of $[0, 1]$. There is an $x \in J$ such that $x - t^* \in D$: then $t = k(x) \in k(J)$. We see that *k maps every subinterval of* $[0, 1]$ *onto all of* $[0, 1]$. Then certainly *k is Darboux continuous.*
(ii) In particular, on every subinterval of $[0, 1]$ the function k takes the values 0 and 1. Therefore, *k is not Riemann integrable over any subinterval of* $[0, 1]$.
(iii) Finally, we show that *on no subinterval of* $[0, 1]$ *does k have an antiderivative.* Indeed, suppose there exists a subinterval $[a - \varepsilon, a + \varepsilon]$ of $[0, 1]$ and a $K : [a - \varepsilon, a + \varepsilon] \to \mathbb{R}$ such that $K'(x) = k(x)$ for all $x \in [a - \varepsilon, a + \varepsilon]$. If $n \in \mathbb{N}$ is so large that $10^{-n} < \varepsilon$, then by the definition of k we have

$$0 = k(x + 10^{-n}) - k(x) = K'(x + 10^{-n}) - K'(x) \qquad (a - \varepsilon \leqslant x \leqslant a)$$

so that $x \mapsto K(x + 10^{-n}) - K(x)$ is constant on $[a - \varepsilon, a]$. Thus, we obtain

$$10^n(K(x + 10^{-n}) - K(x)) = 10^n(K(a + 10^{-n}) - K(a)) \qquad (a - \varepsilon \leqslant x \leqslant a)$$

By taking limits we find $K'(x)=K'(a)$, i.e. $k(x)=k(a)$ for all $x \in [a-\varepsilon, a]$, which is absurd.

**Exercise*. Find $f:[0, 1]\to\mathbb{R}$ and $g:\mathbb{R}\to\mathbb{R}$ that map each subinterval of $[0, 1]$ (or $\mathbb{R}$, respectively) onto $\mathbb{R}$.

1

MONOTONE FUNCTIONS

1. Continuity of monotone functions

DEFINITION 1.1. Let $X \subset \mathbb{R}$. A function $f: X \to \mathbb{R}$ is called

increasing if $x \leqslant y$ implies $f(x) \leqslant f(y)$,

strictly increasing if $x < y$ implies $f(x) < f(y)$,

decreasing if $x \leqslant y$ implies $f(x) \geqslant f(y)$,

strictly decreasing if $x < y$ implies $f(x) > f(y)$,

for all $x, y \in X$.

f is called *monotone* if f is increasing or decreasing. If f is either strictly increasing or strictly decreasing, then we call f *strictly monotone*.

In the following exercises we discuss equivalent definitions of monotony.

**Exercise* 1.A. $f: \mathbb{R} \to \mathbb{R}$ is called *increasing at* $p \in \mathbb{R}$ if there exists an $\varepsilon > 0$ such that $f(x) \leqslant f(p) \leqslant f(y)$ for all $x \in (p-\varepsilon, p)$ and $y \in (p, p+\varepsilon)$. Show that $f: \mathbb{R} \to \mathbb{R}$ is increasing if and only if f is increasing at every $p \in \mathbb{R}$.

Exercise 1.B. Let $c, p, q \in \mathbb{R}$. We say that *c is between p and q* if either $p \leqslant c \leqslant q$ or $q \leqslant c \leqslant p$. Consider the following conditions (a), (b) and (c) for a function $f: [0, 1] \to \mathbb{R}$.

(a) If x is between y and z then $f(x)$ is between $f(y)$ and $f(z)$ $(x, y, z \in [0, 1])$.

(b) If $f(x)$ is between $f(y)$ and $f(z)$ then x is between y and z $(x, y, z \in [0, 1])$.

(c) x is between y and z if and only if $f(x)$ is between $f(y)$ and $f(z)$ $(x, y, z \in [0, 1])$.

Express each of these conditions in terms of monotony of f.

Exercise 1.C. Let $f: [0, 1] \to \mathbb{R}$. Then f is monotone if and only if for every connected set $I \subset \mathbb{R}$ the inverse image $f^{-1}(I)$ is connected. (Comment. $f: [0, 1] \to \mathbb{R}$ is Darboux continuous if and only if the image of every connected set is again connected.)

In 1.D–1.I we collect some elementary facts about monotone functions.

Exercise 1.D. Let $f_1, f_2, \ldots$ be monotone functions such that $f := \lim_{n\to\infty} f_n$ exists. Then show that f is monotone. (The set of all monotone functions is closed under pointwise limits.)

Exercise 1.E. In general, the monotone functions on a subset X of $\mathbb{R}$ do not form a vector space. (The sum of two monotone functions need not be monotone.) Let M

denote the linear space of functions generated by the monotone functions. Prove that $f \in M$ if and only if f is the difference of two increasing functions.

**Exercise* 1.F. Let $\mathscr{F}$ be a set of increasing functions on $[0, 1]$. If $g(x) := \sup \{f(x) : f \in \mathscr{F}\}$ is finite for all $x \in [0, 1]$, then the function g is increasing. If $h(x) := \inf \{f(x) : f \in \mathscr{F}\}$ is finite for all x, then h is increasing. (Of course, there is a similar result for decreasing functions.)

Exercise 1.G. (The rising sun function) Let $f : [0, 1] \to \mathbb{R}$ be a bounded function. Then the set of all decreasing functions $g : [0, 1] \to \mathbb{R}$ for which $g \geqslant f$ has a smallest element, $f_\odot$. In fact,

$$f_\odot(x) = \sup\{f(y) : x \leqslant y \leqslant 1\} \qquad (x \in [0, 1])$$

If f is continuous, then so is $f_\odot$. (Comment. Imagine a rising sun on the x-axis at $+\infty$ (see Fig. 2). Then $\{(x, y) \in \mathbb{R}^2 : y \geqslant f_\odot(x)\}$ is illuminated by the sun whereas $\{(x, y) : y < f_\odot(x)\}$ lies in the shadow. The set $\{(x, f(x)) : f(x) = f_\odot(x)\}$ is the collection of those points of the graph of f that receive light from the sun.)

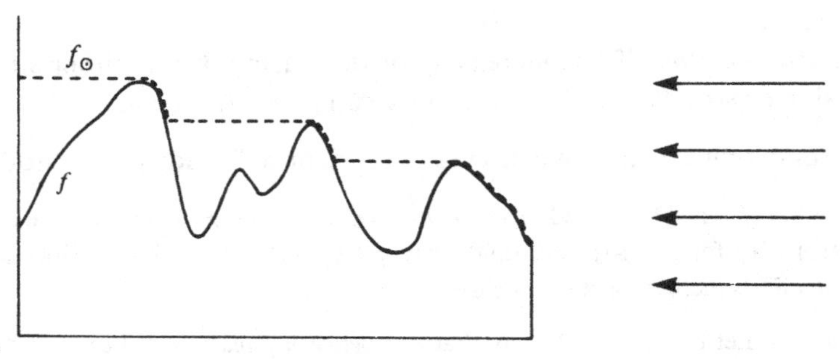

Fig. 2

**Exercise* 1.H. Let $f : [0, 1] \to \mathbb{R}$ be increasing. Show that for each $x \in [0, 1)$

$$f(x+) := \lim_{y \downarrow x} f(y)$$

exists. Also, for each $x \in (0, 1]$

$$f(x-) := \lim_{y \uparrow x} f(y)$$

exists. Recall that $f(0-) = f(0)$ and $f(1+) = f(1)$. Show that the functions $f_l : x \mapsto f(x-)$ and $f_r : x \mapsto f(x+)$ are increasing, $f_l \leqslant f \leqslant f_r$, f_l is left continuous and f_r is right continuous.

**Exercise* 1.I

(i) Let $f : [0, 1] \to [0, 1]$ be monotone and surjective. Show that f is continuous.

(ii) Let $g : [0, 1] \to \mathbb{R}$ be continuous and injective. Show that g is strictly monotone.

Obviously, a monotone function need not be continuous. So a natural question to ask is: How discontinuous can a monotone function be? We may get a first impression by trying to sketch graphs. Functions with only finitely many discontinuities will not give any problem. Fig. 3 represents the graph of an increasing function $f : [0, 1] \to \mathbb{R}$ that is discontinuous at the points $\frac{1}{2}, \frac{3}{4}, \frac{5}{6}, \ldots$ We can get more complicated monotone functions by 'gluing together' graphs of this type countably many times (see Fig. 4). Here the points of discontinuity already form a rather

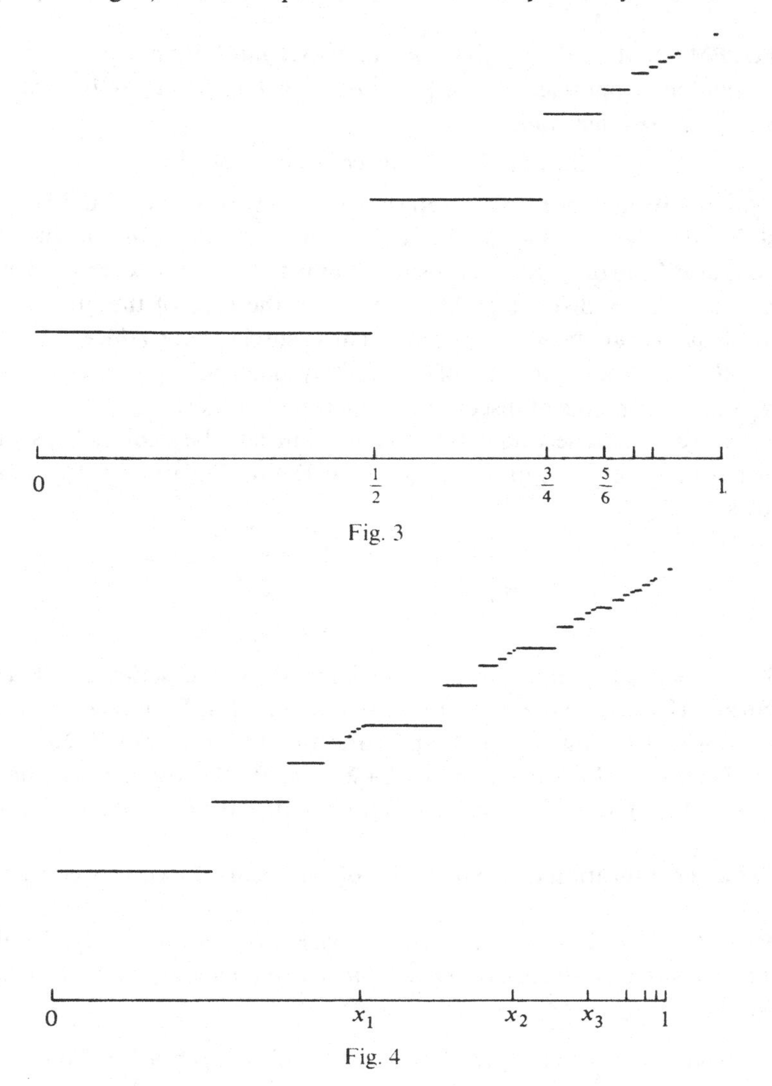

Fig. 3

Fig. 4

complicated (countable) set having x_1, x_2, ... and $1(=\lim_{n\to\infty}x_n)$ as accumulation points.

The reader may begin to feel helpless when we ask for graphs of yet more complicated increasing functions, for example one that is discontinuous at every rational point. Pencil and paper, anyway, seem too rough to make it possible to sketch the graph in such a way that the discontinuities are visible (or even suggested). We continue our investigation by other means:

THEOREM 1.2. *A monotone function $[a, b]\to\mathbb{R}$ has only countably many discontinuities. Conversely, if $S\subset[a, b]$ is countable, then there is a monotone $f:[a, b]\to\mathbb{R}$ such that*

$$S=\{x\in[a, b]: f \text{ is discontinuous at } x\}.$$

Proof. Let f be increasing. By Exercise 1.H, for every x, $f(x+)$ and $f(x-)$ exist. Let $\omega(x):=f(x+)-f(x-)$ (the 'jump' of f at x). Then f is discontinuous at x if and only if $\omega(x)$ is positive. For $n\in\mathbb{N}$ set $S_n:=\{x:\omega(x)\geqslant 1/n\}$. If $p_1,\ldots,p_k$ are k distinct points of S_n, then the sum of the 'jumps' in $p_1,\ldots,p_k$ is at most $f(b)-f(a)$. Thus surely $k/n\leqslant f(b)-f(a)$, i.e. $k\leqslant n(f(b)-f(a))$. So S_n consists of only finitely many points. Consequently, $\bigcup_n S_n$, the set of points of discontinuity of f, is countable.

For a proof of the second part of the theorem, let S be a countable set in $[a, b]$ with an enumeration $(x_1, x_2,\ldots)$. Define functions $f_1, f_2,\ldots$ as follows.

$$f_n(x):=\begin{cases}-n^{-2} & \text{if } x<x_n\\ 0 & \text{if } x=x_n\\ n^{-2} & \text{if } x>x_n\end{cases}$$

and set $f:=\Sigma_{n=1}^{\infty} f_n$. (Since $|f_n(x)|\leqslant n^{-2}$ for all x, the series converges uniformly.) Each f_n is increasing, hence so is f. Let $x\notin S$. Then each function $f_1+\ldots+f_k$ is continuous at x and so is the uniform limit, f. Now let $x\in S$. Then $x=x_i$ for some i and $f=f_i+\Sigma_{j\neq i} f_j$. By the argument we have just used, $\Sigma_{j\neq i} f_j$ is continuous at x. But f_i is not. Then f is not continuous at x.

For a generalization of the first part of Theorem 1.2, see Exercise 1.M.

COROLLARY 1.3. *There exists a strictly increasing function $f:[a, b]\to\mathbb{R}$ that is discontinuous at each rational point and continuous at each irrational point.*

The reader is invited to try and sketch a graph of such a function.

Exercise 1.J. Every monotone function that is discontinuous at each rational point has the following properties. On no subinterval has it an antiderivative. On no subinterval is it Darboux continuous.

Exercise 1.K. Let

$$f(x) := \sum_{n=1}^{\infty} \frac{[nx]}{n!} \qquad (x \in \mathbb{R})$$

where $[nx]$ is the entire part of nx. Determine the set of continuity points of f.

Exercise 1.L. (Extension of monotone functions) Let $\varnothing \neq X \subset \mathbb{R}$ and let f be a bounded increasing function $X \to \mathbb{R}$. Then f can be extended to an increasing function $g : \mathbb{R} \to \mathbb{R}$ as follows.

$$g(x) := \begin{cases} \sup \{f(t) : t \in X, t \leqslant x\} & \text{if } X \cap (-\infty, x) \neq \varnothing \\ \inf \{f(t) : t \in X\} & \text{otherwise.} \end{cases}$$

Exercise 1.M. Let $\varnothing \neq X \subset \mathbb{R}$ and let $f : X \to \mathbb{R}$ be monotone. Then f has at most countably many points of discontinuity.

Exercise 1.N. (The saltus function) Let $f : [a, b] \to \mathbb{R}$ be increasing. Then for each $x \in [a, b]$ we can define

$$g(x) := \sum_{y<x} (f(y+) - f(y-)) + f(x) - f(x-)$$

$$= \sum_{y \leqslant x} (f(y) - f(y-)) - f(x+) + f(x)$$

(Show that the sums converge!) The function g is called *the saltus function* of f. Both g and $f-g$ are increasing and $f-g$ is continuous.

Exercise 1.O. The set of the monotone functions on $[0, 1]$ contains all polynomial functions of degree $\leqslant 1$. These form a two-dimensional vector space. Does the set of all monotone functions contain a three-dimensional vector space?

**Exercise* 1.P. Let $A \subset \mathbb{R}$ be a closed set. Then there exists an increasing $f : \mathbb{R} \to \mathbb{R}$ such that f' is continuous and $A = \{x \in \mathbb{R} : f'(x) = 0\}$. If A contains no interval, then every such f is strictly increasing. (Hint. Let $\phi(x) := \inf\{|x-a| : a \in A\}$. ($\phi(x)$ is the 'distance' of x and A.) Prove that $|\phi(x) - \phi(y)| \leqslant |x-y|$ for all $x, y \in \mathbb{R}$, so that ϕ is continuous. Let f be an antiderivative of ϕ.)

Exercise 1.Q. Let A and B be countable dense subsets of $[0, 1]$, both containing 0 and 1. We construct a strictly increasing continuous bijection $f : [0, 1] \to [0, 1]$ with $f(A) = B$.

(i) Let T be the set of all rational numbers in $[0, 1]$ that can be represented by fractions whose denominators are powers of 2. Show that we may assume $A = T$.

(ii) We first define a strictly increasing bijection $f_0 : T \to B$. Let $(0, 1, b_1, b_2, \ldots)$ be an enumeration of B. We define $f_0(t)$ successively for all elements t of T. We start by setting $f_0(0) := 0$, $f_0(1) := 1$, $f_0(\frac{1}{2}) := b_1$. Next, for $f_0(\frac{1}{4})$ we choose the first entry in the sequence $b_1, b_2, \ldots$ that lies in the interval $(f_0(0), f_0(\frac{1}{2}))$, $f_0(\frac{3}{4})$ is the first entry that lies in $(f_0(\frac{1}{2}), f_0(1))$, $f_0(\frac{1}{8})$ is the first entry lying in $(f_0(0), f_0(\frac{1}{4}))$, etc. Formalized, the procedure is as follows.

For $n \in \mathbb{N}$, let $T_n := \{i2^{-n} : i=0, 1, 2, \ldots, 2^n\}$; then $T_1 \subset T_2 \subset \ldots$ and $\bigcup_n T_n = T$. Let $n \in \mathbb{N}$ and suppose that f_0 has been defined on T_n. To define $f_0(t)$ for $t \in T_{n+1} \setminus T_n$ we set $t' := \max\{s \in T_n : s<t\}$, $t'' := \min\{s \in T_n : s>t\}$, and we let $f_0(t)$ be the first entry of the sequence $b_1, b_2, \ldots$ that lies in the interval $(f_0(t'), f_0(t''))$.

Thus, we can define $f_0 : T \to B$. Show that f_0 is strictly increasing and that $f_0(T) = B$. (Actually, $b_n \in f_0(T_n)$ for all n.)

(iii) Extend f_0 to an increasing function $f: [0, 1] \to [0, 1]$. Prove that f is strictly increasing, continuous and surjective. (For the continuity, observe that $f([0, 1])$ is a dense subset of $[f(0), f(1)]$.)

2. Indefinite integrals of monotone functions (convex functions)

In this section we shall see that the functions $x \mapsto \int_a^x \phi(t)\,dt$, where ϕ is increasing, are just the convex functions that have the value 0 at a (Theorem 2.4).

DEFINITION 2.1. Let I be an interval and let $f: I \to \mathbb{R}$. We shall call f *convex* if for all $x, y \in I$ $(x<y)$ and $0 \leqslant \lambda \leqslant 1$

$$f(\lambda x + (1-\lambda)y) \leqslant \lambda f(x) + (1-\lambda) f(y)$$

Geometrically this means that the chord connecting $(x, f(x))$ and $(y, f(y))$ lies above or on the graph of f (see Fig. 5).

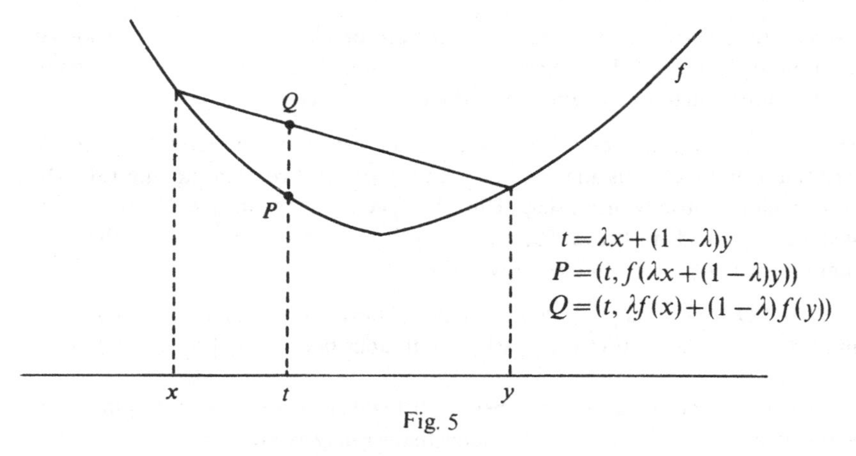

Fig. 5

Exercise 2.A. Show that the function $x \mapsto x^2$ is convex on $\mathbb{R}$.

Exercise 2.B. Show that $f : \mathbb{R} \to \mathbb{R}$ is convex if and only if $V := \{(x, y) \in \mathbb{R}^2 : y \geqslant f(x)\}$ is a convex set (i.e. if $A, B \in V$ and $0 \leqslant \lambda \leqslant 1$, then $\lambda A + (1-\lambda)B \in V$).

From elementary calculus we know that convexity of a function f has to do with positivity of f'', the second derivative of f. We wish to have a precise statement of this kind, even for an f that is not differentiable. Thus, instead of f' we consider the difference quotient of f:

$$\Phi_1 f(x, y) := \frac{f(x)-f(y)}{x-y} \qquad (x \neq y)$$

Instead of f'' we consider the difference quotient of $\Phi_1 f$ (in one of its variables, which one is immaterial) and we get

$$\Phi_2 f(x, y, z) := \left[\frac{f(x)-f(z)}{x-z} - \frac{f(y)-f(z)}{y-z}\right] \Big/ (x-y) \qquad (x \neq y,\ y \neq z,\ x \neq z)$$

$\Phi_2 f$ is a symmetric function of its three variables (i.e. $\Phi_2 f(x, y, z) = \Phi_2 f(x, z, y) = \Phi_2 f(y, z, x) = \dots$). The connection between convexity of f and properties of first and second 'derivatives' is reflected in

THEOREM 2.2. *Let $I \subset \mathbb{R}$ be an interval and let $f: I \to \mathbb{R}$. Then the following conditions are equivalent.*

(α) *f is convex.*

(β) *$\Phi_1 f$ is an increasing function in each variable.*

(γ) *$\Phi_2 f \geqslant 0$.*

Proof. The equivalence of (β) and (γ) follows directly from the definitions. If we substitute $z = \lambda x + (1-\lambda)y$ $(0 < \lambda < 1)$ in the defining formula for $\Phi_2 f$ we obtain after an easy computation

$$(*) \qquad \Phi_2 f(x, y, \lambda x + (1-\lambda)y) = \frac{\lambda f(x) + (1-\lambda) f(y) - f(\lambda x + (1-\lambda)y)}{\lambda(1-\lambda)(x-y)^2}$$

Let f be convex. In order to prove $\Phi_2 f \geqslant 0$, by the symmetry of $\Phi_2 f$, it suffices to consider $\Phi_2 f(x, y, z)$ where $x < z < y$. Then $z = \lambda x + (1-\lambda)y$ for some λ between 0 and 1 and the above formula $(*)$ shows that $\Phi_2 f(x, y, z) \geqslant 0$. Conversely, if $\Phi_2 f \geqslant 0$, then again $(*)$ shows us that f is convex.

The fact that, for a convex function f, $\Phi_1 f$ is increasing, implies the existence of the functions $D_r f$ and $D_l f$ defined by

$$D_r f(x) := \lim_{y \downarrow x} \frac{f(y)-f(x)}{y-x}$$

$$D_l f(x) := \lim_{y \uparrow x} \frac{f(y)-f(x)}{y-x}$$

for all x belonging to the interior of I. For the time being, suppose that I is an open interval and $f: I \to \mathbb{R}$ is convex. From the existence of $D_r f$ and $D_l f$ it follows that $\lim_{y \downarrow x} f(y) = f(x)$ and $\lim_{y \uparrow x} f(y) = f(x)$ for all $x \in I$.

Thus, f is continuous.

It is easy to see that $D_r f$ and $D_l f$ are increasing functions, that $D_l f \leqslant D_r f$ and that

$$D_r f(x) \leqslant \frac{f(x)-f(y)}{x-y} \leqslant D_l f(y) \qquad (x<y)$$

From this we obtain for each $c \in I$

$$\lim_{x \downarrow c} D_r f(x) \leqslant \lim_{y \downarrow c} D_l f(y) \qquad (\leqslant \lim_{y \downarrow c} D_r f(y))$$

Thus, for all $c \in I$, $\lim_{x \downarrow c} D_r f(x) = \lim_{x \downarrow c} D_l f(x)$. Similarly, $\lim_{x \uparrow c} D_r f(x) = \lim_{x \uparrow c} D_l f(x)$.

Let $S := \{x \in I : D_r f$ is discontinuous at $x\}$. If $x \notin S$, then $D_r f$ and $D_l f$ are continuous at x and $D_r f(x) = D_l f(x)$. Hence, f is differentiable at all points of I that are not in S. From Theorem 1.2 we infer that S is countable.

We collect these results in

LEMMA 2.3. *Let I be an open interval and let $f : I \to \mathbb{R}$ be convex. Then f is continuous. There are at most countably many points at which f is not differentiable.*

Exercise 2.C. Give an example of a convex function on (0, 1) that is not differentiable at $\frac{1}{2}, \frac{1}{3}, \frac{1}{4}, \ldots$

Exercise 2.D. Sketch graphs of convex functions f on (0, 1) for which
(i) $\lim_{x \downarrow 0} f(x) = \infty$, $\lim_{x \uparrow 1} f(x) = \infty$,
(ii) $\lim_{x \downarrow 0} f(x) = \infty$, $\lim_{x \uparrow 1} f(x) = 1$,
(iii) $\lim_{x \downarrow 0} f(x) = 1$, $\lim_{x \uparrow 1} f(x) = \infty$,
respectively.

**Exercise* 2.E. Prove that every convex function f: $(0, 1) \to \mathbb{R}$ is bounded below. (Hint. f is bounded on $[\frac{1}{4}, \frac{3}{4}]$. If $0 < x < \frac{1}{2}$, then one has the inequality $f(x) \geqslant 2f(\frac{1}{2}x + \frac{1}{2}\cdot\frac{3}{4}) - f(\frac{3}{4})$.)

Let f be a convex function defined on an open interval I. Take $x, y \in I$ with $x < y$. Both $D_r f$ and $D_l f$ are increasing, and hence Riemann integrable over the closed interval $[x, y]$. Is there any connection between $\int_x^y D_r f(t)\,dt$, $\int_x^y D_l f(t)\,dt$ and $f(y) - f(x)$?

For any partition $x = x_0 < x_1 < \ldots < x_n = y$ of $[x, y]$ we have

$$f(y) - f(x) = \sum_{i=1}^{n} (f(x_i) - f(x_{i-1})) = \sum_{i=1}^{n} \Phi_1 f(x_i, x_{i-1})(x_i - x_{i-1})$$

Now for each i,

$$D_r f(x_{i-1}) \leqslant \Phi_1 f(x_i, x_{i-1}) \leqslant D_l f(x_i)$$

whence

$$\sum_{i=1}^{n} D_r f(x_{i-1})(x_i - x_{i-1}) \leqslant f(y) - f(x) \leqslant \sum_{i=1}^{n} D_l f(x_i)(x_i - x_{i-1})$$

We can choose the $x_0, \ldots, x_n$ such that the Riemann sums in this formula are close to the lower integral of $D_r f$ and the upper integral of $D_l f$, respectively. As $D_r f$ and $D_l f$ are integrable, we find

$$\int_x^y D_r f(t)\, dt \leqslant f(y) - f(x) \leqslant \int_x^y D_l f(t)\, dt$$

But $D_l f \leqslant D_r f$, so

$$f(y) - f(x) = \int_x^y D_r f(t)\, dt = \int_x^y D_l f(t)\, dt$$

At this stage we may drop the condition $x < y$. We have proved the first half of the following theorem.

THEOREM 2.4. *Let I be an open interval and let $f: I \to \mathbb{R}$ be convex. Then there is an increasing function $\phi: I \to \mathbb{R}$ (e.g. $\phi = D_r f$) with*

$$f(y) - f(x) = \int_x^y \phi(t)\, dt \qquad (x, y \in I).$$

Conversely, if $\phi: I \to \mathbb{R}$ is an increasing function and if $a \in I$, then the function $x \mapsto \int_a^x \phi(t)\, dt$ $(x \in I)$ is convex.

Proof. We only need to prove the second statement. Let $F(x) := \int_a^x \phi(t)\, dt$ $(x \in I)$. If $x, y, z \in I$ and $x < y < z$, then

$$\Phi_1 F(y, x) = (y - x)^{-1} \int_x^y \phi(t)\, dt \leqslant \phi(y) \leqslant (z - y)^{-1} \int_y^z \phi(t)\, dt = \Phi_1 F(y, z)$$

Thus, $\Phi_2 F(x, y, z) \geqslant 0$. Since $\Phi_2 F$ is a symmetric function of its variables we obtain $\Phi_2 F \geqslant 0$. By Theorem 2.2, F is convex.

Exercise 2.F. (Another proof of Lemma 2.3) Let $f: (-1, 1) \to \mathbb{R}$ be convex. By Theorem 2.4 there is a monotone ϕ such that $f(x) = f(0) + \int_0^x \phi(t)\, dt$ for all x. Then f is differentiable at every point where ϕ is continuous (and at no other point).

**Exercise* 2.G. Let I be an open interval and let $f: I \to \mathbb{R}$ be convex. Then either f is monotone or there exists a $\xi \in I$ such that f is increasing on $\{x \in I : x \geqslant \xi\}$ and decreasing on $\{x \in I : x \leqslant \xi\}$.

**Exercise* 2.H. Let $f: (0, 1) \to \mathbb{R}$.

(i) Let f be differentiable. Then f is convex if and only if f' is increasing.

(ii) Let f be twice differentiable. Then f is convex if and only if $f'' \geqslant 0$.

So far we have considered convex functions on open intervals. What can we expect when starting with a convex function f, defined on a closed interval $[a, b]$? On (a, b) the situation has been described above, so we study the local behaviour of f at a and b. By Exercise 2.G, f is monotone on (a, c) for some $c>a$. We now show that f is bounded on $[a, b]$. (It will follow that $\lim_{x\downarrow a} f(x)$ exists.) For any $x \in [a, b]$, $x=\lambda a+(1-\lambda)b$ for some $\lambda \in [0, 1]$ and by the convexity of f, $f(x) \leqslant \lambda f(a)+(1-\lambda)f(b)$. Thus, f is bounded above. By Exercise 2.E, f is bounded below. We see that $\lim_{x\downarrow a} f(x)$ exists. Similarly, one proves the existence of $\lim_{x\uparrow b} f(x)$.

The function f need not be continuous (e.g. take $f(x)=x^2$ for $-1<x<1$, $f(1)=f(-1)=10$).

With all this, the reader will have no problems in proving

THEOREM 2.5. *Let $f:[a, b]\to\mathbb{R}$ be convex. Then f is continuous on (a, b) and has removable discontinuities at a and b. If $f:[a, b]\to\mathbb{R}$ is convex and continuous there is an increasing function ϕ on (a, b) that is (improperly) Riemann integrable over $[a, b]$ and such that for all $x, y \in [a, b]$*

$$f(y)=f(x)+\int_x^y \phi(t)\,\mathrm{d}t$$

Conversely, if $\phi:(a, b)\to\mathbb{R}$ is increasing and (improperly) Riemann integrable over $[a, b]$ then for every $x \in [a, b]$ the function

$$y \mapsto \int_x^y \phi(t)\,\mathrm{d}t \qquad (a \leqslant y \leqslant b)$$

is convex and continuous.

**Exercise* 2.I. Let f be a continuous function on an interval I such that

$$f(\tfrac{1}{2}(x+y)) \leqslant \tfrac{1}{2}(f(x)+f(y))$$

for all $x, y \in I$. Show that f is convex.

Exercise 2.J. Every straight line L in $\mathbb{R}^2$ determines two closed half planes, which we call the *sides* of L. If $a, b \in \mathbb{R}$ and if L is the graph of the function $x \mapsto ax+b$ $(x \in \mathbb{R})$, then these sides are the sets $\{(x, y) \in \mathbb{R}^2 : y \leqslant ax+b\}$ and $\{(x, y) \in \mathbb{R}^2 : y \geqslant ax+b\}$.

Now let f be a continuous function $\mathbb{R}\to\mathbb{R}$ with the property that for every point P of the graph Γ_f of f there exists a straight line L containing P and such that Γ_f lies entirely in one of the sides of L. Show that either f or $-f$ is convex. (We outline a proof: For all $c \in \mathbb{R}$, choose a straight line L_c containing $(c, f(c))$ and such that Γ_f lies in one side of L_c. None of the lines L_c can be vertical and one can define $A:=\{c \in \mathbb{R} : \Gamma_f$ lies above $L_c\}$, $B:=\{c \in \mathbb{R} : \Gamma_f$ lies below $L_c\}$. If $c \in A$ and $x<c<y$, then $\Phi_1 f(x, c) \leqslant \Phi_1 f(y, c)$, so that $\Phi_2 f(x, c, y) \geqslant 0$. Hence, if $A=\mathbb{R}$, then f is convex. Similarly, if $B=\mathbb{R}$, then $-f$ is convex. Now assume $A \neq \mathbb{R}$, $B \neq \mathbb{R}$. If $a \in A$ and $b \in B$,

then L_a and L_b are parallel. It follows that all the lines L_c are parallel. Then A and B are closed sets. Consequently, $A \cap B \neq \emptyset$ and Γ_f is itself a straight line. Then f and $-f$ are both convex.)

Exercise 2.K. Let $f:(-1, 1)\to\mathbb{R}$ be convex and let $\phi:(-1, 1)\to\mathbb{R}$ be an increasing function such that for all $x, y \in (-1, 1)$

$$f(y)-f(x)=\int_x^y \phi(t)\,dt$$

(i) Prove that $D_l f \leqslant \phi \leqslant D_r f$.
(ii) Show that $D_l f$ is left continuous (i.e. $D_l f(x)=\lim_{y\uparrow x} D_l f(y)$ for all $x \in (-1, 1)$) and, similarly, that $D_r f$ is right continuous. (Hint. Define $L(x):=\lim_{y\uparrow x} D_l f(y)$. Then L is increasing and $\int_x^y L(t)\,dt=\int_x^y D_l f(t)\,dt$ for all $x, y \in (-1, 1)$. Now use (i).)

3. Differences of monotone functions

The set of the monotone functions on $[a, b]$ is not a vector space. We shall show that the smallest vector space of functions on $[a, b]$ that contains all monotone functions is the set of all functions on $[a, b]$ whose graphs have finite length.

Let $P: a=x_0<x_1<\ldots<x_n=b$ be a partition of $[a, b]$. Then we define for a function $f:[a, b]\to\mathbb{R}$

$$L_P(f):=\sum_{i=1}^{n} \sqrt{\{(x_i-x_{i-1})^2+(f(x_i)-f(x_{i-1}))^2\}}$$

$L_P(f)$ is the length of the polygon with vertices $(x_i, f(x_i))$ $(0\leqslant i\leqslant n)$ and may serve as an approximation of the 'length' of the graph of f (see Fig. 6). In fact, if P' is a partition of $[a, b]$ that is finer than P, then $L_{P'}(f)\geqslant L_P(f)$. The *length of the graph* of f is, by definition,

$$L(f):=\sup\{L_P(f): P \text{ is a partition of } [a, b]\}$$

($L(f)$ may be ∞.)

Let $P: a=x_0<x_1<\ldots<x_n=b$ be a partition of $[a, b]$. Then, since for each i

$$\begin{aligned}|f(x_i)-f(x_{i-1})| &\leqslant \sqrt{\{(x_i-x_{i-1})^2+(f(x_i)-f(x_{i-1}))^2\}}\\ &\leqslant (x_i-x_{i-1})+|f(x_i)-f(x_{i-1})|\end{aligned}$$

we have

$$\sum_{i=1}^{n} |f(x_i)-f(x_{i-1})| \leqslant L_P(f) \leqslant (b-a)+\sum_{i=1}^{n} |f(x_i)-f(x_{i-1})|$$

It follows that the graph of f has finite length if and only if the so-called

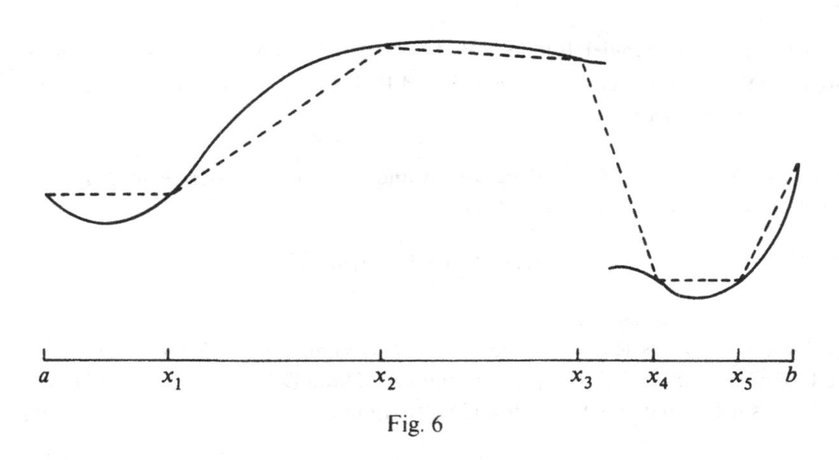

Fig. 6

total variation of f,

$$\operatorname{Var} f := \sup \left\{ \sum_{i=1}^{n} |f(x_i) - f(x_{i-1})| : \right.$$
$$\left. a = x_0 < \ldots < x_n = b \text{ is a partition of } [a, b] \right\}$$

is finite. (Sometimes we shall use $\operatorname{Var}_{[a,b]} f$ rather than $\operatorname{Var} f$.)

DEFINITION 3.1. An $f : [a, b] \to \mathbb{R}$ is called a *function of bounded variation* if its graph has finite length, or, equivalently, if its total variation, $\operatorname{Var} f$, is finite.
The set of all $f : [a, b] \to \mathbb{R}$ that are of bounded variation is called $\mathscr{BV}[a, b]$, or simply $\mathscr{BV}$.

Exercise 3.A. Let $f \in \mathscr{BV}[a, b]$ be continuous, let $\varepsilon > 0$. Then there is a $\delta > 0$ with the following property. If $a = x_0 < x_1 < \ldots < x_n = b$ is a partition of $[a, b]$ and if $x_i - x_{i-1} < \delta$ for every i, then

$$\sum_{i=1}^{n} |f(x_i) - f(x_{i-1})| \geq \operatorname{Var} f - \varepsilon$$

(Hint. Use the uniform continuity of f.)

**Exercise* 3.B. Let $f \in \mathscr{BV}[a, b]$. The function $T : x \mapsto \operatorname{Var}_{[a,x]} f$ $(x \in [a, b])$ (the so-called *indefinite variation* of f) is increasing. If $p, q \in [a, b]$ and if $p < q$, then $T(q) - T(p) = \operatorname{Var}_{[p,q]} f$.

Clearly, a monotone function $f : [a, b] \to \mathbb{R}$ is of bounded variation and $\operatorname{Var} f = |f(b) - f(a)|$. We have

THEOREM 3.2. *$\mathscr{BV}$ is the vector space generated by the set of monotone functions. In fact, every $f \in \mathscr{BV}$ is a difference of two increasing functions.*

Proof. Let $f, g : [a, b] \to \mathbb{R}$, $\lambda \in \mathbb{R}$, $\lambda \neq 0$. Then $\operatorname{Var}(\lambda f) = |\lambda| \operatorname{Var} f$ and $\operatorname{Var}(f+g) \leqslant \operatorname{Var} f + \operatorname{Var} g$. It follows that $\mathscr{BV}$ is a vector space (which contains all monotone functions). Now let $f \in \mathscr{BV}$. Then for $a \leqslant x \leqslant b$, f is also of bounded variation on $[a, x]$. Let $T(x)$ be the total variation of f on $[a, x]$. Then T is an increasing function on $[a, b]$ (Exercise 3.B). We have $f = T - (T - f)$. We show now that $T - f$ is also increasing. Let $p, q \in [a, b]$, $p < q$. By Exercise 3.B,

$$f(q) - f(p) \leqslant \operatorname*{Var}_{[p,q]} f \leqslant T(q) - T(p)$$

or $T(p) - f(p) \leqslant T(q) - f(q)$, and the theorem is proved.

**Exercise* 3.C. Let $f : [0, 1] \to \mathbb{R}$ be differentiable with bounded derivative. Then $f \in \mathscr{BV}$. (Corollary. Functions with continuous derivatives are of bounded variation.)

**Exercise* 3.D. Let $f : [0, 1] \to \mathbb{R}$ be differentiable and let f' be Riemann integrable over $[0, 1]$. (This condition is not redundant, as we have seen in the Introduction.) Then $f \in \mathscr{BV}$ and

$$\operatorname{Var} f = \int_0^1 |f'(t)| \, dt$$

(Corollary. If f' is continuous, then T is differentiable, $T' = |f'|$.) The function $h : [0, 1] \to \mathbb{R}$ defined via $h(0) = 0$, $h(x) = x^2 \sin x^{-3/2}$ if $x \neq 0$, is differentiable and of bounded variation, but h' is unbounded. (Compare Exercises 3.C and 4.J.)

Exercise 3.E. Let $f : [0, 1] \to \mathbb{R}$ be differentiable and let $\sqrt{(1 + (f')^2)}$ be Riemann integrable over $[0, 1]$. (In fact, this condition is equivalent to Riemann integrability of f'; see Exercise 12.L.) Then $f \in \mathscr{BV}$ and the length of its graph is

$$\int_0^1 \sqrt{(1 + f'(t)^2)} \, dt$$

Exercise 3.F. Define $g : [0, 1] \to \mathbb{R}$ as follows. $g(0) := 0$ and if $x \neq 0$, then $g(x) := x \sin x^{-1}$. Then g is not of bounded variation. The function $x \mapsto g(x^2)$ is differentiable but not in $\mathscr{BV}$. (Compare Exercise 3.C.) The function g^2 is of bounded variation.

Exercise 3.G. Let $f, g : [0, 1] \to \mathbb{R}$. Let $f(x)$ and $g(x)$ differ for only finitely many points x. Then $f \in \mathscr{BV}$ if and only if $g \in \mathscr{BV}$.

Exercise 3.H. A function $f : [a, b] \to \mathbb{R}$ is said to be *piecewise monotone* if there exists a partition $a = x_0 < x_1 < \ldots < x_n = b$ of $[a, b]$ such that f is monotone on $[x_{i-1}, x_i]$ for each i. A piecewise monotone function is of bounded variation. (Corollary. A convex function defined on a closed bounded interval is of bounded variation.) If f is differentiable and if f' is piecewise monotone, then so is f.

(In Section 13 we shall discover the existence of a differentiable function of bounded variation that is monotone on no interval. Thus, a difference of two monotone functions may fail to have any monotony property.)

Exercise 3.I. Let $f: [0, 1] \to \mathbb{R}$ be Riemann integrable. Then the function $x \mapsto \int_0^x f(t)\,dt$ is continuous and of bounded variation.

Exercise 3.J. Let $f, g \in \mathscr{BV}$. Then fg, $f \vee g$ and $f \wedge g$ are elements of $\mathscr{BV}$. (Comment. $\mathscr{BV}$ is a vector lattice (Riesz space) and an algebra.)

Exercise 3.K. If $f: [0, 1] \to \mathbb{R}$ is Darboux continuous and of bounded variation, then f is continuous. (Hint. $f(x+)$ and $f(x-)$ exist for every $x \in [0, 1]$.)

Exercise 3.L. Write the sine function, restricted to $[0, 15]$, in two different ways as the difference of two increasing functions.

Exercise 3.M. Let $f \in \mathscr{BV}[a, b]$ and let T be its indefinite variation (see Exercise 3.B). Show that T is the smallest among the functions S on $[a, b]$ that have the following properties.
(i) $S(a) = 0$.
(ii) Both $S - f$ and $S + f$ are increasing.

Exercise 3.N. Let $f: [0, 1] \to \mathbb{R}$ be of bounded variation and let T be its indefinite variation. Prove the following.
(i) If f is continuous, then so is T, and vice versa.
(ii) If f has a continuous derivative, then so does T and $T' = |f'|$. (See Exercise 3.D.)
(Comment on (i). A continuous function of bounded variation is the difference of two continuous increasing functions.)

The next four exercises are linked together by the following idea. A function $f: [a, b] \to \mathbb{R}$ is called a *step function* if there exists a partition $a = x_0 < x_1 < \ldots < x_n = b$ of $[a, b]$ such that f is constant on each open subinterval (x_{i-1}, x_i). The collection of the step functions may serve as a starting point for the definition of an integral for certain functions $[a, b] \to \mathbb{R}$, closely related to the Riemann integral. Roughly, the construction is as follows.

First, one defines the integral of a step function in the obvious way. Then by continuity one defines the integral for functions that are limits of uniformly convergent sequences of step functions. These functions are said to be *Cauchy–Bourbaki integrable.*

It is the purpose of Exercises 3.O, 3.P and 3.Q to describe these Cauchy–Bourbaki integrable functions. Together, they yield Theorem 3.3.

THEOREM 3.3. *The following conditions on $f: [a, b] \to \mathbb{R}$ are equivalent.*
(α) *f is the uniform limit of a sequence of step functions.*
(β) *f is the uniform limit of a sequence of functions of bounded variation.*
(γ) *For each $x \in [a, b]$, $f(x+)$ and $f(x-)$ exist.*
(It follows that an f satisfying (γ) is bounded.)
The proof is furnished by 3.O, 3.P and 3.Q. ((α)$\Leftrightarrow$(β) by 3.O and 3.P, (γ)$\Rightarrow$(α) by 3.Q. The proof of (β)$\Rightarrow$(γ) is standard.)

**Exercise* 3.O. Let $f: [a, b] \to \mathbb{R}$ be increasing. Then there is a sequence $f_1, f_2, \ldots$ of step functions such that $\lim_{n\to\infty} f_n = f$ uniformly. (Hint. Take $f_n(x) := (1/n)[nf(x)]$ ($a \leqslant x \leqslant b$, $n \in \mathbb{N}$) where [ ] refers to the entire part function. (The subsequence $f_1, f_2, f_4, f_8, \ldots$ is even increasing.))

**Exercise* 3.P. A step function is of bounded variation.

**Exercise* 3.Q. Let $f: [a, b] \to \mathbb{R}$ and suppose that $f(x+)$ and $f(x-)$ exist for all $x \in [a, b]$. Let $\varepsilon > 0$. Then there is a step function t on $[a, b]$ such that $|f(x) - t(x)| < \varepsilon$ for all $x \in [a, b]$. (Hint. Define $A := \{x \in [a, b]$: there is a step function s on $[a, x]$ such that $|f(y) - s(y)| < \varepsilon$ for each $y \in [a, x]\}$. Show that A is an interval, that $\sup A \in A$ and finally that $\sup A = b$.)

Exercise 3.R. Find a uniformly convergent sequence of elements of $\mathscr{BV}$ whose limit is not in $\mathscr{BV}$. (The space of the Cauchy–Bourbaki integrable functions properly contains $\mathscr{BV}$.)

Exercise 3.S. Let $f: [a, b] \to [a, b]$ and $g: [a, b] \to \mathbb{R}$. Suppose that f and g are of bounded variation. Is $g \circ f$ necessarily of bounded variation?

Exercise 3.T. Let $f: [a, b] \to [c, d]$ be continuous. Assume that for every $y \in [c, d]$ the set $f^{-1}(\{y\})$ has at most 37 elements. Show that then Var $f \leqslant 37(d-c)$. Hint. Let $a = a_0 < a_1 < \ldots < a_m = b$ be a partition of $[a, b]$. For each $i \in \{1, \ldots, m\}$ let I_i be the interval (a_{i-1}, a_i) : then $f(I_i)$ is again an interval. Now show that

$$\sum_{i=1}^{m} |f(a_i) - f(a_{i-1})| \leqslant \sum_{i=1}^{m} \int_c^d \xi_{f(I_i)}(y)\,dy = \int_c^d \sum_{i=1}^{m} \xi_{f(I_i)}(y)\,dy \leqslant 37(d-c)$$

Exercise 3.U. If $f: [a, b] \to \mathbb{R}$ is continuous, then f and $|f|$ have the same total variation.

4. Differentiability of monotone functions

Is there any guarantee that every monotone function is somewhere differentiable? A theorem of the type of Theorem 1.2 might say that a monotone function is differentiable except at points belonging to a set E which is 'small' in some sense. Differentiability is a stronger property than continuity, so all countable sets would have to be 'small'. Theorem 4.10 will show that the following notion of 'smallness' suits our purpose.

DEFINITION 4.1 Let $E \subset \mathbb{R}$. We call E a *null set* (or a *negligible set*) if for every $\varepsilon > 0$ there exist intervals $I_1, I_2, \ldots$ covering E and such that the sum of their lengths is at most ε. To avoid laborious circumlocutions, we call $\Sigma_{i=1}^{\infty} L(I_i)$ the *total length* of the intervals $I_1, I_2, \ldots$, without requiring these intervals to be pairwise disjoint. (The length of an interval I is denoted by $L(I)$.)

**Exercise* 4.A. Every countable subset of $\mathbb{R}$ is a null set. (In Example 4.2 we shall present an uncountable null set.) Every subset of a null set is a null set. Unions of countably many null sets are null sets.

**Exercise* 4.B. The interval $[0, 1]$ is not a null set. In fact, if $I_1, I_2, \ldots$ are intervals such that $\Sigma_{i=1}^{\infty} L(I_i) < 1$, then they do not cover $[0, 1]$. (Suppose they do; use the Heine–Borel theorem to derive a contradiction.)

**Exercise* 4.C. Let $E \subset \mathbb{R}$. We say that E has *zero content* if for every $\varepsilon > 0$ E can be covered by finitely many intervals whose total length is less than ε.

(i) Prove that E has zero content if and only if there exists a bounded closed interval $[a, b]$, containing E, such that ξ_E is Riemann integrable over $[a, b]$ and its integral is 0. (In general, for a subset E of a closed interval $[a, b]$ one may define the *content* of E to be the integral of ξ_E when ξ_E is Riemann integrable over $[a, b]$.)

(ii) Prove that a set has zero content if and only if its closure is a bounded null set. (Apply the Heine–Borel covering theorem.)

(iii) Give an example of a bounded null set that does not have zero content.

(iv) Let $f: [a, b] \to [0, \infty)$ be Riemann integrable and assume that its integral is 0. Prove that $\{x : f(x) \neq 0\}$ is a null set. (Show that for every $\varepsilon > 0$ the set $\{x : f(x) \geqslant \varepsilon\}$ has zero content.)

EXAMPLE 4.2. (The Cantor set) We construct an uncountable null set. This is done by deleting certain intervals from $[0, 1]$. First, divide $[0, 1]$ into three equal parts and remove the middle interval

$$I_{11} := (\tfrac{1}{3}, \tfrac{2}{3})$$

(More generally, given an interval $[a, b]$, we call $(a + \frac{1}{3}(b-a), b - \frac{1}{3}(b-a))$ the *middle interval* of $[a, b]$.) Divide the remaining two closed intervals $[0, \frac{1}{3}]$ and $[\frac{2}{3}, 1]$ again into three equal parts and remove the middle intervals, viz.

$$I_{12} := (\tfrac{1}{9}, \tfrac{2}{9}) \quad \text{and} \quad I_{22} := (\tfrac{7}{9}, \tfrac{8}{9})$$

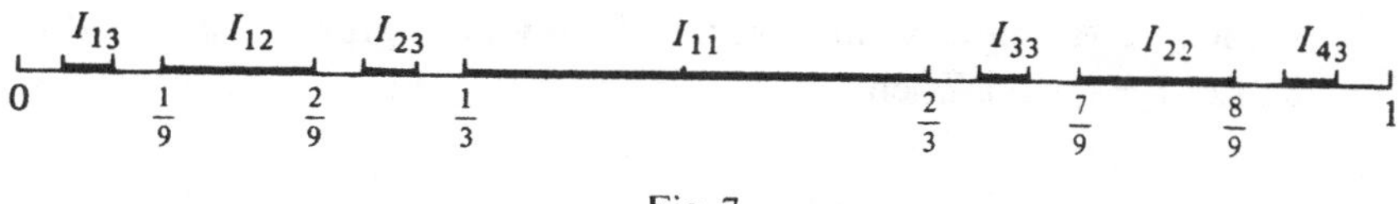

Fig. 7

(See Fig. 7.) Continuing in this way, after $n-1$ steps we have deleted $1+2+\ldots+2^{n-2}=2^{n-1}-1$ disjoint open intervals I_{jk} $(1\leqslant k\leqslant n-1,$ $1\leqslant j\leqslant 2^{k-1})$ and we have left 2^{n-1} closed intervals, each of length $3^{-(n-1)}$. The nth step then consists of leaving out the middle interval of each of these closed intervals $I_{1n},\ldots,I_{2^{n-1}n}$, thus obtaining 2^n closed intervals of length 3^{-n}. We define $\mathbb{D}$, *the Cantor set*, by

$$\mathbb{D}:=[0,1]\setminus\bigcup_{k=1}^{\infty}\bigcup_{j=1}^{2^{k-1}}I_{jk}$$

(the symbol $\setminus$ denoting complementation). Obviously, $\mathbb{D}$ is closed. For each n, $\mathbb{D}$ is contained in

$$[0,1]\setminus\bigcup_{k=1}^{n}\bigcup_{j=1}^{2^{k-1}}I_{jk}$$

which is a union of 2^n intervals with total length $(\frac{2}{3})^n$. Therefore, $\mathbb{D}$ is a null set. On the other hand, $\mathbb{D}$ is uncountable. This will follow from Exercise 4.D.

**Exercise* 4.D. An element of $[0, 1]$ belongs to $\mathbb{D}$ if and only if in base 3 it has a representation $0.x_1x_2\ldots$ with $x_i\in\{0,2\}$ for every i. Every element of $\mathbb{D}$ has exactly one such representation.

By Theorem 1.2 a monotone function has only countably many discontinuities. We prove now that a monotone function may be non-differentiable at uncountably many points. In fact, we prove much more:

THEOREM 4.3. *Let $E\subset\mathbb{R}$ be a null set. Then there exists a continuous increasing function on $\mathbb{R}$ that is differentiable at no point of E.*

Proof. For each $n\in\mathbb{N}$ we can cover E with a countable system S_n of open intervals having total length less than 2^{-n}. Then $S:=S_1\cup S_2\cup\ldots$ is also a countable system of open intervals $I_1, I_2,\ldots$ covering E. Its total length is less than 1. It is easy to see that our system S has the following property.

Let $x\in E$, let U be a neighbourhood of x and let $n\in\mathbb{N}$. Then there exist $i_1,\ldots,i_n$ with $i_1<i_2<\ldots<i_n$ and such that $x\in I_{i_k}\subset U$ for each $k\in\{1,\ldots,n\}$.

('Each element of E lies in infinitely many intervals I_i of arbitrarily small length.')

For each $i \in \mathbb{N}$ define a function f_i as follows. (Its graph is sketched in Fig. 8.) Let $I_i = (p_i, q_i)$. Then

$$f_i(x) := \begin{cases} 0 & \text{if } x < p_i \\ x - p_i & \text{if } p_i \leqslant x \leqslant q_i \\ q_i - p_i & \text{if } x > q_i \end{cases}$$

Clearly, f_i is continuous and increasing, and $0 \leqslant f_i(x) \leqslant q_i - p_i = L(I_i)$ for each $x \in \mathbb{R}$.

Now set $f := \sum_{i=1}^{\infty} f_i$. For all x and n,

$$0 \leqslant f(x) - \sum_{i=1}^{n} f_i(x) \leqslant \sum_{i>n} L(I_i)$$

Thus the convergence is uniform and f is continuous. Of course, f is increasing.

Let $x \in E$. We show that f is not differentiable at x. Let U be a neighbourhood of x. We are done if we can show that for each $n \in \mathbb{N}$ there exists $y \in U$ such that $(f(y) - f(x))/(y - x) \geqslant n$. To this end, choose $i_1, \ldots, i_n$ with $i_1 < i_2 < \ldots < i_n$ and with $x \in I_{i_k} \subset U$ for each $k \in \{1, \ldots, n\}$. Take $y \in I_{i_1} \cap \ldots \cap I_{i_n}$, $y \neq x$. Since the difference quotients of every f_i are nonnegative, we have

$$\frac{f(x) - f(y)}{x - y} \geqslant \sum_{k=1}^{n} \frac{f_{i_k}(x) - f_{i_k}(y)}{x - y} = \sum_{k=1}^{n} 1 = n$$

and the theorem is proved.

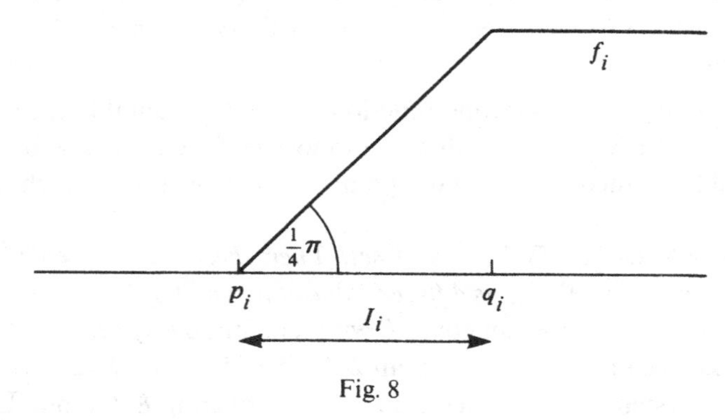

Fig. 8

Exercise 4.E. Let $E \subset \mathbb{R}$ be a null set. Then there exists a strictly increasing continuous function that is differentiable at no point of E.

Theorem 4.3 describes the worst possible situation. Indeed, we shall prove a famous theorem due to Lebesgue stating that the set of points where a monotone function is not differentiable is a null set.

In this connection we introduce some terminology. For $f:[0, 1]\to\mathbb{R}$, instead of

$$\{x\in[0, 1]: f \text{ is not differentiable at } x\} \text{ is a null set}$$

we often say

$$f \text{ is differentiable almost everywhere on } [0, 1]$$

In general, let $X\subset\mathbb{R}$. If for some of the elements x of X a statement $P(x)$ is defined, then we say that P holds *almost everywhere* on X (or that $P(x)$ holds for *almost every* $x\in X$) if the set of all points x for which $P(x)$ is either undefined or false is a null set. We use a.e. as an abbreviation for 'almost every(where)'. Typical expressions of this kind are: 'almost every real number is irrational'; 'a monotone function on $[0, 1]$ is continuous a.e. on $[0, 1]$'; 'if f is a Riemann integrable function on $[0, 1]$, $f\geqslant 0$ and $\int_0^1 f(t)\,dt=0$, then $f=0$ a.e. on $[0, 1]$'; '$\xi_{\mathbb{D}}\leqslant 0$ a.e.'; '$\xi_{\mathbb{D}}=1$ a.e. on $\mathbb{D}$'; '$\xi_{\mathbb{D}}=0$ a.e. on $\mathbb{D}$'.

Before we can start with the actual proof of Lebesgue's theorem we need some preliminaries.

DEFINITION 4.4. Let $a\in\mathbb{R}$. If g is a function defined on some interval $(a, a+\varepsilon)$, we set $\overline{\lim}_{y\downarrow a}g(y):=\inf_{0<\delta<\varepsilon}(\sup\{g(y): a<y<a+\delta\})$, admitting the values ∞ and $-\infty$. In a similar way we define $\underline{\lim}_{y\downarrow a}g(y)$, and, for a function h that is defined on an interval $(a-\varepsilon, a)$, the expressions $\overline{\lim}_{y\uparrow a}h(y)$ and $\underline{\lim}_{y\uparrow a}h(y)$.

Now let I be an interval and let $f: I\to\mathbb{R}$. The four *Dini derivatives* of f are the mappings D_r^+f, D_r^-f, D_l^+f and D_l^-f, with values in the set $\mathbb{R}\cup\{\infty\}\cup\{-\infty\}$ and defined in the interior of I by the following formulas.

$$D_r^+f(x):=\overline{\lim_{y\downarrow x}}\frac{f(y)-f(x)}{y-x}\qquad D_r^-f(x):=\underline{\lim_{y\downarrow x}}\frac{f(y)-f(x)}{y-x}$$

$$D_l^+f(x):=\overline{\lim_{y\uparrow x}}\frac{f(y)-f(y)}{y-x}\qquad D_l^-f(x):=\underline{\lim_{y\uparrow x}}\frac{f(y)-f(x)}{y-x}$$

Clearly, a function $f: I\to\mathbb{R}$ is differentiable at an interior point x of I if and only if all Dini derivatives at x are equal and finite. (The relations between the Dini derivatives and $D_r f$, $D_l f$ are also evident.) For any f we obviously have $D_r^-f\leqslant D_r^+f$ and $D_l^-f\leqslant D_l^+f$. If f is increasing, then the four Dini derivatives are nonnegative.

4.5. Consider the following two statements:

(α) *Every monotone function defined on a bounded closed interval I is differentiable almost everywhere on I.*

(β) *If f is an increasing function on a bounded closed interval I, then $D_l^- f \geqslant D_r^+ f$ a.e. on I and $D_r^+ f < \infty$ a.e. on I.*

(α) is the theorem of Lebesgue which we want to prove. To this end it suffices to show that (β) is valid. Indeed, for the moment let us suppose (β) is true. Let $f : [a, b] \to \mathbb{R}$ be monotone. We may assume that f is increasing. The function $g : x \mapsto -f(-x)$, defined on $[-b, -a]$, is increasing. From the definition of the Dini derivatives it follows at once that

$$D_l^- g(x) = D_r^- f(-x) \qquad (-b < x < -a)$$
$$D_r^+ g(x) = D_l^+ f(-x) \qquad (-b < x < -a)$$

Applying (β) to the function g we see that $D_r^- f(-x) \geqslant D_l^+ f(-x)$ for almost every $x \in (-b, -a)$, i.e. $D_r^- f \geqslant D_l^+ f$ a.e. on $[a, b]$. Of course, (β) yields $D_l^- f \geqslant D_r^+ f$ a.e. on $[a, b]$. Combining these two inequalities with the two trivial ones we obtain (the union of finitely many null sets being a null set)

$$D_l^- f \leqslant D_l^+ f \leqslant D_r^- f \leqslant D_r^+ f \leqslant D_l^- f \text{ a.e. on } [a, b]$$

which implies that the four Dini derivatives are equal a.e. Since one of them is finite a.e. (observe that $D_r^+ f \geqslant 0$), they all are. Thus, f is differentiable a.e.

Hence, in order to prove Lebesgue's theorem it suffices to have (β). To get an idea of the technique that we need to prove (β), let us start with an increasing $f : [a, b] \to \mathbb{R}$ that is continuous. If we want to show, for example, that $D_r^+ f < \infty$ a.e. it seems wise to consider the set

$$A := \left\{ x \in (a, b) : D_r^+ f(x) = \overline{\lim_{y \downarrow x}} \frac{f(y) - f(x)}{y - x} = \infty \right\}$$

For each $c > 0$, A is contained in

$$A_c := \left\{ x \in (a, b) : \frac{f(y) - f(x)}{y - x} > c \text{ for some } y > x \right\}$$

$$(*) \qquad = \{x \in (a, b) : f(y) - cy > f(x) - cx \text{ for some } y > x\}$$

A_c is open. We are done if we can show that $\lim_{c \to \infty} L(A_c) = 0$, where $L(A_c)$ is the sum of the lengths of the components of A_c.

The proof is based upon the following lemma.

LEMMA 4.6. (The lemma of the rising sun) *Let $g : [a, b] \to \mathbb{R}$ be continuous. Then the set*

$$E := \{x \in (a, b) : g(\xi) > g(x) \text{ for some } \xi > x\}$$

is open. If (α, β) is a component of E then $g(\alpha) \leqslant g(\beta)$. (In fact, we have $g(\alpha) = g(\beta)$ except possibly if $\alpha = a$.)

Proof. It is clear that E is open. Let (α, β) be a component of E. Let $x \in (\alpha, \beta)$.

We prove that $g(x) \leqslant g(\beta)$. (It will follow that $g(\alpha) = g(\alpha+) \leqslant g(\beta)$.) The set

$$V := \{t : x \leqslant t \leqslant \beta,\ g(t) \geqslant g(x)\}$$

has a largest element, x_1, say. It is enough to show that $x_1 = \beta$. Let us suppose $x_1 < \beta$. Since $x_1 \in V$ and $\beta \notin V$, we then have

$$(1) \qquad g(\beta) < g(x) \leqslant g(x_1)$$

Since $\alpha < x_1 < \beta$ we have $x_1 \in E$, which means that there is a $\xi > x_1$ such that

$$(2) \qquad g(\xi) > g(x_1)$$

Now we distinguish two possibilities for ξ.

(*a*) $x_1 < \xi \leqslant \beta$. Then, since $\xi \notin V$, $g(\xi) < g(x)$. Together with (1) and (2) this leads to $g(x_1) < g(x_1)$, an absurdity.

(*b*) $\xi > \beta$. Since $\beta \notin E$ and $\xi > \beta$ we must have $g(\xi) \leqslant g(\beta)$, which, combined with (1) and (2), leads again to $g(x_1) < g(x_1)$.

So our assumption that $x_1 < \beta$ is false and we infer that $x_1 = \beta$. (The reader may prove that $g(\alpha) = g(\beta)$ if $\alpha \neq a$.)

Exercise 4.F. We indicate another proof of Lemma 4.6. Let $g_\odot$ be 'the rising sun function' (see Exercise 1.G) of g, i.e the smallest decreasing function on $[a, b]$ that is $\geqslant g$. Then (Exercise 1.G) $g_\odot$ is continuous. With E as in Lemma 4.6 we have $E = \{x \in (a, b) : g_\odot(x) > g(x)\}$. If (α, β) is a component of E, then $g_\odot$ is constant on (α, β), $g_\odot(\beta) = g(\beta)$ and, if $\alpha \neq a$, then $g_\odot(\alpha) = g(\alpha)$.

Exercise 4.G. (We put the sun at the negative x-axis.) Let $g : [a, b] \to \mathbb{R}$ be continuous. Then

$$F := \{x \in (a, b) : g(\xi) > g(x) \text{ for some } \xi < x\}$$

is open. If (α, β) is a component of F, then $g(\alpha) \geqslant g(\beta)$. (In fact, $g(\alpha) = g(\beta)$ except possibly if $\beta = b$.)

For a continuous increasing function $f : [a, b] \to \mathbb{R}$ the set A_c (see formula (*)) is

$$\{x \in (a, b) : f(\xi) - c\xi > f(x) - cx \text{ for some } \xi > x\}$$

so we may apply Lemma 4.6 to the function $x \mapsto f(x) - cx$. For every component (α, β) of A_c we then obtain $f(\alpha) - c\alpha \leqslant f(\beta) - c\beta$, i.e. $\beta - \alpha \leqslant (1/c)(f(\beta) - f(\alpha))$. It is now easy to prove that $L(A_c) \leqslant (1/c)(f(b) - f(a))$. Therefore, $\lim_{c\to\infty} L(A_c) = 0$, which is what we were trying to prove.

But this reasoning depends on the continuity of f. We want to prove Lebesgue's theorem without any continuity assumption. Thus, we need a slightly refined version of Lemma 4.6:

LEMMA 4.7. *Let $g : [a, b] \to \mathbb{R}$ be such that $g(x-)$ and $g(x+)$ exist for all $x \in [a, b]$ and such that*

$$g(x-) \leqslant g(x) \leqslant g(x+) \qquad (x \in [a, b])$$

Then the sets $E := \{x \in (a, b) : g(\xi) > g(x+)$ *for some* $\xi > x\}$ *and* $F := \{x \in (a, b) : g(\xi) > g(x-)$ *for some* $\xi < x\}$ *are open.*

If (α, β) *is a component of* E, *then* $g(\alpha+) \leqslant g(\beta+)$.

If (γ, δ) *is a component of* F, *then* $g(\gamma-) \geqslant g(\delta-)$.

Proof. We prove the statement about E, leaving the rest to the reader. Let $x \in E$. Then there is a $\xi > x$ such that $g(\xi) > g(x+)$. Set $\delta := \frac{1}{2}(g(\xi) - g(x+))$. As $g(x-) \leqslant g(x+)$ and $g(x) \leqslant g(x+)$, for all z in some open neighbourhood of x we have $g(z) \leqslant g(x+) + \delta$. It follows that for all z in that neighbourhood we also have $g(z+) \leqslant g(x+) + \delta$ and, consequently, $g(z+) < g(\xi)$, whence $z \in E$. We infer that E is open.

Further, notice that $E = \{x \in (a, b) :$ there is a $\xi > x$ such that $g(\xi+) > g(x+)\}$. Let (α, β) be a component of E and let $\alpha < x < \beta$. We prove $g(x+) \leqslant g(\beta+)$. The set $V := \{t : x \leqslant t \leqslant \beta,\ g(t+) \geqslant g(x+)\}$ has a largest element, x_1. (If $t_1 < t_2 < \ldots$ are in V and $\lim_{n\to\infty} t_n = t$, then $g(t-) = \lim g(t_n+) \geqslant g(x+)$, so certainly $g(t+) \geqslant g(x+)$, whence $t \in V$.) For the proof of the identity $x_1 = \beta$ we can copy the corresponding part of the proof of Lemma 4.6, replacing g by the function $x \mapsto g(x+)$.

LEMMA 4.8. *Let* $f : [a, b] \to \mathbb{R}$ *be increasing. Then*

$$A := \{x \in (a, b) : D_r^+ f(x) = \infty\}$$

is a null set.

Proof. Let $B := \{x \in (a, b) : f$ is continuous at $x\}$. Since $[a, b] \setminus B$ is countable, hence a null set, it suffices to show that $B \cap A$ is a null set. For every $c > 0$ we have

$$B \cap A \subset \left\{x \in (a, b) : \frac{f(y) - f(x+)}{y - x} > c \text{ for some } y > x\right\}$$

(with $g(x) := f(x) - cx$)

$$= \{x \in (a, b) : g(y) > g(x+) \text{ for some } y > x\}$$

By Lemma 4.7 the latter set is the union of countably many pairwise disjoint intervals $(\alpha_1, \beta_1), (\alpha_2, \beta_2), \ldots$ such that $g(\alpha_n+) \leqslant g(\beta_n+)$ for each n. Hence, $B \cap A$ is contained in the union of these (α_n, β_n) while for each n we have $f(\beta_n+) - f(\alpha_n+) \geqslant c(\beta_n - \alpha_n)$. Since f is increasing,

$$\sum_n (\beta_n - \alpha_n) \leqslant \frac{1}{c} \sum_n (f(\beta_n+) - f(\alpha_n+)) \leqslant \frac{1}{c}(f(b) - f(a))$$

By taking c large enough we can make the sum of the lengths of the intervals as small as we wish. Consequently, $B \cap A$ is a null set.

LEMMA 4.9. *Let* $0 < c < C$. *Then for every increasing* $f : [a, b] \to \mathbb{R}$,

$$V_f := \{x \in (a,b) : D_l^- f(x) < c < C < D_r^+ f(x)\}$$

is a null set.

Proof. As in the proof of Lemma 4.8, we only look at $B \cap V_f$ and prove this set to be a null set.

$$B \cap V_f \subset \left\{x \in (a,b) : \frac{f(y) - f(x-)}{y-x} < c \text{ for some } y < x\right\}$$
$$= \{x \in (a,b) : f(y) - cy > f(x-) - cx \text{ for some } y < x\}$$

We apply Lemma 4.7. $B \cap V_f$ is contained in the union of pairwise disjoint intervals (α_i, β_i) such that $f(\alpha_i-) - c\alpha_i \geqslant f(\beta_i-) - c\beta_i$ for each i, hence $f(\beta_i-) - f(\alpha_i-) \leqslant c(\beta_i - \alpha_i)$ for each i.

For every i, let f_i be the restriction of f to $[\alpha_i, \beta_i]$. Then

$$B \cap V_f \cap (\alpha_i, \beta_i) \subset \left\{x \in (\alpha_i, \beta_i) : \frac{f(y) - f(x+)}{y-x} > C \text{ for some } y \in (x, \beta_i)\right\}$$
$$= \{x \in (\alpha_i, \beta_i) : f_i(y) - Cy > f_i(x+) - Cx \text{ for some } y \in (x, \beta_i)\}$$

Applying 4.7 again, we see that $B \cap V_f \cap (\alpha_i, \beta_i)$ is contained in a union of intervals $(\alpha_{i1}, \beta_{i1}), (\alpha_{i2}, \beta_{i2}), \ldots$ (subsets of (α_i, β_i)) such that

$$f_i(\alpha_{ij}+) - C\alpha_{ij} \leqslant f_i(\beta_{ij}+) - C\beta_{ij} \text{ for each } j \in \mathbb{N}$$

The intervals $(\alpha_{ij}, \beta_{ij})$ $(i, j \in \mathbb{N})$ are mutually disjoint and they cover $B \cap V_f$. We estimate the sum of their lengths:

$$\sum_{i,j} (\beta_{ij} - \alpha_{ij}) \leqslant \frac{1}{C} \sum_{i,j} (f_i(\beta_{ij}+) - f_i(\alpha_{ij}+)) \leqslant \frac{1}{C} \sum_i (f(\beta_i-) - f(\alpha_i+))$$
$$\leqslant \frac{1}{C} \sum_i (f(\beta_i-) - f(\alpha_i-)) \leqslant \frac{c}{C} \sum_i (\beta_i - \alpha_i) \leqslant \frac{c}{C}(b-a)$$

So far we have shown: *If f is an increasing function on $[a, b]$, then $B \cap V_f$ is contained in the union of countably many pairwise disjoint intervals $I_1, I_2, \ldots$ for which*

$$\sum_i L(I_i) \leqslant \frac{c}{C}(b-a)$$

For every $k \in \mathbb{N}$ we apply this statement to the function g_k, which is the restriction of f to the closure of I_k. We find that $B \cap V_{g_k}$ is contained in the union of countably many disjoint intervals whose total length is at most $(c/C)L(I_k)$. Now $B \cap V_f \subset \bigcup_k (B \cap V_{g_k})$, so $B \cap V_f$ is contained in a union of countably many intervals $J_1, J_2, \ldots$ such that

$$\sum_i L(J_i) \leqslant \sum_k \frac{c}{C} L(I_k) \leqslant \left(\frac{c}{C}\right)^2 (b-a)$$

And so on. We see that V_f is a null set.

THEOREM 4.10 (Lebesgue) *A monotone function $f: [a, b] \to \mathbb{R}$ is differentiable almost everywhere on $[a, b]$.*
Proof. Let f be increasing. The set

$$\{x \in (a, b) : D_l^- f(x) < D_r^+ f(x)\}$$

is contained in

$$\bigcup_{c,C \in \mathbb{Q}} \{x \in (a, b) : D_l^- f(x) < c < C < D_r^+ f(x)\}$$

which is a countable union of null sets (Lemma 4.9), hence a null set. Thus, $D_l^- f \geqslant D_r^+ f$ a.e. Together with Lemma 4.8 this implies statement (β) of 4.5. The theorem follows.

COROLLARY 4.11. *Functions of bounded variation are differentiable almost everywhere.*
Proof. Theorems 3.2 and 4.10.

COROLLARY 4.12. *A piecewise monotone function is differentiable almost everywhere.*
Proof. Obvious. (Observe that Exercise 3.H shows that a piecewise monotone function is of bounded variation.)

A function f, defined on a subset X of $\mathbb{R}$, is said to *satisfy a Lipschitz condition* if there exists a number C such that

$$|f(x) - f(y)| \leqslant C|x - y| \qquad (x, y \in X)$$

Such a C is called a *Lipschitz constant* for f.

COROLLARY 4.13. *If $f: [a, b] \to \mathbb{R}$ satisfies a Lipschitz condition, then f is differentiable almost everywhere on $[a, b]$.*
Proof. Let C be a Lipschitz constant for f. Then for every partition $a = x_0 < \ldots < x_n = b$ of $[a, b]$,

$$\sum_{i=1}^{n} |f(x_i) - f(x_{i-1})| \leqslant C(b - a)$$

so f is of bounded variation. Apply Corollary 4.11.

As an application of Lebesgue's theorem we prove

THEOREM 4.14. (Fubini) *Let $f_1, f_2, \ldots$ be increasing functions on $[a, b]$ such that*

$$s(x) := \sum_{n=1}^{\infty} f_n(x)$$

is finite for every $x \in [a, b]$. *Then* $s = \Sigma_{n=1}^{\infty} f_n$ *uniformly and* $s' = \Sigma_{n=1}^{\infty} f'_n$ *almost everywhere on* $[a, b]$. *In particular, we have* $\lim_{n\to\infty} f'_n = 0$ *almost everywhere on* $[a, b]$.

Proof. Without loss of generality we may assume $f_n(a) = 0$ for all n. (Put $f_n(x) = f_n(x) - f_n(a) + f_n(a) = g_n(x) + f_n(a)$ for $x \in [a, b]$. It is easy to see that the theorem, once proved for the g_n, then also holds for the f_n.) Then $f_n \geqslant 0$. Set

$$s_n := f_1 + \ldots + f_n \qquad (n \in \mathbb{N})$$

Then for $x \in [a, b]$ and $n \in \mathbb{N}$ we have

$$0 \leqslant s(x) - s_n(x) = \sum_{i>n} f_i(x) \leqslant \sum_{i>n} f_i(b) = s(b) - s_n(b)$$

whence $\lim_{n\to\infty} s_n = s$ uniformly.

Both s and s_n are increasing, hence differentiable a.e. Thus, if we let $E := \{x \in (a, b) : s,\ s_1,\ s_2, \ldots$ are differentiable at $x\}$, then $[a, b] \setminus E$ is a countable union of null sets, hence a null set. For each n, $s_{n+1} - s_n$ and $s - s_n$ are increasing, so $(s_{n+1} - s_n)' \geqslant 0$ and $(s - s_n)' \geqslant 0$ on E. Therefore we have

$$s'_1(x) \leqslant s'_2(x) \leqslant \ldots \leqslant s'(x) \qquad (x \in E)$$

It follows that $\lim_{n\to\infty} s'_n(x)$ exists for all $x \in E$. But then

$$\lim_{n\to\infty} f'_n(x) = \lim_{n\to\infty} (s'_n(x) - s'_{n-1}(x)) = 0$$

for all $x \in E$. We have proved: *if* $f_1, f_2, \ldots$ *satisfy the conditions of the theorem, then* $\lim_{n\to\infty} f'_n = 0$ *a.e.*

To show that $\lim_{n\to\infty} s'_n$ is, in fact, a.e. equal to s' it suffices to find a subsequence that converges to s' a.e. Choose positive integers $n_1 < n_2 < \ldots$ such that $s(b) - s_{n_k}(b) \leqslant 2^{-k}$ ($k \in \mathbb{N}$). Define $t_k := s - s_{n_k}$ ($k \in \mathbb{N}$). Then each t_k is increasing and $\Sigma_{k=1}^{\infty} t_k(x)$ exists for all $x \in [a, b]$ ($0 \leqslant t_k(x) = s(x) - s_{n_k}(x) \leqslant s(b) - s_{n_k}(b) \leqslant 2^{-k}$). Thus, the sequence $t_1, t_2, \ldots$ satisfies the condition of the theorem. We infer that $\lim_{n\to\infty} t'_n = 0$ a.e., and we are done.

Exercise 4.H. Even if s and all the f_n of Theorem 4.14 are differentiable everywhere on $[a, b]$, it is not certain that $s' = \Sigma_{n=1}^{\infty} f'_n$ everywhere on $[a, b]$. (Hint. Define $f_n(x) := n^{-1} \arctan nx - (n+1)^{-1} \arctan (n+1)x \quad (-1 \leqslant x \leqslant 1)$.)

Exercise 4.I. There exist strictly increasing functions $f : [a, b] \to \mathbb{R}$ with $f' = 0$ a.e. (Take f as in Exercise 1.K and apply Fubini's theorem.) Later on we shall even make a *continuous* strictly increasing function whose derivative vanishes a.e. (15.10).

**Exercise* 4.J. Let $f : [a, b] \to \mathbb{R}$ be of bounded variation and let the function $T : [a, b] \to \mathbb{R}$ be its indefinite variation. As we know, both f and T are differentiable a.e. We want to prove that $T' = |f'|$ a.e. on $[a, b]$.

If f has a continuous derivative, then, by Exercise 3.D, we know that $T'=|f'|$ everywhere. Show that, if both f and T are differentiable at $x_0 \in [a, b]$, then $T'(x_0) \geqslant |f'(x_0)|$. In order to prove the general statement we shall not use the above observations but follow another direction.
(i) Let $a=a_0<\ldots<a_n=b$ be a partition of $[a, b]$. Define $g:[a, b]\to\mathbb{R}$ inductively as follows. Let $x \in [a_0, a_1]$.

$$\text{If } f(a_0)\leqslant f(a_1), \text{ then } g(x):=f(x)-f(a_0).$$
$$\text{If } f(a_0)>f(a_1), \text{ then } g(x):=f(a_0)-f(x).$$

Let $k<n$ and suppose g has been defined on $[0, a_k]$. Take $x \in (a_k, a_{k+1}]$.

$$\text{If } f(a_k)\leqslant f(a_{k+1}), \text{ then } g(x):=f(x)+g(a_k)-f(a_k).$$
$$\text{If } f(a_k)>f(a_{k+1}), \text{ then } g(x):=-f(x)+g(a_k)+f(a_k).$$

(Draw a picture.)

Show that on each $[a_{k-1}, a_k]$ $(1\leqslant k\leqslant n)$ either $g-f$ or $g+f$ is constant, so that T is the indefinite variation of g and $|g'|=|f'|$ a.e. Furthermore $g(b)=\Sigma_k|f(a_k)-f(a_{k-1})|$, so that $T(b)-g(b)=\operatorname{Var} f-\Sigma_k|f(a_k)-f(a_{k-1})|$.
(ii) Now, for every $n \in \mathbb{N}$, find a function g_n on $[a, b]$ such that $T-g_n$ is increasing, $T(a)-g_n(a)=0$, $T(b)-g_n(b)\leqslant 2^{-n}$ and $|g_n'|=|f'|$ a.e. Apply Fubini's theorem (4.14) to deduce that $T'=|f'|$ a.e.

We close this section with three exercises about the Cantor set.

**Exercise* 4.K. The Cantor set has no isolated points (i.e. if $x \in \mathbb{D}$ and $\varepsilon>0$, then there is a $y \in \mathbb{D}$ for which $0<|x-y|<\varepsilon$).

Exercise 4.L. For $f: \mathbb{D}\to\mathbb{R}$ we define differentiability in the usual way. (Here one needs the fact that $\mathbb{D}$ has no isolated points; see 4.K.) There exists a monotone $f: \mathbb{D}\to\mathbb{R}$ that is nowhere differentiable.

Exercise 4.M. Although $\mathbb{D}$ is a null set, in some sense it is not 'very' small. Show that every element of $[0, 2]$ can be written as the sum of two elements of $\mathbb{D}$. (Every element of $[0, 2]$ is of the form $2\Sigma_{i=1}^{\infty} \alpha_i 3^{-i}$ for certain $\alpha_1, \alpha_2, \alpha_3, \ldots \in \{0, 1, 2\}$.)

Notes to Section 4

The proofs presented above of the theorems of Lebesgue and Fubini are taken from the book *Functional Analysis* by F. Riesz & B. Sz.-Nagy (Frederick Ungar, New York 1955). By using some Lebesgue integration theory one can devise shorter proofs. (See Saks (no date), Ch. 4, §5, also for historical notes and further references, or Hewitt & Stromberg (1965), Th. 17.12.) Other elementary proofs are given by, among others, L. A. Rubel (*Coll. Math.* **10** (1963), 277–9) and A. Rajchman & S. Saks (*Fund. Math.* **4** (1923), 204–13).

The lemma of the rising sun (Lemma 4.6) is due to F. Riesz. References to other applications are given in Saks (no date), Ch. 4, §12.

2

SUBSETS OF $\mathbb{R}$

5. Small sets

In Chapter 1 we have seen that monotone functions are continuous or differentiable, except for points belonging to a 'small' set. In the first case this meant 'countable set', in the second it was 'null set'. The usefulness of either notion of 'smallness' lies partly in the following. If $\mathscr{S}$ is either the class of the countable subsets of $\mathbb{R}$ or the class of the null sets, then

(a) if $A \in \mathscr{S}$ and $B \subset A$, then $B \in \mathscr{S}$,

(b) if $A_1, A_2, \ldots \in \mathscr{S}$, then $\bigcup_i A_i \in \mathscr{S}$,

(c) if $A \in \mathscr{S}$, then for all $x \in \mathbb{R}$, $A + x \in \mathscr{S}$,

(d) $\mathbb{R} \notin \mathscr{S}$.

(Here $A + x := \{a + x : a \in A\}$.)

These facts guarantee that, if a property $P(x)$ holds for all $x \in \mathbb{R}$ except for x belonging to a set in $\mathscr{S}$, then $P(x)$ must hold at least for some x, and, when proving such a thing, we may disregard countably many sets of $\mathscr{S}$. (See, for instance, our proof of Lebesgue's theorem (4.10).)

In this section we introduce another type of 'small set', satisfying (a), (b), (c) and (d). 'Countable set' being defined in terms of cardinality and 'null set' in terms of measure, our new notion will be a purely topological one.

5.1. First, let us start with any collection $\mathscr{S}_0$ of subsets of $\mathbb{R}$ that satisfies (c) and try to extend it to a collection $\mathscr{S}$ that has the four properties (a), (b), (c), (d). If such an $\mathscr{S}$ can be made, it will have to contain all countable unions of elements of $\mathscr{S}_0$. Thus we are led to define

$$\mathscr{S}_b := \left\{ \bigcup_{i=1}^{\infty} A_i : A_1, A_2, \ldots \in \mathscr{S}_0 \right\}$$

This $\mathscr{S}_b$ has properties (b) and (c). But we see that $\mathscr{S}$ will also have to contain all subsets of elements of $\mathscr{S}_b$. Therefore, we put

$$\mathscr{S}_a := \{A : A \text{ is a subset of an element of } \mathscr{S}_b\}$$

$\mathscr{S}_a$ has the properties (a), (b) and (c).

Every $\mathscr{S}$ that has properties (*a*), (*b*), (*c*), (*d*) and contains $\mathscr{S}_0$ must contain $\mathscr{S}_a$. Hence, if $\mathbb{R} \in \mathscr{S}_a$, then the desired extension of $\mathscr{S}_0$ is not possible. But if $\mathbb{R} \notin \mathscr{S}_a$, then $\mathscr{S}_a$ has property (*d*). Hence, if $\mathbb{R} \notin \mathscr{S}_a$, we may choose $\mathscr{S} := \mathscr{S}_a$.

From the definition of $\mathscr{S}_a$ it is obvious that $\mathbb{R} \in \mathscr{S}_a$ if and only if $\mathbb{R} \in \mathscr{S}_b$. Thus: *if $\mathbb{R}$ can be covered by countably many sets belonging to $\mathscr{S}_0$, then the extension is impossible. Otherwise, $\mathscr{S}_a$ satisfies the requirements.*

A dense subset of $\mathbb{R}$ may intuitively be regarded as 'big' and we may call a set 'small' if its complement is dense. This notion of smallness, however, does not meet our purposes. ($\mathbb{R}$ is the union of $\mathbb{Q}$ and $\mathbb{R} \setminus \mathbb{Q}$.) The situation becomes more promising when we look at *open* dense sets (such as $\mathbb{R} \setminus \{0\}$, $\mathbb{R} \setminus \mathbb{D}$), because of the following theorem.

THEOREM 5.2. *Let $U_1, U_2, \ldots$ be open dense subsets of $\mathbb{R}$. Then $\bigcap_{i=1}^{\infty} U_i$ is dense in $\mathbb{R}$.*

Proof. Let I be an interval in $\mathbb{R}$. We prove that there is a $\xi \in I$ that lies in every U_i. There is an interval $J_1 \subset U_1 \cap I$. Choose a closed interval $[p_1, q_1] \subset J_1$. Then $(p_1, q_1) \cap U_2$ contains an interval J_2. Choose a closed interval $[p_2, q_2] \subset J_2$. Then $(p_2, q_2) \cap U_3$ contains an interval J_3, etc. We obtain a sequence of closed intervals $[p_1, q_1] \supset [p_2, q_2] \supset \ldots$ such that $[p_i, q_i] \subset U_i$ for each i. By Cantor's theorem the intersection of these intervals contains an element ξ. Then $\xi \in [p_1, q_1] \subset J_1 \subset I$ and ξ is contained in every U_i.

Let us define

$$\mathscr{S}_0 := \{A \subset \mathbb{R} : \mathbb{R} \setminus A \text{ is open and dense}\}$$

or, equivalently,

$$\mathscr{S}_0 := \{B \subset \mathbb{R} : B \text{ is closed and has empty interior}\}$$

Starting with this collection of sets we can construct $\mathscr{S}_b$ and $\mathscr{S}_a$ as we did earlier in this section. By Theorem 5.2, $\mathbb{R}$ is not the union of countably many elements of $\mathscr{S}_0$, and by the conclusion of 5.1 it follows that $\mathscr{S}_a$ satisfies (*a*, (*b*), (*c*) and (*d*). We define:

DEFINITION 5.3. A subset A of $\mathbb{R}$ is called *meagre* (*a set of the first category, small in the sense of Baire*) if A is contained in a countable union of closed sets each of which has an empty interior.

The fact that the collection of meagre sets has property (*d*) (trivially, it satisfies (*a*), (*b*) and (*c*)) is often referred to as the *category theorem of Baire* and is worth stating explicitly.

THEOREM 5.4. (Baire) $\mathbb{R}$ *is not meagre.*

**Exercise* 5.A. Show that: (i) Countable sets are meagre. (ii) $\mathbb{D}$ is meagre. (iii) $\mathbb{R} \setminus \mathbb{Q}$ is not meagre. (iv) No interval is meagre.
A meagre set has empty interior. The converse is false. However, a closed set is meagre if and only if its interior is empty. If U is an open set, then $\overline{U} \setminus U$ is meagre. If A is a meagre set, then for every $x \in \mathbb{R}$ the set $\{xa : a \in A\}$ is meagre.

**Exercise* 5.B. A subset A of $\mathbb{R}$ is called *nowhere dense* if its closure has empty interior. Show that a subset of $\mathbb{R}$ is meagre if and only if it is a union of countably many nowhere dense sets.

**Exercise* 5.C. A subset A of $\mathbb{R}$ is of the *second category* if A is not of the first category. Show that:
If A is meagre, then $\mathbb{R} \setminus A$ is of the second category. The converse is false.
If A is meagre, then $\mathbb{R} \setminus A$ is dense. The converse is false.
If $\mathbb{R} \setminus A$ is both dense and of the second category, must A be meagre?

Now let us compare the two notions 'meagre' and 'null'.

If A is a closed null set, then A does not contain any interval, so A is nowhere dense. Hence, a null set which is the union of countably many closed sets is meagre. This leads to the question whether every null set is meagre. Conversely, one may ask if every meagre set is null. The following example shows that in general there is no connection between the two concepts.

Let $(r_1, r_2, \ldots)$ be an enumeration of $\mathbb{Q}$. Define

$$U_1 := \bigcup_{i=1}^{\infty} (r_i - 2^{-i}, r_i + 2^{-i})$$

Then U_1 is open and dense, so its complement is nowhere dense. But this complement is not a null set. In fact, suppose that one could cover it with intervals $I_1, I_2, \ldots$ having total length $\leqslant 1$. U_1 itself is covered by intervals of total length 2, as is clear from its definition. It follows that $\mathbb{R}$ can be covered with countably many intervals of total length $\leqslant 3$, an absurdity (see Exercise 4.B). Thus, *we have an example of a closed nowhere dense set that is not a null set.*

Next, define

$$U_2 := \bigcup_{i=1}^{\infty} (r_i - 2^{-2i}, r_i + 2^{-2i})$$
$$\vdots$$
$$U_n := \bigcup_{i=1}^{\infty} (r_i - 2^{-ni}, r_i + 2^{-ni})$$
$$\vdots$$

We have $U_1 \supset U_2 \supset \ldots \supset \mathbb{Q}$. Now $U := \bigcap_{i=1}^{\infty} U_i$ is a null set (since U_n is the union of intervals with total length $\leqslant 2^{-n+2}$). But $\mathbb{R} \setminus U$ is meagre (as each U_n is open and dense). Thus, we have:

THEOREM 5.5. *There exist nowhere dense sets that are not null. There exist null sets that are not meagre. More than that : $\mathbb{R}$ can be written as the union of a null set and a meagre set.*

Remark 1. Although the notions 'null set' and 'meagre set' turn out to be radically different, from another point of view they are very similar. Assuming the continuum hypothesis, Sierpiński and Erdös have proved the existence of a bijection $f: \mathbb{R} \to \mathbb{R}$ such that E is a null set if and only if $f(E)$ is meagre while E is meagre if and only if $f(E)$ is null. For a proof, see Oxtoby (1971), 19.3.

Remark 2. One can easily define other notions of 'small' that satisfy the conditions (*a*), (*b*), (*c*), (*d*) mentioned at the beginning of this section. For example, let us call a set A 'unimportant' if A is the union of countably many sets with zero content (see Exercise 4.C). Every 'unimportant' set is null. The converse is false (use Theorem 5.5). More examples of this kind can be found in Exercises 5.E and 5.H.

Exercise 5.D. Can $\mathbb{R}$ be written as the union of a null set and a nowhere dense set?

Exercise 5.E. Let $\mathcal{T}$ denote the collection of all subsets of $\mathbb{R}$ that are both meagre and null. Show that this $\mathcal{T}$ has properties (*a*), (*b*), (*c*), (*d*), that $\mathcal{T}$ contains all countable sets and also certain uncountable sets. Show that $\mathcal{T}$ does not coincide with the collection of all meagre sets or with the collection of all null sets.

We shall need a refinement of Theorem 5.2. Let X be a closed subset of $\mathbb{R}$ and let $Y \subset X$. We say that Y is *dense in* X if every open interval that meets X also meets Y, or, equivalently, if $\bar{Y} = X$.

THEOREM 5.6. *Let $X \subset \mathbb{R}$ be closed and let $U_1, U_2, \ldots$ be open subsets of $\mathbb{R}$ such that $U_i \cap X$ is dense in X for each i. Then the set $(\bigcap_{i=1}^{\infty} U_i) \cap X$ is dense in X.*

Proof. (Similar to the proof of Theorem 5.2) Let $x \in X$ and let I be an open interval that contains x. Then there is a closed interval $[p_1, q_1]$, contained in $I \cap U_1$, such that (p_1, q_1) has a nonempty intersection with X. There is a closed interval $[p_2, q_2] \subset (p_1, q_1) \cap U_2$ such that (p_2, q_2) has a nonempty intersection with X, etc. Thus,

$$[p_1, q_1] \cap X \supset [p_2, q_2] \cap X \supset \ldots$$

and

$$[p_i, q_i] \cap X \subset U_i \text{ for each } i$$

We can force $\lim_{i\to\infty}(q_i - p_i) = 0$. There is a $\xi \in \bigcap_{i=1}^{\infty} [p_i, q_i]$. Now every $[p_i, q_i]$ intersects X, while $\lim_{i\to\infty}(q_i - p_i) = 0$. Consequently, $\xi \in \bar{X} = X$. We have

$$\xi \in [p_1, q_1] \cap X \cap \left(\bigcap_{i=1}^{\infty} U_i\right) \subset I \cap X \cap \left(\bigcap_{i=1}^{\infty} U_i\right)$$

COROLLARY 5.7. (Cantor–Bendixson) *Every nonempty closed countable subset of* $\mathbb{R}$ *has isolated points.*

Proof. Let X be a nonempty closed countable subset of $\mathbb{R}$. If X does not have isolated points, then $\mathbb{R} \setminus \{x\} \cap X$ is dense in X for every $x \in X$. The sets $\mathbb{R} \setminus \{x\}$ $(x \in X)$ form a countable collection of open subsets of $\mathbb{R}$, so by Theorem 5.6 $\bigcap_{x \in X} (\mathbb{R} \setminus \{x\}) \cap X$ is dense in X. But $\bigcap_{x \in X} (\mathbb{R} \setminus \{x\}) \cap X = \varnothing$.

Exercise 5.F. Can $\mathbb{R}$ be written as the disjoint union of countably many closed intervals of finite length? (Hint. Apply Corollary 5.7 to the set of the end points.)

Exercise 5.G. Let X be a countable subset of $\mathbb{R}$ without isolated points. Then $\bar{X} \setminus X$ is dense in $\bar{X}$. (Hint. Consider for each $x \in X$ the set $(\mathbb{R} \setminus \{x\}) \cap \bar{X}$.)

Exercise 5.H. (Another example of smallness) We call a subset X of $\mathbb{R}$ 'tiny', say, if for every $\varepsilon > 0$ there exists a sequence $I_1, I_2, \ldots$ of intervals, covering X and such that $L(I_i) \leqslant \varepsilon^i$ for each i.

(i) The collection of all tiny sets has the properties (a), (b), (c), (d) mentioned at the beginning of this section. (To prove (b), cover A_n with intervals $I_{n1}, I_{n2}, \ldots$ with $L(I_{ni}) \leqslant \varepsilon_n{}^i$, where $\varepsilon_n = \varepsilon^{2^n}$.)

(ii) The Cantor set (which is both meagre and null) is not tiny. (Suppose there exists a covering of $\mathbb{D}$ by intervals $I_1, I_2, \ldots$ with $L(I_i) < 3^{-i}$ for each i. For every $x \in \mathbb{D}$ there exists a sequence $\alpha_1(x), \alpha_2(x), \ldots$ of 0s and 2s with $x = \sum_{n=1}^{\infty} \alpha_n(x) 3^{-n}$. Prove that, for every i, the function α_i is constant on I_i. Using a diagonal argument, construct an $x \in \mathbb{D}$ with $x \notin \bigcup_i I_i$.)

(iii) All countable sets are tiny, but not every tiny set is countable. (The set U, constructed in 5.5, is (tiny and) uncountable, since its complement is meagre.)

Exercise 5.I. In the Introduction we constructed a function k on $[0, 1]$ with the property that $k(I) = [0, 1]$ for every subinterval I of $[0, 1]$. Prove that $k = 0$ a.e. (Let $T^* := \{t^* : 0 \leqslant t \leqslant 1\}$. As k vanishes everywhere outside $\bigcup_{d \in D} (T^* + d)$ and D is countable, it suffices to show that T^* is null. Prove that for every $n \in \mathbb{N}$ T^* is contained in a union of 10^n intervals of length $10^{-(1+2+\ldots+n)}$ each.)

**Exercise* 5.J. (A dual form of Theorem 5.6) Let $X \subset \mathbb{R}$ be closed, $X \neq \varnothing$. Let $C_1, C_2, \ldots$

be closed subsets of $\mathbb{R}$ such that $X \subset \bigcup_i C_i$. Then there exist an $i \in \mathbb{N}$ and an interval I such that $\varnothing \neq I \cap X \subset C_i$.

Notes to Section 5

We have mentioned several forms of 'smallness'. The following notion has played a role in the development of ideas around the continuum hypothesis. (See Sierpiński 1956.) A subset E of $\mathbb{R}$ is said to have property (L) if its intersection with every meagre set is countable. The collection of all sets with this property satisfies the conditions (*a*)–(*d*) of Section 5. Using the continuum hypothesis, Lusin constructed an uncountable set with property (L). It was proved by F. Rothberger (*Fund. Math.* **30** (1938), 215–17) that, conversely, the existence of such 'Lusin sets' entails the continuum hypothesis.

6. F_σ-sets and G_δ-sets

DEFINITION 6.1. A subset A of $\mathbb{R}$ is called an F_σ (or an *F_σ-set*) if it is a countable union of closed sets, a G_δ (or a *G_δ-set*) if it is a countable intersection of open sets.

THEOREM 6.2.
(i) *Every closed set is an F_σ. A union of countably many F_σ-sets is an F_σ. If A_1, A_2 are F_σ-sets, then so is $A_1 \cap A_2$.*
(ii) *Every open set is a G_δ. An intersection of countably many G_δ-sets is a G_δ. If B_1, B_2 are G_δ-sets, then so is $B_1 \cup B_2$.*
(iii) *A set is an F_σ if and only if its complement is a G_δ.*
(iv) *An interval is both an F_σ and a G_δ. Every open set is an F_σ; every closed set is a G_δ.*
(v) *Every countable set is an F_σ.*
Proof. (i), (ii) and (iii) are clear. The open interval (p, q) (where $-\infty < p < q < \infty$) can be written as the union of the countably many closed intervals $[p+n^{-1}, q-n^{-1}]$ $(n \in \mathbb{N})$, so (p, q) is an F_σ. A similar proof works for unbounded open intervals. Every interval is a union of an open interval and a set which contains at most two points. Thus, every interval is an F_σ. The complement of an interval is a union of at most two intervals, hence is an F_σ. It follows that every interval is a G_δ.

Every open set is a union of countably many open intervals with rational end points. Therefore, every open set is an F_σ. By complementation, every closed set is a G_δ.

If $A \subset \mathbb{R}$ is countable, then A is an F_σ because $A = \bigcup_{a \in A}\{a\}$.

To see that not every set is a G_δ we first give a criterion for meagreness in terms of G_δ-sets.

THEOREM 6.3. *A subset of* $\mathbb{R}$ *is meagre if and only if its complement contains a* G_δ*-set that is dense in* $\mathbb{R}$.

Proof. Let A be meagre. Then $A = \bigcup_{i=1}^{\infty} P_i \subset \bigcup_{i=1}^{\infty} \overline{P_i}$, where each P_i is nowhere dense. Then $\mathbb{R} \setminus \overline{P_i}$ is a dense open set. The complement of A contains $\bigcap_{i=1}^{\infty} \mathbb{R} \setminus \overline{P_i}$, a countable intersection of open dense sets, which is dense by Theorem 5.2.

Conversely, if $A \subset \mathbb{R}$ and $\mathbb{R} \setminus A \supset \bigcap_{i=1}^{\infty} U_i$ where each U_i is open and dense, then A is contained in $\bigcup_{i=1}^{\infty} \mathbb{R} \setminus U_i$, a countable union of nowhere dense sets. Thus, A is meagre.

EXAMPLE 6.4. $\mathbb{Q}$ *is not a* G_δ. (If $\mathbb{Q}$ were a G_δ, then $\mathbb{R} \setminus \mathbb{Q}$ would be meagre according to Theorem 6.3.) Thus, the set of all irrational numbers is not an F_σ. Since $\mathbb{Q} = \bigcup_{q \in \mathbb{Q}} \{q\}$, it follows that $\mathbb{Q}$ is a union of countably many G_δ-sets. Hence, the collection of the G_δ-sets is not closed under countable unions (and the collection of the F_σ-sets is not closed under countable intersections).

EXAMPLE 6.5. Let A be the intersection of an open and a closed set. It follows from Theorem 6.2 that A is both an F_σ and a G_δ. To see that the converse does not hold, consider the set $B := \{1, \frac{1}{2}, \frac{1}{3}, \dots\}$. Since B is the intersection of $\bar{B}$ and $(0, \infty)$, B is both an F_σ and a G_δ. Then so is its complement, which we call A. Suppose $A = P \cap Q$ where P is open and Q is closed. Then $B = S \cup T$ for some closed S and some open T. Now from $T \subset B$ it follows that $T = \varnothing$. Thus, $B = S$, but B is not closed. We have a contradiction.

**Exercise* 6.A. Show that the intersection of countably many dense G_δ-sets is itself a dense G_δ-set.

Exercise 6.B. Give an example of a subset of $\mathbb{R}$ that is neither an F_σ nor a G_δ.

Exercise 6.C. Let E be a subset of $\mathbb{R}$ that is a union of intervals. Then E is a union of countably many pairwise disjoint intervals, so E is an F_σ. (See the proof for open E, given in Appendix A, Theorem A.9.)

Exercise 6.D. Let $f_1, f_2, \dots$ be a sequence of continuous functions on $\mathbb{R}$. Then $\{x \in \mathbb{R} : \liminf_{n \to \infty} f_n(x) > 0\}$ is an F_σ. (Hint. For $n, m \in \mathbb{N}$ consider the set $\{x : f_k(x) \geq 1/n \text{ for all } k \geq m\}$.)

Exercise 6.E. (On countable G_δ-sets)
(i) An example: let (p_1, q_1), (p_2, q_2), ... be the components of $[0, 1]\setminus\mathbb{D}$. Let $D_m := \{\frac{1}{2}(p_i+q_i) : i \in \mathbb{N}\}$. Then D_m is a countable G_δ.
(ii) A countable set whose closure is an interval cannot be a G_δ.
(iii) The closure of a countable G_δ is nowhere dense. (Use (ii).)
(iv) For a set $X \subset \mathbb{R}$, let X_{is} be the set of all isolated points of X. If X is a countable G_δ, then X_{is} is dense in $\bar{X}$. (Hint. Let $U_1, U_2, \ldots$ be open sets whose intersection is X. Let $Y := X \setminus X_{is}$. Consider the set $\bar{X} \cap \bigcap_{i=1}^{\infty} U_i \cap \bigcap_{y\in Y} \mathbb{R}\setminus\{y\}$.)
(v) With q_i defined as in (i), $D_r := \{q_i : i \in \mathbb{N}\}$ is a countable subset of $\mathbb{D}$ that is not a G_δ. (Use (iv).)
(vi) The converse of (iv) is false. (Take $X := D_m \cup D_r$.)
(vii) Let $X \subset \mathbb{R}$. If every point of X is isolated, then X is a countable G_δ. (For every $x \in X$, choose a positive number ε_x such that $(x-\varepsilon_x, x+\varepsilon_x) \cap X = \{x\}$. Show that the intervals $(x-\frac{1}{2}\varepsilon_x, x+\frac{1}{2}\varepsilon_x)$ are pairwise disjoint.)
(viii) Let $X \subset \mathbb{R}$. If $\bar{X}$ is countable, then X (or, in fact, any subset of $\bar{X}$) is a countable G_δ. The converse is false (consider D_m).

Exercise 6.F. Let $f : [0, 1] \to \mathbb{R}$ be continuous. Define $\phi : [0, 1] \to [0, 1]$:

$$\phi(x) := \inf\{y \in [0, 1] : f(y) = f(x)\} \qquad (x \in [0, 1])$$

Show that $\phi([0, 1])$ is a G_δ. (Hint. Show that for every $n \in \mathbb{N}$ the set $\{a \in [0, 1]$: there is an $x \in [0, 1]$, $x \leqslant a - 1/n$ with $f(x) = f(a)\}$ is closed and that the union of these sets is the complement of $\phi([0, 1])$ relative to $[0, 1]$.)

Exercise 6.G. Let $A \subset \mathbb{R}$. A *two-sided accumulation point* of A is an $x \in \mathbb{R}$ with the property that for every $\varepsilon > 0$ the intervals $(x-\varepsilon, x)$ and $(x, x+\varepsilon)$ both intersect A.
(i) Assuming that all two-sided accumulation points of A are elements of A, prove that A is a G_δ. (Show that every element of $\bar{A}\setminus A$ is an end point of a component of $\mathbb{R}\setminus\bar{A}$, so that $\bar{A}\setminus A$ is countable.)
(ii) Let D_r be as in Exercise 6.E. Set $A := \mathbb{D}\setminus D_r$. Then A contains all of its two-sided accumulation points, but A is not an F_σ.

Exercise 6.H
(i) If $f : \mathbb{R} \to \mathbb{R}$ is monotone, then $f(\mathbb{R})$ is a G_δ. (Use the preceding exercise.)
(ii) Show that there does not exist a monotone $f : \mathbb{R} \to \mathbb{R}$ with $f(\mathbb{R}) = \mathbb{R}\setminus\mathbb{Q}$. (First prove that such an f would have to be continuous.)
(iii) Show that for every nonempty closed set $A \subset \mathbb{R}$ there is a monotone $f : \mathbb{R} \to \mathbb{R}$ with $f(\mathbb{R}) = A$. (Start by defining $f(x) := x$ for all $x \in A$.)

**Exercise* 6.I. An F_σ-set is meagre if and only if its interior is empty. Every null F_σ-set is meagre.

**Exercise* 6.J. Every null set is contained in a null G_δ-set.

7. Behaviour of arbitrary functions

To illustrate the use of 'small' sets, F_σ-sets and G_δ-sets in the theory of functions we now give examples of theorems valid for arbitrary functions. (In the following chapters we will meet many more applications.)

7.1. The set of maxima and minima of an arbitrary function. Let $f:\mathbb{R}\to\mathbb{R}$. We say that s is a *local maximum* of f if there exist an $x\in\mathbb{R}$ and a $\delta>0$ such that $f(x)=s$ while $f(y)\leqslant s$ for $x-\delta<y<x+\delta$. Similarly one can define *local minimum*. If s is a local maximum or a local minimum we say that s is a *local extremum* of f. Now we have the following, somewhat surprising, fact.

THEOREM 7.2. *Let $f:\mathbb{R}\to\mathbb{R}$ be any function. Then*

$$\{a\in\mathbb{R} : a \text{ is a local extremum of } f\}$$

is a countable set.

Proof. We show that the set S_+ of all local maxima of f is countable. For every $s\in S_+$, choose an interval I_s with rational end points such that $s=\max\{f(x) : x\in I_s\}$. If $s,t\in S_+$ and $I_s=I_t$, then $s=t$. The set $\{I_s : s\in S_+\}$ is countable because of the countability of $\mathbb{Q}\times\mathbb{Q}$. Hence, S_+ is also countable.

Exercise 7.A. Prove a converse of Theorem 7.2: if $S\subset\mathbb{R}$ is countable, then there exists a (continuous) function $f:\mathbb{R}\to\mathbb{R}$ such that S is the set of the local extrema of f.

Exercise 7.B. There exists a function $f:\mathbb{R}\to\mathbb{R}$, constant on no interval, such that for every $x\in\mathbb{R}$, $f(x)$ is a local extremum. Such an f cannot be continuous.

Exercise 7.C. Let $f:\mathbb{R}\to\mathbb{R}$ and $a\in\mathbb{R}$. We say that f *has a local maximum* (*a local strict maximum*) *at* a if there exists a $\delta>0$ such that $f(y)\leqslant f(a)$ ($f(y)<f(a)$) for $a-\delta<y<a+\delta$, $y\neq a$. Prove that the set $\{a\in\mathbb{R} : f \text{ has a local strict maximum at } a\}$ is countable.

7.3. It is interesting to compare, for a differentiable $f:\mathbb{R}\to\mathbb{R}$, the sets S and T where

$$S:=\{s\in\mathbb{R} : s \text{ is a local extremum of } f\}$$
$$T:=\{f(x) : f'(x)=0\}$$

By an elementary theorem from calculus we have $S\subset T$. How big can T be with respect to S? In other words, if $f'(x)=0$ for some x, how 'probable' is it that f has a local maximum or minimum at x? The following example shows that S can be empty while T is uncountable.

**Exercise* 7.D. Let

$$\phi(x) := \min\{|x-a| : a \in \mathbb{D}\} \qquad (x \in \mathbb{R})$$

Then ϕ is continuous ($|\phi(x)-\phi(y)| \leqslant |x-y|$ for all $x, y \in \mathbb{R}$) and nonnegative, $\phi(x)=0$ if and only if $x \in \mathbb{D}$. Let f be the indefinite integral of ϕ:

$$f(x) := \int_0^x \phi(t)\,\mathrm{d}t \qquad (x \in \mathbb{R})$$

Then f is strictly increasing and has no local extrema. On the other hand, $f'(x)=0$ if and only if $x \in \mathbb{D}$, hence $T = \{f(x) : f'(x)=0\} = f(\mathbb{D})$, so T is uncountable since f is injective.

One may ask, nevertheless, whether the set T is 'small' in some sense or another. Indeed, in Corollary 21.5 we shall prove that for every differentiable f the set $T = \{f(x) : f'(x)=0\}$ is a null set.

7.4. The set of continuity points of an arbitrary function. For an $f : \mathbb{R} \to \mathbb{R}$ we consider the set $C_f := \{x \in \mathbb{R} : f$ is continuous at $x\}$. Of course, C_f may be empty (choose $f := \xi_{\mathbb{Q}}$). Not every subset of $\mathbb{R}$ can serve as a C_f. In fact, we have:

THEOREM 7.5. *Let $f : \mathbb{R} \to \mathbb{R}$. Then the set of points of continuity of f is a G_δ.*

(See also Exercise 7.H.)

Proof. For $n \in \mathbb{N}$, let

$$U_n := \{a \in \mathbb{R} : \text{there is a } \delta > 0 \text{ such that, if } x, y \in (a-\delta, a+\delta), \text{ then } |f(x)-f(y)| < 1/n\}$$

Each U_n is open. It is easy to see that f is continuous at a if and only if $a \in U_n$ for every n. Thus, the set of points of continuity of f is $\bigcap_n U_n$, which is a G_δ.

COROLLARY 7.6. *There is no function $\mathbb{R} \to \mathbb{R}$ that is continuous at each rational point and discontinuous at each irrational point.*

Proof. $\mathbb{Q}$ is not a G_δ.

Exercise 7.E. Let $(q_1, q_2, \ldots)$ be an enumeration of $\mathbb{Q}$. Define $f : \mathbb{R} \to \mathbb{R}$ as follows. $f(x) := 0$ if x is irrational, $f(q_n) := 1/n$ ($n \in \mathbb{N}$). Show that $\{x : f$ is discontinuous at $x\} = \mathbb{Q}$ (compare Corollary 7.6).

Exercise 7.F. Let $U \subset \mathbb{R}$ be open. Show that there exists an $f : \mathbb{R} \to \mathbb{R}$ which is discontinuous at each point of U and continuous at each point of $R \setminus U$. (Hint. First try the case where U is an open interval.)

Exercise 7.G. Let S be an F_σ which does not contain any interval. Then S is a union of closed, nowhere dense sets $C_1, C_2, \ldots$ Show that the function $\Sigma_{n=1}^{\infty} 2^{-n}\xi_{C_n}$ is continuous at every point of $\mathbb{R} \setminus S$ and discontinuous at every point of S.

Exercise 7.H. With the help of the preceding exercises, prove the following theorem. *If $T \subset \mathbb{R}$ is a G_δ, then there exists a function $\mathbb{R} \to \mathbb{R}$ that is continuous at each point of T and discontinuous at each point of $\mathbb{R} \setminus T$.*

Exercise 7.I. Let $f: \mathbb{R} \to \mathbb{R}$. Show that the set

$$\{a \in \mathbb{R} : \lim_{x \to a} f(x) \text{ exists}\}$$

is a G_δ.

Exercise 7.J. Let $f: \mathbb{R} \to \mathbb{R}$. Show that $\{a \in \mathbb{R} : \lim_{x \to a} f(x) = 3\}$ and $\{a \in \mathbb{R} : \lim_{x \to a} f(x) = \infty\}$ are G_δ-sets.

Exercise 7.K. Let L be the set of all functions $f: [0, 1] \to \mathbb{R}$ that have the property that $\lim_{x \to a} f(x)$ exists for all $a \in [0, 1]$. Show that
(i) L is a vector space. Each $f \in L$ is bounded.
(ii) For each $f \in L$, define $f^c(x) := \lim_{y \to x} f(y)$ $(x \in [0, 1])$. f^c is continuous.
(iii) '$f^c = 0$' is equivalent to 'there exist $x_1, x_2, \ldots$ in $[0, 1]$ and $\alpha_1, \alpha_2, \ldots$ in $\mathbb{R}$ with $\lim_{n \to \infty} \alpha_n = 0$, such that $f(x_n) = \alpha_n$ for every n, and $f = 0$ elsewhere'.
(iv) Describe the general form of an element of L. Show that every $f \in L$ is Riemann integrable.

THEOREM 7.7. *Let f be any function on $\mathbb{R}$. Then there exist only countably many points x of $\mathbb{R}$ for which f is not continuous at x but $f(x+)$ exists.*
Proof. We may assume f to be bounded. (Otherwise, consider $\arctan \circ f$ instead of f.) Set $L := \{x \in \mathbb{R} : f(x+) \text{ exists}\}$. For $n \in \mathbb{N}$, let U_n be as in the proof of Theorem 7.5. As $\bigcap_n U_n$ is the set of continuity points of f, we have to prove that $L \setminus \bigcap_n U_n$ is countable. It suffices to show that, for each $n \in \mathbb{N}$, $L \setminus U_n$ is countable.

Take $n \in \mathbb{N}$. Let $a \in L \setminus U_n$. There exists a $\delta > 0$ such that $|f(x) - f(a+)| < 1/2n$ for all $x \in (a, a+\delta)$. If now $x, y \in (a, a+\delta)$, then $|f(x) - f(y)| < 1/n$. Hence, $(a, a+\delta) \subset U_n$. Thus, every element a of $L \setminus U_n$ is the left end point of an open interval J_a that does not intersect $L \setminus U_n$. Consequently, if $a, b \in L \setminus U_n$ and $a \neq b$, then $J_a \cap J_b = \varnothing$. It follows that $\{J_a : a \in L \setminus U_n\}$ must be countable, so $L \setminus U_n$ is itself countable.

COROLLARY 7.8. *If f is any function on $\mathbb{R}$, then the set*

$$\{x \in \mathbb{R} : \lim_{y \to x} f(y) = \infty\}$$

is countable. (See also Exercise 7.J.)

Proof. Apply Theorem 7.7 to the function $\arctan \circ f$.

COROLLARY 7.9. *A left continuous function on $\mathbb{R}$ has only countably many discontinuities.*

7.10. The set of points where a continuous function is differentiable. Let $f:\mathbb{R}\to\mathbb{R}$ be a continuous function. We now consider the set

$$S:=\{a\in\mathbb{R} : f \text{ is not differentiable at } a\}.$$

Take $c\in\mathbb{R}$. For each $n, k\in\mathbb{N}$ put

$$U_{n,k}:=\left\{a\in\mathbb{R} : \text{there is an } x\in\mathbb{R} \text{ such that } 0<|x-a|<\frac{1}{n} \text{ and } \frac{f(x)-f(a)}{x-a}>c-\frac{1}{k}\right\}$$

$U_{n,k}$ is open. It is easy to see that $\overline{\lim}_{x\to a}(f(x)-f(a))/(x-a)\geqslant c$ if and only if a is contained in all of the sets $U_{n,k}$. Thus, for every $c\in\mathbb{R}$ the set

$$\left\{a : \overline{\lim_{x\to a}}\frac{f(x)-f(a)}{x-a}\geqslant c\right\}$$

is a G_δ. Similarly we can prove that $\{a : \underline{\lim}_{x\to a}(f(x)-f(a))/(x-a)\leqslant d\}$ is a G_δ for every $d\in\mathbb{R}$. Now the set S of points where f is not differentiable can be written as

$$\bigcup_{c,d\in\mathbb{Q}}\left\{a : \underline{\lim_{x\to a}}\frac{f(x)-f(a)}{x-a}\leqslant d<c\leqslant\overline{\lim_{x\to a}}\frac{f(x)-f(a)}{x-a}\right\}$$

and therefore is a countable union of G_δ-sets. We have found the following result. (For a more general version, see Theorem 20.12.)

THEOREM 7.11. *Let $f:\mathbb{R}\to\mathbb{R}$ be a continuous function. Then the set of points where f is differentiable is a countable intersection of F_σ-sets.*

In Lemma 7.12 we shall see that not every set is a countable intersection of F_σ-sets, which saves Theorem 7.11 from being trivial.

In Theorem 4.3 we constructed, given any null set E, a monotone function $f:\mathbb{R}\to\mathbb{R}$, differentiable at no point of E. Combining this with the results we have just obtained we see that, if E does not happen to be a countable union of G_δ-sets, then f is bound to be nondifferentiable at points outside of E.

There exist continuous functions that are nowhere differentiable (see Example 7.16).

Exercise 7.L. Let $f:[0, 1]\to\mathbb{R}$ be continuous. Then the set of points where f is differentiable is an intersection of countably many F_σ-sets. (Theorem 7.11 has an analogue for functions on $[0, 1]$, or, for that matter, for functions on any interval.)

Exercise 7.M

(i) Let $f:\mathbb{R}\to\mathbb{R}$ be increasing and let $F(x):=\int_0^x f(t)\,dt\ (x\in\mathbb{R})$. Show that, for all $x\in\mathbb{R}$,

$$\lim_{h\downarrow 0}\frac{F(x+h)-F(x)}{h}=f(x+),\quad \lim_{h\uparrow 0}\frac{F(x+h)-F(x)}{h}=f(x-)$$

(ii) Give an example of a continuous $g:\mathbb{R}\to\mathbb{R}$ for which $\{a : g \text{ is not differentiable at } a\}=\mathbb{Q}$.

(iii) For any countable subset S of $\mathbb{R}$ there exists a continuous function $h:\mathbb{R}\to\mathbb{R}$ such that $S=\{a : h \text{ is not differentiable at } a\}$.

A set that is an intersection of countably many F_σ-sets is sometimes called an $F_{\sigma\delta}$-set. Unions of countably many $F_{\sigma\delta}$-sets are $F_{\sigma\delta\sigma}$-sets, etc. Theorem 7.11 raises the question of whether all subsets of $\mathbb{R}$ are $F_{\sigma\delta}$-sets. Fortunately for the theorem, the answer is negative:

LEMMA 7.12. *There exist subsets of* $\mathbb{R}$ *that are not* $F_{\sigma\delta}$*-sets.*

Proof. If a set S has cardinality $\leqslant c$, then so has the set of all sequences of elements of S (Appendix B, Theorem B.10). Hence, the following sets have cardinality $\leqslant c$: the set of all bounded open intervals with rational end points; the set of all open subsets of $\mathbb{R}$; the set of all closed subsets of $\mathbb{R}$; the set of all F_σ-sets; the set of all $F_{\sigma\delta}$-sets. Thus, none of these sets coincides with the set of *all* subsets of $\mathbb{R}$. (See Appendix B, Theorem B.4.)

(Note that, similarly, not all subsets of $\mathbb{R}$ are $F_{\sigma\delta\sigma}$-sets.)

7.13. The set of points where a continuous function is increasing. Let $f:\mathbb{R}\to\mathbb{R}$ be continuous, $a\in\mathbb{R}$. Then f is *increasing at a* if there exists a $\delta>0$ such that $x\in(a-\delta, a)$ implies $f(x)\leqslant f(a)$, and $x\in(a, a+\delta)$ implies $f(a)\leqslant f(x)$. In a similar way one defines the term *decreasing at a.* Exercise 1.A shows that, if f is increasing at each point of an interval (p, q), then the restriction of f to (p, q) is increasing. Thus, if f is increasing at every point of $(0, 1)\cup(1, 2)$, then for any $\alpha\in(0, 1)$ and $\beta\in(1, 2)$ we have $f(\alpha)\leqslant\lim_{x\uparrow 1} f(x)=f(1)=\lim_{x\downarrow 1} f(x)\leqslant f(\beta)$, so f is also increasing at 1. It follows that the set

$$\{a\in\mathbb{R} : f \text{ is not increasing at } a\}$$

does not have isolated points.

THEOREM 7.14. *Let* $f:\mathbb{R}\to\mathbb{R}$ *be continuous. Then the set* $\{a\in\mathbb{R} : f$ *is not increasing at* $a\}$ *is a* G_δ *without isolated points.*

Proof. For $n \in \mathbb{N}$, let $U_n := \{a \in \mathbb{R} :$ there is an $x \in (a - n^{-1}, a)$ for which $f(x) > f(a)\}$ and $V_n := \{a \in \mathbb{R} :$ there is an $x \in (a, a + n^{-1})$ for which $f(x) < f(a)\}$. Then U_n and V_n are open for each $n \in \mathbb{N}$, hence so is $W_n := U_n \cup V_n$. Let $a \in \mathbb{R}$. f is not increasing at a if and only if a lies in every W_n. The theorem follows.

COROLLARY 7.15. *There is no continuous function* $f : \mathbb{R} \to \mathbb{R}$ *for which* $\{a \in \mathbb{R} : f$ is increasing at $a\} = \mathbb{R} \setminus \mathbb{Q}$.
Proof. $\mathbb{Q}$ is not a G_δ.

Exercise 7.N. Let $f : \mathbb{R} \to \mathbb{R}$ be continuous. Show: if $\{a \in \mathbb{R} : f$ is not increasing at $a\}$ is countable, then f is monotone (see Exercise 6.E(iv)).

Let $f : \mathbb{R} \to \mathbb{R}$ be continuous and let $f(0) = 0$, $f(1) = 1$. Question: does there exist a ξ between 0 and 1 such that f is increasing at ξ? For differentiable f the answer is yes: by the mean value theorem there exists $\xi \in (0, 1)$ such that $f'(\xi) = f(1) - f(0) > 0$ and then it is an easy matter to show that f is increasing at ξ. In general, we may analyse the situation roughly as follows. When t runs from 0 to 1, $f(t)$ runs continuously from 0 to 1. The net result is that the value of $f(t)$ has increased by 1. This must have happened 'somewhere on the way', so we may guess that f is increasing at many points of $[0, 1]$.

However, what we had in mind when thinking about the graph of a 'general' continuous function turns out to be insufficient: we do not have enough imagination to get a clear idea of the graph of the following function.

EXAMPLE 7.16. (Weierstrass' continuous function that is nowhere increasing and nowhere decreasing) *Let* $p \in \mathbb{R}$, $q \in \mathbb{N}$ *such that* $p > 1$, q *is even and* $q > (4\pi + 1)p$. *Then the function* $f : \mathbb{R} \to \mathbb{R}$ *defined by*

$$f(x) := \sum_{n=0}^{\infty} p^{-n} \sin 2\pi q^n x \qquad (x \in \mathbb{R})$$

has the following properties.

(*a*) f *is continuous and bounded,*

(*b*) f *is increasing at no point of* $\mathbb{R}$,

(*c*) f *is decreasing at no point of* $\mathbb{R}$,

(*d*) f *is differentiable at no point of* $\mathbb{R}$.

Proof. Since $p > 1$, the series defining f is uniformly convergent, so f is continuous and bounded.

Now let $c \in \mathbb{R}$. For $k \in \mathbb{N}$, set $I_k := [c - q^{-k}, c)$, $J_k := (c, c + q^{-k}]$. The function $x \mapsto \sin 2\pi q^k x$ has period q^{-k}, so both on I_k and on J_k it takes all values between -1 and 1. Thus, we can choose

$$\begin{aligned} &x_1 \in I_k \text{ such that } \sin 2\pi q^k x_1 = 1, \\ &x_2 \in I_k \text{ such that } \sin 2\pi q^k x_2 = -1, \\ &y_1 \in J_k \text{ such that } \sin 2\pi q^k y_1 = -1, \\ &y_2 \in J_k \text{ such that } \sin 2\pi q^k y_2 = 1. \end{aligned}$$

If $n > k$, then q^{n-k} is even, so $\sin q^{n-k}\alpha = 0$ for every $\alpha \in \mathbb{R}$ for which $|\sin \alpha| = 1$. In particular,

$$\sin 2\pi q^n x = 0 \qquad (n > k,\ x \in \{x_1, x_2, y_1, y_2\})$$

We find

$$f(x) = \sum_{n=0}^{k} p^{-n} \sin 2\pi q^n x \qquad (x \in \{x_1, x_2, y_1, y_2\})$$

Define

$$s_m(x) := \sum_{n<m} p^{-n} \sin 2\pi q^n x \qquad (m \in \mathbb{N},\ x \in \mathbb{R}).$$

For any $\alpha, \beta \in [c - q^{-k}, c + q^{-k}]$ we have the following inequalities.

$$\begin{aligned} |s_k(\alpha) - s_k(\beta)| &= \left| \sum_{n=0}^{k-1} p^{-n} \sin 2\pi q^n \alpha - \sum_{n=0}^{k-1} p^{-n} \sin 2\pi q^n \beta \right| \\ &\leqslant \sum_{n=0}^{k-1} p^{-n} \cdot |\sin 2\pi q^n \alpha - \sin 2\pi q^n \beta| \\ &\leqslant \sum_{n=0}^{k-1} p^{-n} \cdot 2\pi q^n |\alpha - \beta| = 2\pi \cdot \frac{(q/p)^k - 1}{(q/p) - 1} \cdot |\alpha - \beta| \\ &< \tfrac{1}{2} \left(\frac{q}{p}\right)^k \cdot |\alpha - \beta| \leqslant \left(\frac{q}{p}\right)^k \cdot q^{-k} = p^{-k} \end{aligned}$$

Hence,

$$\begin{aligned} f(x_1) - f(y_1) &= s_k(x_1) - s_k(y_1) + p^{-k}(1+1) > -p^{-k} + 2p^{-k} = p^{-k} > 0, \\ f(x_2) - f(y_2) &= s_k(x_2) - s_k(y_2) + p^{-k}(-1-1) < p^{-k} - 2p^{-k} = -p^{-k} < 0. \end{aligned}$$

We can construct such x_1, x_2, y_1, y_2 for each k. It follows that f is neither increasing nor decreasing at c. To see that f is not differentiable at c, observe that, with x_2 and y_2 as above,

$$\frac{f(y_2) - f(x_2)}{y_2 - x_2} > \frac{p^{-k}}{y_2 - x_2} \geqslant \frac{1}{2} \cdot \frac{p^{-k}}{q^{-k}} = \frac{1}{2}\left(\frac{q}{p}\right)^k \geqslant \frac{1}{2}(4\pi + 1)^k$$

and apply the following exercise.

**Exercise* 7.O. Let $f:\mathbb{R}\to\mathbb{R}$ be differentiable at $c\in\mathbb{R}$. Let $\varepsilon>0$. Then there exists a $\delta>0$ such that

$$\left|\frac{f(y)-f(x)}{y-x}-f'(c)\right|<\varepsilon$$

for all x, y with $c-\delta<x\leqslant c\leqslant y<c+\delta$ and $x\neq y$.

Exercise 7.P. We construct a continuous nowhere differentiable function on $[0, 1]$ which is different from Weierstrass' example. Let $0<\alpha\leqslant\frac{2}{3}$. For $n=0, 1, 2, \ldots$ let $T_n:=\{m3^{-n}: m=0, 1, \ldots, 3^n\}$. Set $T:=T_0\cup T_1\cup T_2\cup\ldots$.
(i) Show that there exists a (unique) function $f: T\to\mathbb{R}$ having the properties

$$f(0)=0,\quad f(1)=1$$

$$\left.\begin{aligned} f((3m+1)3^{-n})&=(1-\alpha)f(m3^{-n+1})+\alpha f((m+1)3^{-n+1})\\ f((3m+2)3^{-n})&=\alpha f(m3^{-n+1})+(1-\alpha)f((m+1)3^{-n+1})\end{aligned}\right\}\ (n\in\mathbb{N},\ 0\leqslant m<3^{n-1})$$

(Draw pictures for the cases $\alpha=\frac{1}{3}, \frac{1}{2}, \frac{2}{3}$.)
(ii) Show that, if $x\in T$ and $m3^{-n}<x<(m+1)3^{-n}$ for certain n and $m<3^n$, then $f(x)$ lies between $f(m3^{-n})$ and $f((m+1)3^{-n})$. Prove that $|f(m3^{-n})-f((m+1)3^{-n})|\leqslant\alpha^n$. Deduce that for every $\varepsilon>0$ there exists a $\delta>0$ such that $|f(x)-f(y)|\leqslant\varepsilon$ as soon as $x, y\in T$ and $|x-y|\leqslant\delta$. Infer that f can (uniquely) be extended to a continuous function on $[0, 1]$. We use the letter f again to indicate the extension.
(iii) Let $c\in[0, 1]$. If $c\neq 1$, set

$$p_n:=\max\{t\in T_n: t\leqslant c\},\quad q_n:=\min\{t\in T_n: t>c\}$$

For $c=1$, put $p_n:=1-3^{-n}$, $q_n:=1$. In any case we have $p_n, q_n\in T_n$, $q_n=p_n+3^{-n}$ and $p_1\leqslant p_2\leqslant\ldots\leqslant c\leqslant\ldots\leqslant q_2\leqslant q_1$. Define

$$\lambda_n:=(q_n-p_n)^{-1}(f(q_n)-f(p_n))$$

Trivially, $\lambda_0=1$. By Exercise 7.O, if f is differentiable at c, then $\lim\lambda_n=f'(c)$.
Show that for every n either $\lambda_{n+1}=3\alpha\lambda_n$ or $\lambda_{n+1}=3(1-2\alpha)\lambda_n$.
(iv) Prove that f is (*continuous and*) *nowhere differentiable* if $\alpha=\frac{2}{3}$.
(v) Now consider the case $\frac{1}{3}<\alpha<\frac{1}{2}$. Show that f is *continuous and strictly increasing, and yet* $f'=0$ *almost everywhere on* $[0, 1]$ (use Theorem 4.10).
(vi) The function f obtained for $\alpha=\frac{1}{2}$ is called *the Cantor function*: we denote it by Φ. Show that Φ is constant on each of the components of $[0, 1]\setminus\mathbb{D}$ (see Example 4.2). Deduce that Φ maps the null set $\mathbb{D}$ *onto* $[0, 1]$.

Notes to Section 7

In connection with Weierstrass' example (7.16) and Exercise 7.P we wish to point out the exposé 'The Theory and Construction of non-differentiable Functions' by A. N. Singh. It has appeared in a collection of four papers, entitled *Squaring the Circle* (ed. E. W. Hobson), published by the Chelsea Publishing Company (New York) in 1969.

There is a considerable number of theorems implying that 'few' continuous functions are differentiable. Most of these are of the following type.

The sup-norm induces a metric on $\mathscr{C}[0, 1]$, enabling one to define a concept of meagreness for subsets of $\mathscr{C}[0, 1]$, generalizing Definition 5.3. Thus, if $n \in \mathbb{N}$, then the polynomial functions on $[0, 1]$ of order $\leqslant n$ are easily seen to form a closed subset of $\mathscr{C}[0, 1]$ whose interior is void, so that the set of *all* polynomial functions on $[0, 1]$ is meagre in $\mathscr{C}[0, 1]$.

As a metric space, $\mathscr{C}[0, 1]$ is complete. Hence, $\mathscr{C}[0, 1]$ itself is not meagre (Oxtoby 1971, Theorem 9.1) and meagre subsets of $\mathscr{C}[0, 1]$ may be regarded as being small. In this sense, one may say that it is rare for a continuous function on $[0, 1]$ to be a polynomial.

This observation is hardly surprising. More interesting is the fact that 'most' continuous functions on $[0, 1]$ are nowhere differentiable (i.e. there exists a meagre set $S \subset \mathscr{C}[0, 1]$ such that all elements of $\mathscr{C}[0, 1] \setminus S$ are nowhere differentiable. It can be shown that 'most' continuous functions f on $[0, 1]$ have the properties:

for all $x \in [0, 1]$, either $D_r^+ f(x) = \infty$ or $D_r^- f(x) = -\infty$,

for almost all $x \in [0, 1]$, $D_r^+ f(x) = \infty$ *and* $D_r^- f(x) = -\infty$,

for all $x \in [0, 1]$, $D^+ f(x) = \infty$ and $D^- f(x) = -\infty$.

For these and similar results, see V. Jarník (*Fund. Math.* **21** (1933), 48–58), P. Kostyrko (*Coll. Math.* **25** (1972), 265–7), S. Mazurkiewicz (*Studia Math.* **3** (1931), 92–4 and 114–18), W. Orlicz (*Fund. Math.* **34** (1947), 45–60); further references are given by A. M. Bruckner (*Am. Math. Monthly* **80** (1973), 679–83).

Propositions of similar structure can be proved for other spaces of functions than $\mathscr{C}[0, 1]$. Thus, for 'most' bounded functions f on $[0, 1]$ the set $S_f := \{x : D^+ f(x) < \infty\}$ is a countable but dense subset of $[0, 1]$. (V. Jarník, *Fund. Math.* **23** (1934), 1–8, where it is also shown that for all $f \in \mathscr{C}[0, 1]$ the set S_f is uncountable.)

3

CONTINUITY

8. Continuous functions

By $\mathscr{C}(I)$, or simply $\mathscr{C}$, we denote the set of all continuous functions defined on the interval I. We collect well-known facts about continuity in the following theorem.

THEOREM 8.1. *Let I be an interval.*
(i) *If $f, g \in \mathscr{C}$ and $\alpha, \beta \in \mathbb{R}$, then $\alpha f+\beta g \in \mathscr{C}$ and $fg \in \mathscr{C}$. ($\mathscr{C}$ is an algebra.)*
(ii) *If f, g $\mathscr{C}$, then $f \vee g \in \mathscr{C}$ and $f \wedge g \in \mathscr{C}$. ($\mathscr{C}$ is a lattice.)*
(iii) *If $f_1, f_2, \ldots \in \mathscr{C}$ and $f := \lim_{n\to\infty} f_n$ uniformly, then $f \in \mathscr{C}$. ($\mathscr{C}$ is uniformly closed.)*
(iv) *If $f \in \mathscr{C}$ and if $f(x) \neq 0$ for all $x \in I$, then $1/f \in \mathscr{C}$.*
(v) *If $f \in \mathscr{C}$ and if $g : f(I) \to \mathbb{R}$ is continuous, then $g \circ f \in \mathscr{C}$.*
(vi) *If $f \in \mathscr{C}$, then $f(I)$ is an interval or, if f is constant, a singleton. Moreover, if f is injective, then f is strictly monotone and $f^{-1} : f(I) \to I$ is continuous. If I is closed and bounded, then so is $f(I)$. In particular, f is bounded on I and has a largest and a smallest value.*
(vii) *If $f \in \mathscr{C}[a, b]$, then there exist polynomial functions P_n on $[a, b]$ ($n \in \mathbb{N}$) such that $\lim_{n\to\infty} P_n = f$ uniformly. (Approximation theorem of Weierstrass)*
(viii) *If $f \in \mathscr{C}$, then f has an antiderivative F, f is Riemann integrable over $[a, b]$ for every $a, b \in I$ with $a<b$ and*

$$\int_a^b f(x)\,\mathrm{d}x = F(b) - F(a)$$

(Fundamental theorem of calculus)

Less known may be the fact that continuity of f can be expressed in topological properties of the graph of f. Recall that a set $S \subset \mathbb{R}^2$ is called *arcwise connected* if for every $v, w \in S$ there is a continuous $\phi : [0, 1] \to S$ such that $\phi(0) = v$ and $\phi(1) = w$.

THEOREM 8.2. *Let $f : [a, b] \to \mathbb{R}$ and let $\Gamma_f := \{(x, f(x)) : x \in [a, b]\}$ be*

the graph of f. Then the following conditions are equivalent.
(α) *f is continuous.*
(β) *Γ_f is compact.*
(γ) *Γ_f is arcwise connected.*
Proof. (α)$\Rightarrow$(γ). Let $(p, f(p))$, $(q, f(q)) \in \Gamma_f$ and let $a \leqslant p < q \leqslant b$. Let $\sigma : [0, 1] \to [p, q]$ be an increasing continuous function for which $\sigma(0) = p$, $\sigma(1) = q$, and let $\psi : [p, q] \to \Gamma_f$ be the map $x \mapsto (x, f(x))$. Then $\phi := \psi \circ \sigma$ is a continuous map $[0, 1] \to \Gamma_f$ with $\phi(0) = (p, f(p))$ and $\phi(1) = (q, f(q))$.

(γ)$\Rightarrow$(β). There is a continuous $\phi : [0, 1] \to \Gamma_f$ such that $\phi(0) = (a, f(a))$ and $\phi(1) = (b, f(b))$. For $x \in [0, 1]$, let $\phi_1(x)$ and $\phi_2(x)$ be defined by $\phi(x) = (\phi_1(x), \phi_2(x))$. Clearly, $a \leqslant \phi_1(x) \leqslant b$ for all x, and by the (Darboux) continuity of ϕ_1 we see that ϕ_1 maps $[0, 1]$ onto $[a, b]$. But then $\phi : [0, 1] \to \Gamma_f$ is surjective. It follows that Γ_f is compact.

(β)$\Rightarrow$(α). Let $x_1, x_2, \ldots \in [a, b]$ with $\lim_{n\to\infty} x_n = x$. We prove that $\lim_{n\to\infty} f(x_n) = f(x)$. The sequence $f(x_1), f(x_2), \ldots$ is bounded (Γ_f being bounded). Let $\alpha := \limsup_{n\to\infty} f(x_n) = \lim_{i\to\infty} f(x_{n_i})$ for some subsequence $x_{n_1}, x_{n_2}, \ldots$ of $x_1, x_2, \ldots$ Then $(x, \alpha) = \lim_{i\to\infty} (x_{n_i}, f(x_{n_i})) \in \Gamma_f$, so that $\alpha = f(x)$. Similarly we can prove that $\beta = f(x)$, where $\beta := \liminf_{n\to\infty} f(x_n)$. Thus the sequence $f(x_1), f(x_2), \ldots$ converges to $f(x)$.

Exercise 8.A. Construct a countable set $\{f_1, f_2, \ldots\}$ of continuous functions on $[a, b]$ such that $f_i \leqslant 1$ for all i while $\sup\{f_i : i \in \mathbb{N}\}$ is not continuous. (Compare Theorem 8.1(ii).)

**Exercise* 8.B. Construct a sequence $f_1, f_2, \ldots$ of continuous functions on $[a, b]$ such that $f(x) := \lim_{n\to\infty} f_n(x)$ exists for every $x \in [a, b]$ but such that f is not continuous. (Compare Theorem 8.1(iii).)

**Exercise* 8.C. (A slight generalization of Theorem 8.1(viii))
(i) Let $f : [a, b] \to \mathbb{R}$ be a Riemann integrable function having an antiderivative F. Show that

$$F(b) - F(a) = \int_a^b f(x)\,dx$$

(Hint. For a partition $a = x_0 < x_1 < \ldots < x_n = b$ of $[a, b]$, write $F(b) - F(a) = \Sigma_{i=1}^n (F(x_i) - F(x_{i-1}))$.)

(ii) Give an example of a Riemann integrable $f : [a, b] \to \mathbb{R}$ having an antiderivative but not being continuous on $[a, b]$.

Exercise 8.D. Find an example of a discontinuous $f : [0, 1] \to \mathbb{R}$ with a closed graph (see Exercises 8.E and 9.B).

Exercise 8.E. Find an example of a discontinuous $f : [0, 1] \to \mathbb{R}$ with a connected graph (see Exercises 8.D and 9.B).

**Exercise* 8.F. (Dini) Let $g, f_1, f_2, \ldots$ be continuous functions on $[a, b]$ such that $f_1(x) \geqslant f_2(x) \geqslant \ldots$ and $\lim_{n\to\infty} f_n(x) = g(x)$ for all $x \in [a, b]$. Then $\lim_{n\to\infty} f_n = g$ uniformly. Give an example of a decreasing sequence of continuous functions, tending pointwise (but not uniformly) to a limit function that is not continuous.

Exercise 8.G. (On Lipschitz conditions of order α) Let $\alpha > 0$ and $f: [0, 1] \to \mathbb{R}$. Then, by definition, $f \in \mathrm{Lip}_\alpha$ if there exists a number C such that for all $x, y \in [0, 1]$ we have $|f(x) - f(y)| \leqslant C|x - y|^\alpha$. (The case $\alpha = 1$ has occurred in Corollary 4.13.) Show the following:
(i) If $f \in \mathrm{Lip}_\alpha$ for some $\alpha > 0$, then f is continuous.
(ii) If $\alpha > 1$, then Lip_α contains only the constant functions.
(iii) If $0 < \alpha \leqslant \beta$, then $\mathrm{Lip}_\alpha \supset \mathrm{Lip}_\beta$.
(iv) Let $0 < \alpha \leqslant 1$. Then $x \mapsto x^\alpha$ is in Lip_α but not in Lip_β for any $\beta > \alpha$.
(v) If $f \in \mathrm{Lip}_1$, then f is of bounded variation.
(vi) There exists an $f \in \mathrm{Lip}_{1/2}$ which is not of bounded variation.
(vii) There exists a continuous monotone function that is in no Lip_α.

Exercise 8.H. For every $\alpha \in (0, 1)$ there exists an $f \in \mathrm{Lip}_\alpha$ that is nowhere differentiable. (Compare Corollary 4.13.) To prove this statement, take p, q, f as in Example 7.16. Let $C := 2\pi pq/(q-p) + 2p/(p-1)$ and α such that $p = q^\alpha$. For $x, y \in [0, 1]$ with $x \neq y$ there is a $k \in \mathbb{N}$ for which $q^{-k} < |x - y| \leqslant q^{-k+1}$. Writing $f(t) = \Sigma_{n<k} p^{-n} \sin 2\pi q^n t + \Sigma_{n \geqslant k} p^{-n} \sin 2\pi q^n t$, show that $|f(x) - f(y)| \leqslant Cp^{-k} \leqslant C|x - y|^\alpha$.

Notes to Section 8

Various generalizations of continuity have been proposed. Thus, S. Mazurkiewicz and H. Auerbach have considered functions $f: \mathbb{R} \to \mathbb{R}$ for which

$$\lim_{h \downarrow 0} (f(x+h) + f(x-h)) = 2f(x) \qquad (x \in \mathbb{R})$$

(see H. Auerbach, *Fund. Math.* **8** (1926), 49–55, E. Stein & A. Zygmund, *Studia Math.* **23** (1963), 249–83, H. Fried, *Fund. Math.* **29** (1937), 134–7).

S. Kempisty (*Fund. Math.* **19** (1932), 185–97) and R. E. Dresler & K. R. Stromberg (*Am. Math. Monthly* **81** (1974), 67–8) study 'quasicontinuous' functions, i.e. functions $f: \mathbb{R} \to \mathbb{R}$ with the property that there exist compact sets $K_1, K_2, \ldots$ such that for each n the restriction of f to K_n is continuous while $\mathbb{R} \setminus \bigcup_n K_n$ is a null set. (Quasicontinuity turns out to be the same as Lebesgue measurability, as follows from Lusin's theorem (20.5).)

K. R. Kellum (*Coll. Math.* **31** (1974), 125–8) calls a function $f: \mathbb{R} \to \mathbb{R}$ 'almost continuous' if every open subset of $\mathbb{R}^2$ that contains the graph of f also contains the graph of a continuous function; he shows that almost continuous functions are Darboux continuous.

Another variant of continuity is the 'approximate continuity' for which we refer to the notes to Section 21.

9. Darboux continuous functions

In elementary analysis one learns that continuous functions and derivatives are Darboux continuous. (For derivatives we have given a proof in the Introduction.) We recall the definition.

DEFINITION 9.1. Let $f:[a, b]\to\mathbb{R}$. f is called *Darboux continuous* if for any p, q with $a\leqslant p<q\leqslant b$ and any $c\in\mathbb{R}$ between $f(p)$ and $f(q)$ there is an s between p and q such that $f(s)=c$.

In this section we collect some facts about the class $\mathscr{DC}$ of the Darboux continuous functions on $[a, b]$. First we formulate an equivalent definition. In Appendix A.8 we obtain the well-known classification of the connected subsets of $[a, b]$: they are of the types (p, q), $[p, q)$, $(p, q]$ or $[p, q]$ where $a\leqslant p\leqslant q\leqslant b$.

THEOREM 9.2. *Let* $f:[a, b]\to\mathbb{R}$. *The following conditions are equivalent.*
(α) f *is Darboux continuous.*
(β) *For any connected set* $C\subset[a, b]$, $f(C)$ *is connected.*
(γ) *For any closed subinterval* $[p, q]$ *of* $[a, b]$, $f([p, q])$ *is connected.*
Proof. $(\alpha)\Rightarrow(\beta)$. Let C be a connected subset of $[a, b]$ and let $f(\alpha)$ and $f(\beta)$ be elements of $f(C)$ ($\alpha, \beta\in C$, $\alpha\leqslant\beta$). Then $[\alpha, \beta]\subset C$. For every c between $f(\alpha)$ and $f(\beta)$ there is an s between α and β such that $f(s)=c$. Hence, $f(C)$ contains the closed interval with end points $f(\alpha)$ and $f(\beta)$. Thus, $f(C)$ is connected.
$(\beta)\Rightarrow(\gamma)$. Trivial.
$(\gamma)\Rightarrow(\alpha)$. Let $p, q\in[a, b]$, $p<q$ and let c be between $f(p)$ and $f(q)$. Since $f([p, q])$ is connected, we have $c\in f([p, q])$.

The reader will have enough imagination to define Darboux continuity for functions defined on an arbitrary connected subset of $\mathbb{R}$ and to prove a characterization similar to Theorem 9.2.

COROLLARY 9.3. *Let* $f:[a, b]\to\mathbb{R}$, $g: C\to\mathbb{R}$ *where* C *is connected and* $f([a, b])\subset C$. *If* f *and* g *are Darboux continuous, then so is* $g\circ f$. *In particular we have*
(i) *If* f *is Darboux continuous, then so are* $|f|, f^2, \lambda f$ *(all* $\lambda\in\mathbb{R}$*).*

(ii) *If f is Darboux continuous and $f(x) \neq 0$ for all $x \in [a, b]$, then $1/f$ is Darboux continuous.*

The class $\mathscr{DC}$ is not a vector space: let $f(x)=g(x)=\sin x^{-1}$ for $x \neq 0$ and $f(0)=0$, $g(0)=1$. Then both f and g are Darboux continuous but $f-g$ is not.

As in the previous section we may ask: can we, from the topological structure of the graph of a function, see whether it is Darboux continuous? We have a partial answer. (Compare Theorem 8.2 and, for the 'converse', Exercise 9.C.)

THEOREM 9.4. *Let $f: [a, b] \to \mathbb{R}$. If the graph Γ_f of f is connected, then f is Darboux continuous.*

Proof. (See Fig. 9.) If f were not Darboux continuous we would have a $[p, q] \subset [a, b]$ (with, say, $f(p) < f(q)$) and a c between $f(p)$ and $f(q)$ for which $c \neq f(x)$ for all $x \in [p, q]$. Define the sets

$$A := \{(x, y) \in \mathbb{R}^2 : x < p\} \cup \{(x, y) \in \mathbb{R}^2 : x < q, y < c\}$$

$$B := \{(x, y) \in \mathbb{R}^2 : x > q\} \cup \{(x, y) \in \mathbb{R}^2 : x > p, y > c\}$$

Then A and B are open, $A \cap B = \varnothing$, $\Gamma_f \subset A \cup B$, $(p, f(p)) \in A$, $(q, f(q)) \in B$. It follows that Γ_f is not connected.

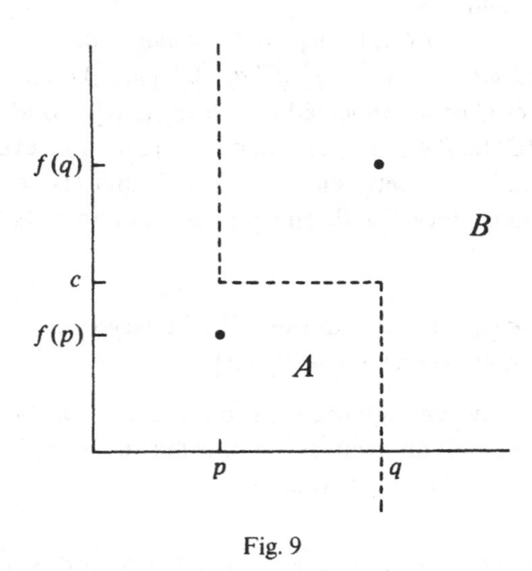

Fig. 9

Exercise 9.A. Let $f(x)=\sin x^{-1}$ for $x \neq 0$, $f(0)=1$. Is $f+h$ Darboux continuous for all continuous functions $h: \mathbb{R} \to \mathbb{R}$? (See Exercise 9.G.)

**Exercise* 9.B. Let $f:[a,b]\to\mathbb{R}$. Consider the conditions (δ) and (ε).
(δ) Γ_f is closed and connected.
(ε) f is Darboux continuous and Γ_f is closed.
Show that both (δ) and (ε) are equivalent to (α), (β) and (γ) of Theorem 8.2 (i.e. equivalent to continuity of f).

Exercise 9.C. (A Darboux continuous g with a disconnected graph)
(i) By an exercise in the Introduction we know of the existence of an $f:\mathbb{R}\to\mathbb{R}$ such that $f(I)=\mathbb{R}$ for every interval I. Define

$$g(x):=\begin{cases} f(x) & \text{if } f(x)\neq x \\ x+1 & \text{if } f(x)=x \end{cases}$$

Prove that g is Darboux continuous but that $x\mapsto g(x)-x$ is not. Since $g(x)\neq x$ for every x we have that $A:=\{(x,y)\in\mathbb{R}^2 : y>x\}$ and $B:=\{(x,y)\in\mathbb{R}^2 : y<x\}$ are disjoint open sets covering Γ_g. As both have nonempty intersection with Γ_g, the graph of g is not connected.
(ii) Show that the graphs of the functions g and $x\mapsto g(x)-x$ are homeomorphic. Apparently, Darboux continuity of a function is *not* a topological property of its graph (see the preamble to Theorem 9.4).

As we did in the case of monotone functions we now look for a description of the linear space generated by the set of the Darboux continuous functions. The answer is as surprising as it is simple: it is the space of *all* functions.

THEOREM 9.5. *Every $f:\mathbb{R}\to\mathbb{R}$ is the difference of two Darboux continuous functions.*
Proof. Let $f:\mathbb{R}\to\mathbb{R}$. Let $g:\mathbb{R}\to\mathbb{R}$ be a function satisfying $g(I)=\mathbb{R}$ for every interval I. (In the Introduction we made such a g.) Define a function $h:\mathbb{R}\to\mathbb{R}$ as follows.

$$h(x):=\begin{cases} \log|g(x)| & \text{if } g(x)\neq 0 \\ 0 & \text{if } g(x)=0 \end{cases}$$

Let $A:=\{x : g(x)\geqslant 0\}$, $B:=\{x : g(x)<0\}$. For every interval I it is easy to see that $h(A\cap I)=\mathbb{R}$ and $h(B\cap I)=\mathbb{R}$. Define

$$f_1(x):=\begin{cases} h(x) & \text{if } x\in A \\ f(x)-h(x) & \text{if } x\in B \end{cases}$$

For any interval I we have $f_1(I)\supset f_1(A\cap I)=h(A\cap I)=\mathbb{R}$, so f_1 is Darboux continuous. For similar reasons, the function f_2, defined by

$$f_2(x):=\begin{cases} f(x)-h(x) & \text{if } x\in A \\ h(x) & \text{if } x\in B \end{cases}$$

is Darboux continuous. As $f_1+f_2=f$, it follows that f is the sum (or difference) of two Darboux continuous functions.

Exercise 9.D. Is every function $\mathbb{R}\to\mathbb{R}$ a product of two Darboux continuous functions?

Exercise 9.E. Show that there exist Darboux continuous functions that do not have antiderivatives.

Exercise 9.F. Let $f:[a, b]\to\mathbb{R}$ be Darboux continuous. Show that f and $|f|$ have the same points of continuity.

Exercise 9.G. Let $f:[a, b]\to\mathbb{R}$ be such that $f+h\in\mathscr{DC}$ for all $h\in\mathscr{DC}$. Show that f is constant. (Hint. First prove that f is Darboux continuous. Now suppose that $f([a, b])$ contains an interval I. Let $g:\mathbb{R}\to\mathbb{R}$ be as in Exercise 9.C. Show that there exist $p, q\in I$ such that $g(p)>p$ and $g(q)<q$. Define $h:=-g\circ f$. Prove that $h\in\mathscr{DC}$ but $f+h\notin\mathscr{DC}$.)

Exercise 9.H

(i) Let $f:\mathbb{R}\to\mathbb{R}$ be Darboux continuous and such that $f(x+)$ and $f(x-)$ exist for every $x\in\mathbb{R}$. Prove that f is continuous.
(ii) Find a Darboux continuous $g:\mathbb{R}\to\mathbb{R}$ such that $g(x+)$ exists for all $x\in\mathbb{R}$ while g is not continuous.

Exercise 9.I. Every injective Darboux continuous function is continuous.

Exercise 9.J. Every function $f:[0, 1]\to[0, 1]$ is the limit of a sequence of Darboux continuous functions. We outline a proof. Let D and t^* be as in the last part of the Introduction. Let $V:=\{m5^{-n}: m\in\mathbb{Z}, n\in\mathbb{N}\}$; for $i\in\mathbb{N}$, put $V_i:=V+2^{-i}$. Show that the sets $V_1, V_2, \ldots$ are pairwise disjoint subsets of D and that for every $x\in[0, 1]$ there exist at most one $i\in\mathbb{N}$ and one $t\in[0, 1]$ with $x-t^*\in V_i$. Define $f_1, f_2, \ldots:[0, 1]\to[0, 1]$ by

$$f_i(x):=\begin{cases} t & \text{if } t\in[0, 1] \text{ and } x-t^*\in V_i \\ f(x) & \text{if there is no } t\in[0, 1] \text{ with } x-t^*\in V_i \end{cases}$$

Then every f_i is Darboux continuous and $\lim_{i\to\infty} f_i=f$.

Exercise 9.K. $\xi_{\{0\}}$ is not a limit of a uniformly convergent sequence of Darboux continuous functions. (Compare Exercise 9.J.)

Exercise 9.L. Show that there exists a uniformly convergent sequence of Darboux continuous functions whose limit is not Darboux continuous (see Exercises 9.J, 9.K). For the construction of such a sequence, let $g:[0, 1]\to\mathbb{R}$ be such that g maps every subinterval of $[0, 1]$ onto $\mathbb{R}$. Let D be a countable subset of $\{x: g(x)=0\}$ that is dense in $[0, 1]$ and let $(d_1, d_2, \ldots)$ be an enumeration of D. For $n\in\mathbb{N}$, define $f_n:[0, 1]\to\mathbb{R}$ by

$$f_n(x):=\begin{cases} 0 & \text{if } x\in\{d_n, d_{n+1}, \ldots\} \\ k^{-1} & \text{if } x=d_k \text{ and } k<n \\ 1 & \text{if } g(x)=0 \text{ but } x\notin D \\ g(x) & \text{if } g(x)\neq 0 \end{cases}$$

Then show that every f_n is Darboux continuous: their uniform limit is not.

Exercise 9.M. (Another function that maps every interval onto $[0, 1]$) For $x \in \mathbb{R}$ let $0.x_1x_2x_3\ldots$ be the standard dyadic development of $x-[x]$:

$$x_n := [2^n x] - 2[2^{n-1}x] \qquad (n \in \mathbb{N})$$

where $[x]$ is the entire part of x. Define $\phi : \mathbb{R} \to \mathbb{R}$ by

$$\phi(x) := \limsup_{n\to\infty} \frac{1}{n}(x_1 + x_2 + \ldots + x_n) \qquad (x \in \mathbb{R})$$

Show that ϕ maps every interval *onto* $[0, 1]$. (Hint. First, show that $\phi(x)=\phi(y)$ if there exist $p, q \in \mathbb{N}$ with $x_p = y_q$, $x_{p+1} = y_{q+1}$, $x_{p+2} = y_{q+2}$, etc., so that it suffices to show that ϕ maps $[0, 1]$ onto $[0, 1]$. Now let $t \in [0, 1]$, $t \neq 1$. Find an $x \in [0, 1]$ such that $x_1 + \ldots + x_n = [nt]$ for every n and prove that $\phi(x) = t$. Finally, find an x with $\phi(x) = 1$.)

Exercise 9.N. For a function $f: [a, b] \to \mathbb{R}$ we define the *counting function* $N_f : \mathbb{R} \to \{0, 1, 2, \ldots, \infty\}$: for every $y \in \mathbb{R}$, $N_f(y)$ will be the number of elements of the set $\{x : f(x) = y\}$.

Prove the following. If $f: [a, b] \to \mathbb{R}$ is Darboux continuous but not continuous, then there exists an interval I such that $N_f(y) = \infty$ for all $y \in I$.

Exercise 9.O. Let $f: \mathbb{R} \to \mathbb{R}$ be a Darboux continuous function such that for every $y \in \mathbb{R}$ the set $\{x \in \mathbb{R} : f(x) = y\}$ is closed. Show that f is continuous.

Exercise 9.P. Let $0 < \alpha < 1$. Let $f: [0, 1] \to \mathbb{R}$, $f(0) = f(1)$.
(i) Show that, if α is one of the numbers $\frac{1}{2}, \frac{1}{3}, \frac{1}{4}, \ldots$ and if f is continuous, then the graph of f has a horizontal chord of length α, i.e. there exist $s, t \in [0, 1]$ with $f(s) = f(t)$ and $|s - t| = \alpha$.
(ii) The proof you gave probably relies on Darboux continuity. Prove, however, that the given continuity condition on f may not be weakened to Darboux continuity. (Take $\alpha := \frac{1}{2}$ and start with a function on $(0, \frac{1}{2}]$ that maps every subinterval of $(0, \frac{1}{2}]$ onto $\mathbb{R}$.)
(iii) Now let $\alpha \notin \{\frac{1}{2}, \frac{1}{3}, \ldots\}$. Define a continuous function f on $[0, 1]$ with $f(0) = f(1)$ whose graph has *no* horizontal chord of length α. (Choose f such that $f(x+\alpha) = f(x) + 1$ for all $x \in [0, 1-\alpha]$.)

Notes to Section 9

Darboux continuity has not been systematically investigated but many stray results are known. For a bibliography, see A. M. Bruckner & J. G. Ceder (*Jahresber. Deutsch. Math. Vereinigung* **67** (1965), 93–117) and A. M. Bruckner, J. G. Ceder & M. Weiss (*Coll. Math.* **15** (1966), 65–77).

10. Semicontinuous functions

Let I be an interval, let $f: I \to \mathbb{R}$. The definition of 'f is continuous' can be written as 'for every $p \in I$ and every $\varepsilon > 0$ there exists a $\delta > 0$ such that, if $x \in I$ and $|x-p|<\delta$, then $-\varepsilon<f(x)-f(p)$ and $f(x)-f(p)<\varepsilon$. Thus, we may split this definition into two halves:

DEFINITION 10.1. Let $f: I \to \mathbb{R}$.
f is called *lower semicontinuous* if for every $p \in I$ and every $\varepsilon>0$ there is a $\delta>0$ such that $x \in I$, $|x-p|<\delta$ implies $-\varepsilon<f(x)-f(p)$.
f is called *upper semicontinuous* if for every $p \in I$ and every $\varepsilon>0$ there is a $\delta>0$ such that $x \in I$, $|x-p|<\delta$ implies $f(x)-f(p)<\varepsilon$.
We denote by $\mathscr{C}^+(I)$ or $\mathscr{C}^+$ the collection of the lower semicontinuous functions on I, by $\mathscr{C}^-(I)$ or $\mathscr{C}^-$ the collection of the upper semicontinuous functions on I. It is clear from the definitions that $\mathscr{C}^+ \cap \mathscr{C}^-$ is $\mathscr{C}$, the collection of the continuous functions, and that $f \in \mathscr{C}^+$ if and only if $-f \in \mathscr{C}^-$.

It is an easy matter to define semicontinuity for a function on an arbitrary subset of $\mathbb{R}$.

The following theorem gives examples of semicontinuous functions that are not continuous.

THEOREM 10.2 *Let $f, g: I \to \mathbb{R}$, $A \subset I$. Then*
(i) *$f \in \mathscr{C}^+$ if and only if for every $s \in \mathbb{R}$ the set $\{x : f(x)>s\}$ is open in I (i.e. is the intersection of I and an open set).*
$f \in \mathscr{C}^-$ if and only if for every $s \in \mathbb{R}$ the set $\{x : f(x)<s\}$ is open in I.
(ii) *$\xi_A \in \mathscr{C}^+$ if and only if A is open in I.*
$\xi_A \in \mathscr{C}^-$ if and only if A is closed in I (i.e. is the intersection of I and a closed set).
(iii) *if $f, g \in \mathscr{C}^+$ and $\lambda \geqslant 0$, then $f+g \in \mathscr{C}^+$ and $\lambda f \in \mathscr{C}^+$.*
If $f, g \in \mathscr{C}^-$ and $\lambda \geqslant 0$, then $f+g \in \mathscr{C}^-$ and $\lambda f \in \mathscr{C}^-$.
Proof. As $f \in \mathscr{C}^-$ if and only if $-f \in \mathscr{C}^+$, it suffices to prove the statements for $\mathscr{C}^+$.
(i) Let $f \in \mathscr{C}^+$ and $s \in \mathbb{R}$. If $f(t)>s$ for certain $t \in I$, then there is a $\delta>0$ such that $x \in I$, $|x-t|<\delta$ implies $-(f(t)-s)<f(x)-f(t)$, i.e. $f(x)>s$. Hence, $\{x : f(x)>s\}$ is open in I. Conversely, assume that $\{x : f(x)>s\}$ is open in I for every $s \in \mathbb{R}$. Let $t \in I$, $\varepsilon>0$. Then $\{x : f(x)>f(t)-\varepsilon\}$ is open in I and contains t. So there is a $\delta>0$ such that $x \in I$, $|x-t|<\delta$ implies $-\varepsilon<f(x)-f(t)$. Hence, $f \in \mathscr{C}^+$.
(ii) Let $\xi_A \in \mathscr{C}^+$. Then $A=\{x : \xi_A(x)>\frac{1}{2}\}$ is open in I. Conversely, if A is open in I, then $\{x : \xi_A(x)>s\}$ is empty if $s \geqslant 1$, is A if $0 \leqslant s<1$ and is I if $s<0$, so in any case is open in I. Thus, $\xi_A \in \mathscr{C}^+$.
(iii) Let $t \in I$ and $\varepsilon>0$. There is a $\delta_1>0$ such that $x \in I$, $|x-t|<\delta_1$ implies $f(x)-f(t)>-\frac{1}{2}\varepsilon$. There is a $\delta_2>0$ such that $x \in I$, $|x-t|<\delta_2$ implies

$g(x)-g(t)>-\frac{1}{2}\varepsilon$. For $x\in I$ with $|x-t|<\min(\delta_1,\delta_2)$ we then have $f(x)+g(x)-(f(t)+g(t))>-\varepsilon$. Thus, if f, $g\in\mathscr{C}^+$, then also $f+g\in\mathscr{C}^+$. We leave the rest to the reader.

Exercise 10.A. For each $p\in\mathbb{R}$ we define $f_p:\mathbb{R}\to\mathbb{R}$ as follows.

$$f_p(x):=\begin{cases}\sin x^{-1} & \text{if } x\neq 0\\ p & \text{if } x=0\end{cases}$$

For which values of $p\in\mathbb{R}$ is f_p upper (lower) semicontinuous, Darboux continuous?

Exercise 10.B. Is the function $x\mapsto[x]$ $(x\in\mathbb{R})$ upper (lower) semicontinuous? (Recall that $[x]$ is the entire part of x.)

Exercise 10.C. Let $f:\mathbb{R}\to\mathbb{R}$ be upper semicontinuous and $h:\mathbb{R}\to\mathbb{R}$ continuous. Does it follow that $f\circ h$ is semicontinuous? Does it follow that $h\circ f$ is semicontinuous?

Exercise 10.D. If $f\in\mathscr{C}^+$ and $f(x)>0$ for all x, does it follow that $1/f\in\mathscr{C}^-$? If $g\in\mathscr{C}^+$ and $g(x)\neq 0$ for all x, does it follow that $1/g\in\mathscr{C}^-$?

Exercise 10.E. Let f, $g\in\mathscr{C}^+$, $f\geqslant 0$, $g\geqslant 0$. Then $fg\in\mathscr{C}^+$.

Exercise 10.F. If $f_1, f_2, \ldots\in\mathscr{C}^+$ and $f:=\lim_{n\to\infty} f_n$ uniformly, then $f\in\mathscr{C}^+$.

**Exercise* 10.G. Let $f:I\to\mathbb{R}$. Then the following conditions are equivalent.

(α) $f\in\mathscr{C}^+$.

(β) If $a, x_1, x_2, \ldots\in I$ and $a=\lim_{n\to\infty} x_n$, then $f(a)\leqslant\liminf_{n\to\infty} f(x_n)$.

**Exercise* 10.H. Give an example of a semicontinuous function on $[0, 1]$ that is discontinuous at uncountably many points.

Exercise 10.I. Let $f\in\mathscr{C}^+[a, b]$. Then f has an (absolute) minimum on $[a, b]$. (Hint. Consider the proof you know for continuous functions.)

THEOREM 10.3. *Let $S\subset\mathscr{C}^+$.*

(i) *If $h(x):=\sup\{f(x):f\in S\}$ exists for all x, then $h\in\mathscr{C}^+$.*

(ii) *If f, $g\in\mathscr{C}^+$, then $f\wedge g\in\mathscr{C}^+$.*

(*In other words, $\mathscr{C}^+$ is closed under arbitrary suprema and finite infima.*)

Proof. Let $s\in\mathbb{R}$. Then $\{x:h(x)>s\}=\bigcup_{f\in S}\{x:f(x)>s\}$, so $h\in\mathscr{C}^+$ by Theorem 10.2(ii). For the second part of the theorem, take f, $g\in\mathscr{C}^+$ and $s\in\mathbb{R}$. Then $\{x:(f\wedge g)(x)>s\}=\{x:f(x)>s\}\cap\{x:g(x)>s\}$. It follows that $f\wedge g\in\mathscr{C}^+$.

**Exercise* 10.J. Show that $\mathscr{C}^-$ is closed under arbitrary infima and finite suprema.

One may wonder whether the inverse of a semicontinuous bijection $[0, 1]\to[0, 1]$ is necessarily semicontinuous. A counterexample is given by the function whose graph is sketched in Fig. 10.

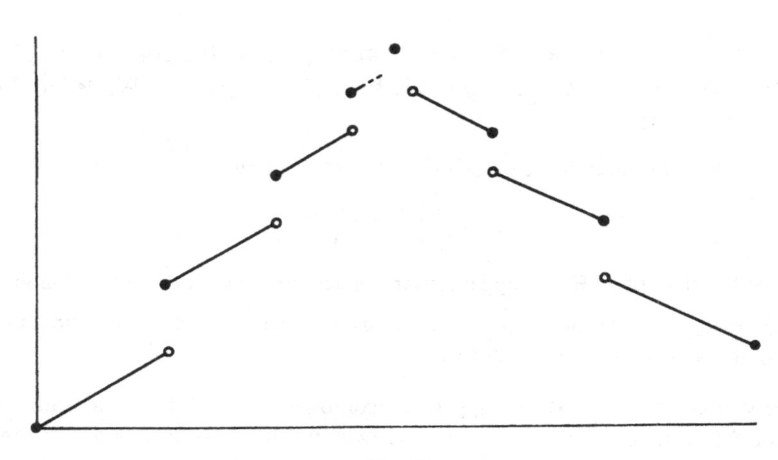

Fig. 10

Let $f: I \to \mathbb{R}$ be bounded below. The set $S := \{g \in \mathscr{C}^+ : g \leqslant f\}$ is non-empty. The function $f^\uparrow$ defined by $f^\uparrow(x) := \sup\{g(x) : g \in S\}$ $(x \in I)$ is the largest $\mathscr{C}^+$-function that is $\leqslant f$ (Theorem 10.3).

DEFINITION 10.4. Let $f: I \to \mathbb{R}$ be a bounded function. We define

$$f^\uparrow(x) := \sup\{g(x) : g \in \mathscr{C}^+, g \leqslant f\} \qquad (x \in I)$$
$$f^\downarrow(x) := \inf\{h(x) : h \in \mathscr{C}^-, h \geqslant f\} \qquad (x \in I)$$

THEOREM 10.5. *Let f and g be bounded functions on an interval I. Then*
(i) $f^\uparrow \leqslant f \leqslant f^\downarrow, f^\uparrow \in \mathscr{C}^+, f^\downarrow \in \mathscr{C}^-$.
If $h \leqslant f$, $h \in \mathscr{C}^+$, then $h \leqslant f^\uparrow$. If $j \geqslant f$, $j \in \mathscr{C}^-$, then $j \geqslant f^\downarrow$.
(ii) *If $f \leqslant g$, then $f^\uparrow \leqslant g^\uparrow$ and $f^\downarrow \leqslant g^\downarrow$.*
(iii) $\inf_{x \in I} f^\downarrow(x) \geqslant \inf_{x \in I} f(x) = \inf_{x \in I} f^\uparrow(x)$.
$\sup_{x \in I} f^\downarrow(x) = \sup_{x \in I} f(x) \geqslant \sup_{x \in I} f^\uparrow(x)$.
(iv) *f is lower semicontinuous if and only if $f = f^\uparrow$.*
f is upper semicontinuous if and only if $f = f^\downarrow$.
f is continuous if and only if $f^\uparrow = f^\downarrow$.
(v) *For every $t \in I$,*

$$f^\uparrow(t) = \lim_{n \to \infty} \inf\{f(x) : x \in I, |x - t| \leqslant 1/n\}$$
$$f^\downarrow(t) = \lim_{n \to \infty} \sup\{f(x) : x \in I, |x - t| \leqslant 1/n\}$$

Proof. We only prove the first statement of (v). For $t \in I$, define $h(t) := \lim_{n \to \infty} \inf\{f(x) : x \in I, |x - t| \leqslant 1/n\}$. We first show that $h \in \mathscr{C}^+$. Let $t \in I$, $\varepsilon > 0$. There is an n such that $\inf\{f(x) : x \in I, |x - t| \leqslant 1/n\} > h(t) - \varepsilon$.

It follows that, if $y \in I$ and $|y-t| \leqslant 1/2n$, then

$$h(y) \geqslant \inf\{f(x) : x \in I, |x-y| \leqslant 1/2n\}$$
$$\geqslant \inf\{f(x) : x \in I, |x-t| \leqslant 1/n\} > h(t)-\varepsilon.$$

Thus, $h \in \mathscr{C}^+$.

Now let $j \in \mathscr{C}^+$, $j \leqslant f$. We prove $j \leqslant h$. (It then follows that $h = f^\uparrow$.) Take $t \in I$, $\varepsilon > 0$. There is an n such that $f(x) \geqslant j(x) \geqslant j(t) - \varepsilon$ for all $x \in I$, $|x-t| \leqslant 1/n$. Then $\inf\{f(x) : x \in I, |x-t| \leqslant 1/n\} \geqslant j(t) - \varepsilon$, whence $h(t) \geqslant j(t) - \varepsilon$. Since the last inequality holds for all $\varepsilon > 0$ we have $h(t) \geqslant j(t)$.

**Exercise* 10.K. Let $f : I \to \mathbb{R}$ be a bounded function. For $t \in I$ and $n \in \mathbb{N}$ define $S(n, t) := \sup\{|f(x) - f(y)| : x, y \in I, \quad |x-t| \leqslant 1/n, \quad |y-t| \leqslant 1/n\} = \sup\{f(x) : x \in I, |x-t| \leqslant 1/n\} - \inf\{f(x) : x \in I, |x-t| \leqslant 1/n\}$. Then $\omega(t) := \lim_{n\to\infty} S(n, t)$ exists for all $t \in I$. It is called the *oscillation of f at t*. f is continuous at t if and only if $\omega(t) = 0$. Show that $\omega = f^\downarrow - f^\uparrow \in \mathscr{C}^-$. Then $\{x : \omega(x) \neq 0\} = \bigcup_n \{x : \omega(x) \geqslant 1/n\}$ is an F_σ. (We have rediscovered Theorem 7.5.)

Exercise 10.L. Let f and g be bounded functions on I. Do we necessarily have $f^{\uparrow\downarrow\uparrow} = f^\uparrow$? Do we necessarily have $f^{\uparrow\downarrow\uparrow\downarrow} = f^{\uparrow\downarrow}$? Must $(f+g)^\uparrow = f^\uparrow + g^\uparrow$?

Exercise 10.M. Let $g, f_1, f_2, \ldots$ be bounded functions on I such that $\lim_{n\to\infty} f_n = g$ uniformly. Then $\lim_{n\to\infty} f_n^\uparrow = g^\uparrow$ uniformly.

Exercise 10.N. Let $f : I \to \mathbb{R}$. Suppose there exist sequences $h_1 \geqslant h_2 \geqslant \ldots$ and $g_1 \leqslant g_2 \leqslant \ldots$ of continuous functions that both tend to f pointwise. Then f is continuous. (Compare Exercise 8.F.)

Exercise 10.O. Let $f : [a, b] \to \mathbb{R}$ be a bounded function whose oscillation (see Exercise 10.K) is everywhere less than 10. Then there is a continuous function $g : [a, b] \to \mathbb{R}$ such that $|f(x) - g(x)| < 10$ for all $x \in [a, b]$.

Exercise 10.P. Let $f : [a, b] \to \mathbb{R}$ be a bounded function with only finitely many discontinuities. Then f is the difference of two bounded lower semicontinuous functions.

**Exercise* 10.Q. Let $f : [a, b] \to \mathbb{R}$ be monotone. Then $f = g - h$ for certain bounded $g, h \in \mathscr{C}^+$. Thus, *any function of bounded variation is the difference of two bounded semicontinuous functions.*

Exercise 10.R. Let $f : [0, 1] \to \mathbb{R}$ be defined as follows. $f(x) := 0$ if x is irrational, while for rational x, $f(x) := 1/n$, where n is the denominator of the irreducible fraction representing x. Compute $f^\uparrow$ and $f^\downarrow$.

Exercise 10.S. Let f be a function on $[a, b]$ which has a local maximum at every point of $[a, b]$. (Example: the restriction of $\xi_{[1,2]}$ to $[0, 2]$.)
(i) Show that f is upper semicontinuous.
(ii) Show that, if f is continuous, then f is constant.
(iii) Show that there exists an interval on which f is constant. (Otherwise, there exist

closed intervals $I_1 \supset I_2 \supset \ldots$ and points $x_i \in I_i$ such that $f(x_1) > f(x_2) > \ldots$ and $f \leqslant f(x_i)$ on I_i for each i.)
(iv) Find a function g on $[0, 1]$ that has a local maximum at every point of $[0, 1]$ and that has infinitely many values.

Exercise 10.T. If a function is semicontinuous and Darboux continuous, is it necessarily continuous?

**Exercise* 10.U. Let $f \in \mathscr{C}^+$ be bounded. Show that the set $\{x : f^{\downarrow}(x) \neq f(x)\}$ is meagre. Give a bounded function g for which $\{x : g^{\downarrow}(x) \neq g^{\uparrow}(x)\}$ is not meagre.

Exercise 10.V. Are the following two statements true?
(i) If $f : \mathbb{R} \to \mathbb{R}$ is bounded and if $g : \mathbb{R} \to \mathbb{R}$ is defined by

$$g(x) := \sup\{f(t) : x-1 < t < x+1\}$$

then g is lower semicontinuous.
(ii) If $f : \mathbb{R} \to \mathbb{R}$ is bounded and if $h : \mathbb{R} \to \mathbb{R}$ is defined by

$$h(x) := \sup\{f(t) : x-1 \leqslant t \leqslant x+1\}$$

then h is lower semicontinuous.

**Exercise* 10.W. Let $f, g \in \mathscr{C}^+$. Let p be such that $f+g$ is continuous at p. Show that both f and g are continuous at p.

**Exercise* 10.X. Let I be an interval. Let $f : I \to \mathbb{R}$ be lower semicontinuous and nowhere negative. Show that f can be extended to a lower semicontinuous function $\mathbb{R} \to \mathbb{R}$ that is nowhere negative.

If $f_1 \leqslant f_2 \leqslant \ldots$ is a sequence of continuous functions, converging pointwise to a function f, then, by Theorem 10.3, $f \in \mathscr{C}^+$. There is also a converse to this statement:

THEOREM 10.6. (Baire) *Let I be an interval. Let $f : I \to \mathbb{R}$ be bounded below. Then the following two conditions are equivalent.*
(α) *f is lower semicontinuous.*
(β) *There is an increasing sequence of continuous functions on I, tending to f pointwise.*
Proof. Only the implication $(\alpha) \Rightarrow (\beta)$ is interesting, so let us assume that f is lower semicontinuous. We may also assume $f \geqslant 0$, and, by Exercise 10.X, $I = \mathbb{R}$. It suffices to construct a countable set $\{f_1, f_2, \ldots\}$ of continuous functions on $\mathbb{R}$ such that $f(x) = \sup\{f_n(x) : n \in \mathbb{N}\}$ for all $x \in \mathbb{R}$. (Then $f = \lim_{n \to \infty} f_1 \vee \ldots \vee f_n$.)
For $q \in \mathbb{Q} \cap (0, \infty)$, let $A_q := \{x \in \mathbb{R} : f(x) > q\}$. Then

$$f = \sup\{q\xi_{A_q} : q \in \mathbb{Q} \cap (0, \infty)\},$$

so it suffices to show that for every q the function $q\xi_{A_q}$ is the supremum of countably many continuous functions on $\mathbb{R}$. Now A_q is an open set. Therefore, we may assume that f is the characteristic function of an open set U. But then U is a union of countably many open intervals $J_1, J_2, \ldots$: then $f=\sup_{n\in\mathbb{N}} \xi_{J_n}$. Thus, we may assume that f actually is the characteristic function of an open interval. Clearly, we lose nothing by considering only the case $f=\xi_{(-1,1)}$. But in this case we have $f=\sup_{n\in\mathbb{N}} f_n$ where each f_n is defined by

$$f_n(x) := \begin{cases} (1-|x|)^{1/n} & \text{if } |x|<1 \\ 0 & \text{if } x\in\mathbb{R},\ |x|\geqslant 1. \end{cases}$$

**Exercise* 10.Y. (Another proof of Theorem 10.6) Let I be an interval and let $f: I\to\mathbb{R}$ be bounded below. For $n\in\mathbb{N}$, let

$$f_n := \sup\{g : I\to\mathbb{R} : g\leqslant f,\ |\Phi_1 g|\leqslant n\}$$

(Here $\Phi_1 g$ is the difference quotient of g; see Section 2.)
(i) Show that each f_n is continuous, $f_n\leqslant f$ for all n, and $f_1\leqslant f_2\leqslant \ldots$
(ii) Show that

$$f_n = \inf\{f(y)+n|x-y| : y\in I\} \qquad (x\in I)$$

and deduce from this equality that $\lim_{n\to\infty} f_n(x)=f^{\uparrow}(x)$ for all $x\in I$.

Exercise 10.Z
(i) A semicontinuous function cannot be discontinuous everywhere. Indeed, if $f:\mathbb{R}\to\mathbb{R}$ is semicontinuous, then the points where f is discontinuous form a meagre set. (Hint. First, suppose that $f\in\mathscr{C}^-(\mathbb{R})$ and that f is bounded. Let ω be as in Exercise 10.K. Take $\varepsilon>0$. Show that it is enough to prove that $\{x\in\mathbb{R} : \omega(x)\geqslant\varepsilon\}$ has empty interior. Now suppose that this set contains an interval (a, b). Choose $c\in(a, b)$ with $f(c)<\inf\{f(x) : a<x<b\}+\frac{1}{2}\varepsilon$. By using the semicontinuity of f at c, deduce that $\omega(c)<\varepsilon$, which is a contradiction.
To remove the boundedness restriction on f, consider $\arctan\circ f$.)
(ii) Combining the above with Theorem 7.5 we see that the points of discontinuity of a semicontinuous function form a meagre F_σ. (For a more general result, see Theorem 11.4.) Use Exercise 7.G to prove the converse: If $S\subset\mathbb{R}$ is a meagre F_σ, then there is an $f\in\mathscr{C}^-(\mathbb{R})$ such that S is the set of all points of discontinuity of f. (You may now skip Exercise 11.H.)

11. Functions of the first class of Baire

The last theorem of the previous section states that a semicontinuous function is the limit of a *monotone* sequence of continuous functions. We define:

DEFINITION 11.1. Let I be an interval. A function $f: I\to\mathbb{R}$ is a *function of* *the first class (of Baire)* if f is the pointwise limit of a sequence of continuous functions on I.

(This definition has an obvious extension for $f: X \to \mathbb{R}$ where X is any subset of $\mathbb{R}$.) Let us denote the first class of Baire by $\mathscr{B}^1$ (or $\mathscr{B}^1(X)$ if there is any possibility of confusion).

By Theorem 10.6, a bounded semicontinuous function on an interval is of the first class. It will follow from Exercise 11.E that the boundedness is irrelevant: on an interval, all semicontinuous functions and all linear combinations of such functions are of the first class. Monotone functions and functions of bounded variation are of the first class (Exercise 10.Q). If $f: \mathbb{R} \to \mathbb{R}$ is differentiable, then f' is of the first class. (For every x, $f'(x) = \lim_{n \to \infty} f_n(x)$ where f_n is defined by $f_n(x) := n(f(x+1/n) - f(x))$.)
Not every function is of the first class (Corollary 11.5).

Exercise 11.A. Show the following: A function of the first class need not be semicontinuous. A function of the first class need not be a derivative. A function of the first class need not be of bounded variation.

Exercise 11.B. Show that a function $[a, b] \to \mathbb{R}$ with only finitely many discontinuities is of the first class (see Theorem 11.8).

**Exercise* 11.C. Let f be a differentiable function on an interval. Show that f' is of the first class. (Be careful!)

**Exercise* 11.D. Let I be an interval and let $f: I \to \mathbb{R}$ be a bounded function of the first class, not the constant function 0. Show that there is a sequence $f_1, f_2, \ldots$ of continuous functions, converging to f, and such that $|f_n(t)| < \sup\{|f(x)| : x \in I\}$ for all $t \in I$ and $n \in \mathbb{N}$.

**Exercise* 11.E. Show that for an $f: \mathbb{R} \to \mathbb{R}$ the conditions $f \in \mathscr{B}^1$ and $\arctan \circ f \in \mathscr{B}^1$ are equivalent.

**Exercise* 11.F. Let $f: [0, 1] \to \mathbb{R}$ be such that its restriction to $(0, 1]$ is of the first class. Show that f itself is also of the first class.

**Exercise* 11.G. Let I be an interval, let $f: I \to \mathbb{R}$. Define $g: \mathbb{R} \to \mathbb{R}$ by

$$g(x) := \begin{cases} f(x) & \text{if } x \in I \\ 0 & \text{if } x \notin I \end{cases}$$

Show that $f \in \mathscr{B}^1(I)$ if and only if $g \in \mathscr{B}^1(\mathbb{R})$.

Exercises 11.E and 11.G serve very practical purposes. In many situations, when we want to prove certain types of functions to be of the first class, 11.E allows us to restrict our attention to bounded functions. Similarly, thanks to 11.G, we may often assume that the interval I we deal with is actually $\mathbb{R}$.

THEOREM 11.2. *Let I be an interval. Let f, $g : I \to \mathbb{R}$ be of the first class. Then*
(i) $f+g \in \mathscr{B}^1$ *and* $\lambda f \in \mathscr{B}^1$ *for every* $\lambda \in \mathbb{R}$. *Furthermore,* $fg \in \mathscr{B}^1$. ($\mathscr{B}^1$ *is an algebra of functions.*)
(ii) $f \vee g \in \mathscr{B}^1$ *and* $f \wedge g \in \mathscr{B}^1$. ($\mathscr{B}^1$ *is a lattice.*)
(iii) *if* $h : \mathbb{R} \to \mathbb{R}$ *is continuous, then* $h \circ f \in \mathscr{B}^1$.
(iv) *If J is an interval and* $\sigma : J \to I$ *is continuous, then* $f \circ \sigma \in \mathscr{B}^1(J)$.
We leave the proof as an easy exercise.

A function f, defined on an interval I, is continuous if and only if for every open set $U \subset \mathbb{R}$, $f^{-1}(U)$ is open in I. In the same vein there is a characterization of functions of the first class. We now present half of this characterization. (For a proof of the other part we have to develop more machinery.)

THEOREM 11.3. *Let I be an interval and let* $f \in \mathscr{B}^1(I)$. *Then for every open set* $U \subset \mathbb{R}$, $f^{-1}(U)$ *is an* F_σ. (*See Theorem* 11.12 *for the converse.*)
Proof. Define $g : \mathbb{R} \to \mathbb{R}$ by setting $g(x) := f(x)$ if $x \in I$ and $g(x) := 0$ if $x \notin I$. Then $f^{-1}(U) = g^{-1}(U) \cap I$. By Exercise 11.G it suffices to prove the theorem for the case $I = \mathbb{R}$.

As U is a union of countably many intervals, we may assume that U itself is an open interval, (a, b), say. Let $f_1, f_2, \ldots$ be a sequence of continuous functions on $\mathbb{R}$ with $\lim_{n\to\infty} f_n = f$. Then

$$\{x : f(x) > a\} = \bigcup_{\substack{\varepsilon \in \mathbb{Q} \\ \varepsilon > 0}} \bigcup_{m \in \mathbb{N}} \bigcap_{\substack{n \in \mathbb{N} \\ n \geq m}} \{y : f_n(y) \geq a + \varepsilon\}$$

so $\{x : f(x) > a\}$ is an F_σ. Similarly, $\{x : f(x) < b\}$ is an F_σ. Then so is their intersection, which is $f^{-1}((a, b))$.

THEOREM 11.4. *Let I be an interval and let* $f \in \mathscr{B}^1(I)$. *Then the points of discontinuity of f form a meagre* F_σ. *In particular the points of continuity form a dense set.*
Proof. By Exercise 11.G we may assume that $I = \mathbb{R}$. For $p, q \in \mathbb{R}$ with $p < q$, set $A_{pq} := \mathbb{R} \setminus (p, q)$. If a is a point of discontinuity of f, there exist $p, q \in \mathbb{Q}$ with

$$\begin{cases} p < f(a) < q \\ \text{there exist } a_1, a_2, \ldots \text{ such that } \lim_{n\to\infty} a_n = a \text{ and } f(a_n) \notin (p, q) \text{ for all } n. \end{cases}$$

Thus, the set of points of discontinuity of f is

$$S := \bigcup_{\substack{p,q \in \mathbb{Q} \\ p < q}} \overline{f^{-1}(A_{pq})} \setminus f^{-1}(A_{pq})$$

Now each $f^{-1}(A_{pq})$ is a G_δ (Theorem 11.3), so each $\overline{f^{-1}(A_{pq})} \setminus f^{-1}(A_{pq})$

is an F_σ: then so is S. Furthermore, for every subset X of $\mathbb{R}$, $\bar{X}\setminus X$ has empty interior. It follows from Exercise 6.I that S is meagre.

COROLLARY 11.5. *$\xi_\mathbb{Q}$ is not a function of the first class.*

Exercise 11.H. Let $A\subset\mathbb{R}$ be a meagre F_σ. Show that there exists a (semicontinuous) function $f:\mathbb{R}\to\mathbb{R}$ of the first class such that the elements of A are just the points of discontinuity of f. (Choose closed sets $K_1\subset K_2\subset\ldots$ whose union is A, and set $f:=\Sigma_{n=1}^{\infty}2^{-n}\xi_{K_n}$.)

Exercise 11.I. Let $f:\mathbb{R}\to\mathbb{R}$ be differentiable. Show that the set of points where f' is continuous is dense in $\mathbb{R}$.

Exercise 11.J. Let $f:\mathbb{R}\to\mathbb{R}$ be such that $f(I)=\mathbb{R}$ for every interval I. Show that f cannot be of the first class. (See, however, Exercise 16.V.)

**Exercise* 11.K
(i) Let $A\subset\mathbb{R}$ be nonempty. Define $d_A:\mathbb{R}\to\mathbb{R}$ by

$$d_A(x):=\inf\{|x-a| : a\in A\}$$

Then $|d_A(x)-d_A(y)|\leqslant|x-y|$ for all $x, y\in\mathbb{R}$. d_A is continuous and for $x\in\mathbb{R}$ we have $d_A(x)=0$ if and only if $x\in\bar{A}$.
(ii) Let $A\subset\mathbb{R}$ be closed, let $U\subset\mathbb{R}$ be open and $A\subset U$. Then there exists a continuous function $f:\mathbb{R}\to[0, 1]$ such that $f(x)=1$ for all $x\in A$ while $f(y)=0$ for $y\notin U$. (Take $f:=d_{\mathbb{R}\setminus U}/(d_A+d_{\mathbb{R}\setminus U})$.)

THEOREM 11.6. *Let $X\subset\mathbb{R}$. Then ξ_X, as a function on $\mathbb{R}$, is of the first class if and only if X is both an F_σ and a G_δ.*
Proof. If X is both an F_σ and a G_δ, then there exist a sequence $A_1, A_2,\ldots$ of closed sets and a sequence $U_1, U_2,\ldots$ of open sets such that $A=A_1\cup A_2\cup\ldots=U_1\cap U_2\cap\ldots$ Choose such sequences with $A_1\subset A_2\subset\ldots$ and $U_1\supset U_2\supset\ldots$ By the preceding exercise, there exist continuous functions $f_1, f_2,\ldots$ with $f_n=1$ on A_n and $f_n=0$ on $\mathbb{R}\setminus U_n$ ($n\in\mathbb{N}$). Now we obtain $\xi_X=\lim_{n\to\infty}f_n\in\mathscr{B}^1$.

Conversely, suppose that ξ_X is the limit of a sequence $g_1, g_2,\ldots$ of continuous functions. Then

$$X=\bigcup_n\bigcap_{m\geqslant n}\{x\in\mathbb{R} : g_m(x)\geqslant\tfrac{1}{2}\}=\bigcap_n\bigcup_{m\geqslant n}\{x\in\mathbb{R} : g_m(x)>\tfrac{1}{2}\}$$

so X is an F_σ and also a G_δ.

Exercise 11.L. (The converse of Theorem 11.4 is false.) Let D_r be as in Exercise 6.E(v) and let $f:\mathbb{R}\to\mathbb{R}$ be its characteristic function. Then f is not of the first class, although its points of discontinuity form a meagre closed set.

The function $\xi_\mathbb{Q}$ (or, for that matter, the function f of Exercise 11.L) is not of the first class but it is the pointwise limit of functions that are of the first class: if $(r_1, r_2, \ldots)$ is an enumeration of $\mathbb{Q}$, then

$$\xi_{\mathbb{Q}} = \lim_{n\to\infty} \xi_{\{r_1,\ldots,r_n\}}$$

Apparently, $\mathscr{B}^1$ is not closed under taking pointwise limits of sequences. (In Section 16 we shall study in more detail the sets $\mathscr{B}^1, \mathscr{B}^2, \ldots$, where for each n, $\mathscr{B}^{n+1}$ is the set of functions that are limits of sequences of elements of $\mathscr{B}^n$.) However, we have

THEOREM 11.7. *Let $X \subset \mathbb{R}$. Then $\mathscr{B}^1(X)$ is uniformly closed, i.e. if $f_1, f_2, \ldots \in \mathscr{B}^1(X)$ and $f := \lim_{n\to\infty} f_n$ uniformly, then $f \in \mathscr{B}^1(X)$.*

Proof. The sequence $f_1, f_2, \ldots$ has a subsequence $g_1, g_2, \ldots$ such that $|f(x)-g_n(x)| \leqslant 2^{-n-1}$ for all $n \in \mathbb{N}$ and $x \in X$. Setting $h_n := g_{n+1} - g_n$ $(n \in \mathbb{N})$ we have $h_n \in \mathscr{B}^1$, $|h_n(x)| \leqslant 2^{-n-2} + 2^{-n-1} < 2^{-n}$ for all n and x, and $f = g_1 + \Sigma_{n=1}^{\infty} h_n$ pointwise. It suffices to prove that $\Sigma_{n=1}^{\infty} h_n \in \mathscr{B}^1$.

Since h_n is an element of $\mathscr{B}^1$ there is a sequence $s_{n1}, s_{n2}, \ldots$ of continuous functions converging to h_n pointwise. We may suppose that $|s_{ni}(x)| \leqslant 2^{-n}$ for all i and x (Exercise 11.D). Then for each i the series $\Sigma_n s_{ni}$ is uniformly convergent, so $t_i := \Sigma_{n=1}^{\infty} s_{ni}$ is continuous for each i. For every $x \in X$ we shall show that

$$\lim_{i\to\infty} t_i(x) = \sum_{n=1}^{\infty} h_n(x)$$

(Then it follows that $\Sigma_{n=1}^{\infty} h_n \in \mathscr{B}^1$.)

Take $x \in X$. Let $\varepsilon > 0$. There is an $N \in \mathbb{N}$ such that for all i

$$\sum_{n=N+1}^{\infty} |s_{ni}(x)| < \tfrac{1}{3}\varepsilon \quad \text{and} \quad \sum_{n=N+1}^{\infty} |h_n(x)| < \tfrac{1}{3}\varepsilon$$

Hence, $|t_i(x) - \Sigma_{n=1}^{\infty} h_n(x)| \leqslant \frac{2}{3}\varepsilon + |\Sigma_{n=1}^{N} (s_{ni}(x) - h_n(x))|$.

As $\lim_{i\to\infty} \Sigma_{n=1}^{N}(s_{ni}(x) - h_n(x)) = 0$, we obtain $|t_i(x) - \Sigma_{n=1}^{\infty} h_n(x)| < \varepsilon$ for sufficiently large values of i.

**Exercise* 11.M. Use Theorems 3.3 and 11.7 to give an alternative proof of the fact that a function of bounded variation is of the first class. Even stronger: *let $f : [a, b] \to \mathbb{R}$ be a function such that $f(x-)$ and $f(x+)$ exist for all $x \in [a, b]$. Then f is of the first class.*

**Exercise* 11.N. The proof of Theorem 11.7 suggests the following. Let $X \subset \mathbb{R}$ and let $\mathscr{A}$ be a uniformly closed linear space of functions on X. Then the functions $X \to \mathbb{R}$ that are limits of sequences of elements of $\mathscr{A}$ form a uniformly closed linear space. Deduce that $\mathscr{B}^2, \mathscr{B}^3, \ldots$ are uniformly closed.

As an application of Theorem 11.7 we prove:

THEOREM 11.8. *Let I be an interval. Every function $I\to\mathbb{R}$ that has only countably many points of discontinuity is of the first class.*
(The converse is false, as the example $\xi_{\mathbb{D}}$ shows.)
Proof. Suppose $f: I\to\mathbb{R}$ has only countably many discontinuities. Without restriction we may assume that f is bounded (Exercise 11.E).

$f^{\downarrow}$ and $f^{\uparrow}$ are semicontinuous and therefore of the first class. Set $\omega:=f^{\downarrow}-f^{\uparrow}$. Then
(i) $0\leqslant f-f^{\uparrow}\leqslant\omega$,
(ii) $\{x:\omega(x)>0\}$ is countable,
(iii) ω is upper semicontinuous and bounded.
It suffices to prove that $f-f^{\uparrow}\in\mathscr{B}^1$. Let $\varepsilon>0$; by Theorem 11.7 we are done if we can construct a $g\in\mathscr{B}^1$ such that $g\leqslant f-f^{\uparrow}\leqslant g+\varepsilon$.

By (ii) and (iii), the set $A:=\{x:\omega(x)\geqslant\varepsilon\}$ is countable and closed. If A' is any subset of A, then A' is countable, so A' is an F_σ and $\mathbb{R}\setminus A'$ is a G_δ; then $A\setminus A'$ is a countable G_δ, whence $\xi_{A\setminus A'}\in\mathscr{B}^1$ by Theorem 11.6. Hence, if $B\subset A$, then $\xi_B\in\mathscr{B}^1$.

Now, for $n\in\mathbb{N}$ put $B_n:=\{x:n\varepsilon\leqslant(f-f^{\uparrow})(x)<(n+1)\varepsilon\}$; let $g:=\Sigma_{n=1}^{\infty}n\varepsilon\xi_{B_n}$. As $f-f^{\uparrow}$ is bounded, $\xi_{B_n}=0$ for large n. Furthermore, each B_n is a subset of A. Consequently, $g\in\mathscr{B}^1$. But $g\leqslant f-f^{\uparrow}\leqslant g+\varepsilon$.

COROLLARY 11.9. *Every left or right continuous function is of the first class of Baire.*
Proof. Apply Theorems 7.7 and 11.8.

THEOREM 11.10 *Let I be an interval. Let $f: I\to\mathbb{R}$ be such that its graph is a closed subset of $\mathbb{R}^2$. Then f is a difference of two semicontinuous functions and therefore is of the first class.*
Proof. We first show that f^2 is lower semicontinuous. Suppose it is not. It follows from Exercise 10.G that there exist $a, x_1, x_2, \ldots\in I$ such that $\lim_{i\to\infty}x_i=a$ while $\lim_{i\to\infty}f(x_i)^2<f(a)^2$. The sequence $f(x_1), f(x_2), \ldots$ is bounded, so we may assume that it converges; let λ be its limit. As $\lim_{i\to\infty}x_i=a$ and the graph of f is closed, it follows that $f(a)=\lambda$, which contradicts the inequality $\lim_{i\to\infty}f(x_i)^2<f(a)^2$.

Thus, f^2 is lower semicontinuous. By the same token, $(f+1)^2$ is lower semicontinuous. Now observe that $f=\frac{1}{2}(f+1)^2-\frac{1}{2}(f^2+1)$.

We proceed to prove the converse of Theorem 11.3. We need the following lemma.

LEMMA 11.11. *Let $A_1, A_2, \ldots, A_N$ be F_σ-sets whose union is $\mathbb{R}$. Then there exist pairwise disjoint F_σ-sets $P_1, \ldots, P_N$ such that $P_1 \cup \ldots \cup P_N = \mathbb{R}$ and $P_i \subset A_i$ for each i.*

Proof. A subset X of $\mathbb{R}$ is *splitting* if there exist pairwise disjoint F_σ-sets $Q_1, \ldots, Q_N$ with $\bigcup_i Q_i = X$ and $Q_i \subset A_i$ for each i. We want to prove that $\mathbb{R}$ is splitting.

(i) Obviously, *every splitting set is an F_σ.*

(ii) It is also obvious that, *if X is splitting and if Z is an F_σ-set that is contained in X, then Z is splitting.*

(iii) *If $X_1, X_2, \ldots$ are splitting G_δ-sets, then their union is splitting.* We prove this as follows. For every $n \in \mathbb{N}$, choose disjoint F_σ-sets Q_{ni} $(i = 1, \ldots, N)$ with $X_n = Q_{n1} \cup \ldots \cup Q_{nN}$ and $Q_{ni} \subset A_i$ for each i. For all n, define

$$X'_n := X_n \setminus \bigcup_{k<n} X_k, \qquad Q'_{ni} := Q_{ni} \cap X'_n$$

Finally, set $Q'_i := \bigcup_n Q'_{ni}$ $(i = 1, \ldots, N)$. If $n, m \in \mathbb{N}$, $i, j \in \{1, \ldots, N\}$ and $i \neq j$, then $Q'_{ni} \cap Q'_{mj} = \varnothing$. (If $n \neq m$, then $Q'_{ni} \cap Q'_{mj}$ is contained in $X'_n \cap X'_m$, which is empty; otherwise, $Q'_{ni} \cap Q'_{mi}$ $(= Q'_{ni} \cap Q'_{nj})$ is contained in $Q_{ni} \cap Q_{nj}$, which is also empty.) It follows that $Q'_1, \ldots, Q'_N$ are pairwise disjoint. They are F_σ-sets because each X_n is both an F_σ and a G_δ while each Q_{ni} is an F_σ. Furthermore, $\bigcup_i Q'_i = \bigcup_{n,i} Q'_{ni} = \bigcup_n X'_n = \bigcup_n X_n$ and $Q'_i \subset A_i$ for each i.

(iv) Now we come to the point. Let Y be the union of all splitting open intervals that have rational end points. Let $Y^c := \mathbb{R} \setminus Y$. By (iii), Y is splitting; we prove $Y = \mathbb{R}$. Suppose $Y \neq \mathbb{R}$. Then Y^c is a nonempty closed subset of $\mathbb{R}$ that is covered by the F_σ-sets $A_1, \ldots, A_N$. Each A_i is a union of countably many closed sets $A_{i1}, A_{i2}, \ldots$ From Exercise 5.J we obtain the existence of an open interval J with rational end points such that $Y^c \cap J$ is not empty but is contained in some A_{ij}, hence in some A_i. Being itself an F_σ-set, $Y^c \cap J$ is trivially splitting. Hence, (iii) implies that $(Y^c \cap J) \cup Y$ is splitting. Then so is J, according to (ii). By the definition of Y, it follows that $J \subset Y$, i.e. $Y^c \cap J = \varnothing$. This is a contradiction.

THEOREM 11.12. (Lebesgue) *Let I be an interval and let $f: I \to \mathbb{R}$. Then f is of the first class if and only if for every open subset U of $\mathbb{R}$ $f^{-1}(U)$ is an F_σ.*

Proof. One half of the theorem has been proved in 11.3. Now let $f: I \to \mathbb{R}$ be such that $f^{-1}(U)$ is an F_σ for every open U: we prove $f \in \mathscr{B}^1$. Without loss of generality, assume $I = \mathbb{R}$ (apply Exercise 11.G). Furthermore, it follows from Exercise 11.E that we may assume that f is bounded, and even $0 \leqslant f \leqslant 1$. Take $M \in \mathbb{N}$. By Theorem 11.7 it suffices to find a $g \in \mathscr{B}^1$ with $|f(x) - g(x)| \leqslant M^{-1}$ for all x. For $i = 0, 1, \ldots, M$, set $A_i :=$

$\{x : (i-1)M^{-1} < f(x) < (i+1)M^{-1}\}$. Every A_i is, by hypothesis, an F_σ and together the A_i cover $\mathbb{R}$. As we have just proved, there exist pairwise disjoint F_σ-sets $P_0, \dots, P_M$, covering $\mathbb{R}$, and such that $P_i \subset A_i$ for each i. Now the complement of each P_i is an F_σ-set, so each P_i is not only an F_σ but also a G_δ. Applying Theorem 11.6, we see that the function

$$g := \sum_{i=0}^{M} \frac{i}{M} \xi_{P_i}$$

is of the first class. Clearly, $g(x) - M^{-1} \leqslant f(x) \leqslant g(x) + M^{-1}$ for each $x \in \mathbb{R}$.

COROLLARY 11.13. *Let $f : I \to \mathbb{R}$. Suppose that there exist functions g_1, $g_2, \dots, h_1, h_2, \dots$ of the first class such that $g_1 \geqslant g_2 \geqslant \dots$, $\lim_{n\to\infty} g_n = f$, $h_1 \leqslant h_2 \leqslant \dots$, $\lim_{n\to\infty} h_n = f$. Then $f \in \mathscr{B}^1$.* (Compare Exercise 10.N.)
Proof. Let $a, b \in \mathbb{R}$. Then

$$\{x : f(x) < b\} = \bigcup_n \{x : g_n(x) < b\}$$

and

$$\{x : f(x) > a\} = \bigcup_n \{x : h_n(x) > a\}$$

so $\{x : f(x) < b\}$ and $\{x : f(x) > a\}$ are F_σ-sets. Then so is their intersection, $\{x : f(x) \in (a, b)\}$. It follows that, for every open $U \subset \mathbb{R}$ the set $f^{-1}(U)$ is an F_σ. Then $f \in \mathscr{B}^1$.

Baire has given the following characterization of the elements of $\mathscr{B}^1$: *a function $f : \mathbb{R} \to \mathbb{R}$ belongs to $\mathscr{B}^1$ if and only if for every nonempty closed subset A of $\mathbb{R}$ the restriction of f to A is continuous at some point of A.* (We give a proof in Appendix C.)

Exercise 11.O. For a subset X of $\mathbb{R}$ we denote by X° the interior of X. Let S be a subset of $\mathbb{R}$ that is both a G_δ and an F_σ. Show that $S^\circ \cup (\mathbb{R} \setminus S)^\circ$ is a dense subset of $\mathbb{R}$. (Prove that $S^\circ \cup (\mathbb{R} \setminus S)^\circ = \{x \in \mathbb{R} : \xi_S \text{ is continuous at } x\}$.)

Exercise 11.P
(i) If $f : \mathbb{R} \to \mathbb{R}$ has a closed graph, then the points of discontinuity of f form a closed set with empty interior.
(ii) Conversely, if A is a closed subset of $\mathbb{R}$ with empty interior, then there exists an $f : \mathbb{R} \to \mathbb{R}$ whose graph is closed and such that A is just the set of all points of discontinuity of f. (Hint. Let ϕ be as in Exercise 1.P. Define $f(x) := 0$ for $x \in A$, $f(x) := \phi(x)^{-1}$ for $x \in \mathbb{R} \setminus A$.)

We close this section with a collection of exercises that deal with the functions of the first class in general and are not particularly related to the theorem of Lebesgue that we have just proved.

Exercise 11.Q. Let $f:\mathbb{R}\to\mathbb{R}$ be such that for every integer n the restriction of f to $[n, n+1]$ is of the first class. Then f itself is of the first class.

Exercise 11.R. Let $f\in\mathscr{B}^1(\mathbb{R})$ be such that $f(x)=0$ for all $x\in\mathbb{Q}$. Show that $\{x : f(x)\neq 0\}$ is meagre.

Exercise 11.S. Let $f\in\mathscr{B}^1(\mathbb{R})$ and let g be a continuous function on the set $f(\mathbb{R})$. Show that $g\circ f\in\mathscr{B}^1(\mathbb{R})$. (Warning. If $f_1, f_2, \ldots$ are continuous functions and if $\lim_{n\to\infty} f_n=f$, then g may not be defined on $f_n(\mathbb{R})$!)

Exercise 11.T. Let I be an interval, let $f: I\to\mathbb{R}$. For $n\in\mathbb{N}$, define

$$f_n(x):=\begin{cases}-n & \text{if } x\in I \text{ and } f(x)\leqslant -n\\ f(x) & \text{if } x\in I \text{ and } -n<f(x)<n\\ n & \text{if } x\in I \text{ and } f(x)\geqslant n\end{cases}$$

If each f_n is of the first class, then so is f.

Exercise 11.U

(i) Let $X\subset\mathbb{R}$ be countable. Then every function $X\to\mathbb{R}$ is an element of $\mathscr{B}^1(X)$. (Hint. For $x_1, \ldots, x_n\in X$ ($x_i\neq x_j$ when $i\neq j$) and $\alpha_1, \ldots, \alpha_n\in\mathbb{R}$ there is a continuous function $f: X\to\mathbb{R}$ such that $f(x_i)=\alpha_i$ for each i.)

(ii) The following statement is *false*. For every countable set $X\subset\mathbb{R}$ and every $g\in\mathscr{B}^1(X)$ there exists an $f\in\mathscr{B}^1(\mathbb{R})$ such that g is the restriction of f to X. (Let $X:=\mathbb{Q}\cup(\mathbb{Q}+\pi)$, $g(x):=0$ if $x\in\mathbb{Q}$, $g(x):=1$ if $x\in\mathbb{Q}+\pi$.)

Exercise 11.V. For $n\in\mathbb{N}$ let $S_n:=\{t2^{-n} : t \text{ is an odd integer}\}$. Observe that the sets $S_1, S_2, \ldots$ are pairwise disjoint and that for every infinite subset T of $\mathbb{N}$, $\bigcup_{n\in T} S_n$ is dense in $\mathbb{R}$.

Let $b, a_1, a_2, \ldots$ be real numbers. Define $f:\mathbb{R}\to\mathbb{R}$ by

$$f(x):=\begin{cases}a_n & \text{if } n\in\mathbb{N},\ x\in S_n\\ b & \text{if } x\in\mathbb{R}\setminus\bigcup\limits_{n\in\mathbb{N}} S_n\end{cases}$$

Show that $f\in\mathscr{B}^1(\mathbb{R})$ if and only if $\lim_{n\to\infty} a_n=b$.

Exercise 11.W

(i) Let $\phi:\mathbb{R}\to\mathbb{R}$. Prove that the following two conditions are equivalent.

(α) If $f\in\mathscr{B}^1(\mathbb{R})$, then $\phi\circ f\in\mathscr{B}^1(\mathbb{R})$.

(β) ϕ is continuous.

(Hint. The previous exercise may be useful.)

(ii) Find a $\psi:\mathbb{R}\to\mathbb{R}$ that is not continuous but has the property that $f\circ\psi\in\mathscr{B}^1(\mathbb{R})$ for every $f\in\mathscr{B}^1(\mathbb{R})$. (See also Exercise 11.X.)

Exercise 11.X. Let $\psi:\mathbb{R}\to\mathbb{R}$. Show that the following conditions are equivalent.

(α) For every $f\in\mathscr{B}^1(\mathbb{R})$ we have $f\circ\psi\in\mathscr{B}^1(\mathbb{R})$.

(β) For every F_σ-set $X\subset\mathbb{R}$, $\psi^{-1}(X)$ is an F_σ.

(γ) For every open $U\subset\mathbb{R}$, $\psi^{-1}(U)$ is both an F_σ and a G_δ.

Exercise 11.Y. Let $f:\mathbb{R}\to\mathbb{R}$, let Γ_f be the graph of f.
(i) Show that

$$\Gamma_f=\bigcap_{\substack{a,b\in\mathbb{Q}\\ a<b}}\{(x,y)\in\mathbb{R}^2 : a\leqslant f(x)\leqslant b \text{ or } y>b \text{ or } y<a\}$$

Deduce that, if $f\in\mathscr{B}^1$, then Γ_f is a G_δ-subset of $\mathbb{R}^2$.
(ii) Let $(q_1, q_2, \ldots)$ be an enumeration of $\mathbb{Q}$. The following formula defines an $f:\mathbb{R}\to\mathbb{R}$ that is not of the first class although its graph is a G_δ.

$$\begin{cases} f(q_n):=n^{-1} \text{ for all } n\in\mathbb{N}\\ f(x):=-1 \text{ if } x\in\mathbb{R}\setminus\mathbb{Q}\end{cases}$$

Challenge 11.Z. Let $f:\mathbb{R}\to\mathbb{R}$, let Γ_f be the graph of f. If Γ_f is an F_σ-subset of $\mathbb{R}^2$, then Γ_f is also a G_δ-subset of $\mathbb{R}^2$.

12. Riemann integrable functions

Let $f:[a,b]\to\mathbb{R}$ be a bounded function. An *overestimate* of f on $[a,b]$ is a number $\Sigma_{i=1}^n h_i(x_i-x_{i-1})$ where $a=x_0<x_1<\ldots<x_n=b$ is a partition of $[a,b]$ and where, for each i, $h_i\geqslant f(x)$ if $x\in(x_{i-1},x_i)$. The set of all overestimates of f on $[a,b]$ is bounded below. Its infimum is called the *upper Riemann integral* of f over $[a,b]$. Similarly, one defines *underestimates* of f on $[a,b]$, and the supremum of the set of all underestimates is called the *lower Riemann integral* of f over $[a,b]$. f is said to be *Riemann integrable over* $[a,b]$ if its upper and lower Riemann integrals over $[a,b]$ are the same. Their common value is then called the *Riemann integral of f over* $[a,b]$ and denoted $\int_a^b f(x)\,dx$. The set of all Riemann integrable functions (on $[a,b]$) is indicated by $\mathscr{R}[a,b]$ or, simply, $\mathscr{R}$. Recall that $\mathscr{C}\subset\mathscr{R}$.

**Exercise* 12.A. Let $f:[a,b]\to\mathbb{R}$ be bounded. Then f is Riemann integrable over $[a,b]$ if and only if for each $\varepsilon>0$ there exist an overestimate v and an underestimate w of f on $[a,b]$ such that $v-w<\varepsilon$.

**Exercise* 12.B
(i) $\mathscr{R}$ is a vector space.
(ii) If $f\in\mathscr{R}$, then $f^2\in\mathscr{R}$. (Use the fact that $f^2(x)-f^2(y)\leqslant 2M|f(x)-f(y)|$ where $M:=\sup\{|f(x)| : x\in[a,b]\}$.
(iii) $f, g\in\mathscr{R}$ implies $fg\in\mathscr{R}$. (Hint. $(f+g)^2-(f-g)^2=4fg$.)
(iv) $f\in\mathscr{R}$ implies $f\vee 0\in\mathscr{R}$.
(v) $f, g\in\mathscr{R}$ implies $f\vee g\in\mathscr{R}$ and $f\wedge g\in\mathscr{R}$. (Hint. $f\vee g=(f-g)\vee 0+g$.)
Thus, $\mathscr{R}$ is an algebra and a vector lattice.

**Exercise* 12.C. Let $f_1, f_2, \ldots\in\mathscr{R}[a,b]$, $f:=\lim_{n\to\infty} f_n$ uniformly. Then f is Riemann integrable over $[a,b]$. ($\mathscr{R}[a,b]$ is uniformly closed.) Moreover,

$$\int_a^b f(x)\,dx=\lim_{n\to\infty}\int_a^b f_n(x)\,dx$$

Exercise 12.D. Functions of bounded variation on $[a, b]$ are Riemann integrable. Convex functions on $[a, b]$ are Riemann integrable.

Exercise 12.E. A (bounded) Darboux continuous function need not be Riemann integrable. (In Example 14.3 and also in Example 13.2 we shall see that even a bounded derivative may fail to be Riemann integrable.)

**Exercise* 12.F. Let $f : [a, b] \to \mathbb{R}$ be Riemann integrable. If $f=0$ a.e., then $\int_a^b f(x)\,dx=0$. (Compare Exercise 4.C(iv).) More than that: if $f=0$ on a dense set, then $\int_a^b f(x)\,dx=0$.

The reason for talking about Riemann integrability in a chapter concerning continuity lies in the following.

THEOREM 12.1. *Let $f : [a, b] \to \mathbb{R}$ be bounded. Then the following conditions are equivalent.*

(α) *f is Riemann integrable.*

(β) *The set of points of discontinuity of f is a null set.* ('*f is continuous almost everywhere*'.)

Proof. $(\alpha)\Rightarrow(\beta)$. Let $f^{\downarrow}, f^{\uparrow}$ be as in Definition 10.4. If $a \leqslant a' < b' \leqslant b$ and if $h \in \mathbb{R}$ is such that $f \leqslant h$ on (a', b'), then $f^{\downarrow} \leqslant h$ on (a', b'). Hence, every overestimate of f is an overestimate of $f^{\downarrow}$. Similarly, every underestimate of f is an underestimate of $f^{\uparrow}$. Consequently, for every $\delta > 0$ there exist an overestimate S of $f^{\downarrow}$ and an underestimate s of $f^{\uparrow}$ with $S-s<\delta$. Then $S-s$ is an overestimate of $f^{\downarrow}-f^{\uparrow}$. As $f^{\downarrow}-f^{\uparrow} \geqslant 0$ it follows that $f^{\downarrow}-f^{\uparrow}$ is Riemann integrable and that $\int_a^b (f^{\downarrow}(x)-f^{\uparrow}(x))dx=0$. Then by Exercise 4.C, $f^{\downarrow}-f^{\uparrow}=0$ a.e. on $[a, b]$. But f is continuous at every point where $f^{\downarrow}-f^{\uparrow}$ vanishes (Exercise 10.K).

$(\beta)\Rightarrow(\alpha)$. Let $\varepsilon > 0$. We construct a partition $a=t_0<t_1<\ldots<t_n=b$ of $[a, b]$ such that

$$(*) \qquad \sum_{j=1}^{n} (t_j-t_{j-1}) \cdot \sup\{f(x)-f(y) : x, y \in (t_{j-1}, t_j)\} \leqslant \varepsilon$$

(The integrability of f follows.) There is a number M with $|f(x)-f(y)| \leqslant M$ for all $x, y \in [a, b]$. The set E of all points of discontinuity of f can be covered by open intervals $I_1, I_2, \ldots$ whose total length is at most $\varepsilon/2M$. We call an interval J 'useful' if either J is contained in an I_i or $|f(x)-f(y)| \leqslant \varepsilon/2(b-a)$ for all $x, y \in J \cap [a, b]$. Every point of $[a, b]$ lies in a useful open interval. It follows from the Heine–Borel theorem that there exists a partition $a=t_0<t_1<\ldots<t_n=b$ of $[a, b]$ such that each of the intervals (t_{j-1}, t_j) is useful. It is now easy to see that (*) is true.

Exercise 12.G. Give new proofs of the statements of Exercises 12.B, 12.C and 12.D by applying Theorem 12.1.

**Exercise* 12.H. Let $f:[a, b]\to\mathbb{R}$ be bounded. The expression 'f is continuous a.e. on $[a, b]$' (Theorem 12.1) should not be misunderstood. Show that neither of the following conditions implies Riemann integrability of f.
(i) There exists a continuous $g:[a, b]\to\mathbb{R}$ such that $f=g$ a.e. on $[a, b]$.
(ii) There exists a null set $X\subset[a, b]$ such that the restriction of f to $[a, b]\setminus X$ is continuous.

**Exercise* 12.I
(i) Let $A\subset[0, 1]$. Show that ξ_A is Riemann integrable if and only if $\bar{A}\setminus A^\circ$ is a null set. (A° is the interior of A.)
(ii) Show that there exists a bounded upper semicontinuous function on $[0, 1]$ that is not Riemann integrable. (Hint. Let U_2 be as in the introduction to Theorem 5.5 and set $P:=[0, 1]\setminus U_2$. Prove that P is a closed set with empty interior and that P is not null. Apply (i) to show that $\xi_P\notin\mathscr{R}$.)
(iii) It follows that a bounded function on $[a, b]$ that is of the first class may fail to be Riemann integrable. Now show that a Riemann integrable function may not be of the first class (e.g. ξ_{D_r}, where D_r is as in Exercise 11.L).

Exercise 12.J
(i) Define $f_0:[0,1]\to[0,1]$ as follows. For $x\in[0,1]\setminus\mathbb{Q}$, set $f_0(x):=0$. If $x\in[0,1]\cap\mathbb{Q}$, let $f_0(x):=1/q$ if $q\in\mathbb{N}$, $p\in 0, 1, 2, \ldots, x=p/q$ and p, q are relatively prime. Show that f_0 is Riemann integrable and that $\int_0^1 f_0(x)\,dx=0$.
(ii) Show that although f_0 and $\xi_{(0,1]}:[0, 1]\to\mathbb{R}$ are both Riemann integrable, the composite function $\xi_{(0,1]}\circ f_0$ is not.
(iii) Show that if $f:[0, 1]\to[a, b]$ is Riemann integrable and if $g:[a, b]\to\mathbb{R}$ is continuous, then $g\circ f$ is Riemann integrable over $[0, 1]$.
(iv) Finally, we give a continuous $f:[0, 1]\to[0, 1]$ and a Riemann integrable $g:[0, 1]\to\mathbb{R}$ such that $g\circ f$ is *not* Riemann integrable. Let P be as in Exercise 12.I. Define

$$f(x):=\inf\{|x-p| : p\in P\} \qquad (x\in[0, 1])$$

Then f is continuous (see Exercise 7.D). Take $g:=\xi_{\{0\}}$. Of course, $g\in\mathscr{R}[0, 1]$. Show that $g\circ f\notin\mathscr{R}[0, 1]$.

**Exercise* 12.K. Let $f:[a, b]\to\mathbb{R}$ be bounded. Let $f^\downarrow$, $f^\uparrow$ be as in the proof of Theorem 12.1. Set $\omega:=f^\downarrow-f^\uparrow$.
(i) Show that the following conditions are equivalent.
(α) $f\in\mathscr{R}$.
(β) $\omega\in\mathscr{R}$ and $\int_a^b\omega(x)\,dx=0$.
(For the implication (β)$\Rightarrow$(α), use Exercise 4.C(iv).)
(ii) If $f\in\mathscr{R}$, then $f^\downarrow\in\mathscr{R}$, $f^\uparrow\in\mathscr{R}$ and $\int_a^b f^\downarrow(x)\,dx=\int_a^b f^\uparrow(x)\,dx=\int_a^b f(x)\,dx$.
(iii) Conversely, suppose that $f^\downarrow\in\mathscr{R}$, $f^\uparrow\in\mathscr{R}$ and that $\int_a^b f^\downarrow(x)\,dx=\int_a^b f^\uparrow(x)\,dx$. Show that $f\in\mathscr{R}$.

Exercise 12.L

(i) Let $f:[a,b]\to\mathbb{R}$ be Darboux continuous. Prove that f is Riemann integrable if and only if $|f|$ is Riemann integrable. (Hint. Exercise 9.F.)

(ii) Let $f:[a,b]\to\mathbb{R}$ be differentiable. Then f' is Riemann integrable if and only if $(1+f'^2)^{1/2}$ is Riemann integrable. (Compare Exercise 3.E.)

Exercise 12.M. Let $f:[a,b]\to\mathbb{R}$ be Cauchy–Bourbaki integrable (see Theorem 3.3). Show that f is Riemann integrable. Give an example of a Riemann integrable function that is not Cauchy–Bourbaki integrable.

Exercise 12.N. Let f be a bounded function on $[a,b]$ such that $f(x+)$ exists for all (or almost all) $x\in[a,b]$. Show that $f\in\mathscr{R}[a,b]$.

**Exercise* 12.O. Let $f:[a,b]\to\mathbb{R}$ be Riemann integrable. Consider the function $F: x\mapsto\int_a^x f(t)dt$ $(x\in[a,b])$. Prove that $F'(x)=f(x)$ for every point of continuity x of f. In particular, F is differentiable a.e. on $[a,b]$ and $F'=f$ a.e. on $[a,b]$.

Exercise 12.P. Let $f:[a,b]\to\mathbb{R}$. Suppose that there exist $g_1,g_2,\ldots,h_1,h_2,\ldots\in\mathscr{R}[a,b]$ such that $g_1\geqslant g_2\geqslant\ldots$, $h_1\leqslant h_2\leqslant\ldots$ and $\lim_{n\to\infty}g_n=f$, $\lim_{n\to\infty}h_n=f$. Prove that $f\in\mathscr{R}[a,b]$. (Compare Exercise 10.N, Corollary 11.13.)

Exercise 12.Q. (A partial converse to Theorem 12.1) Given a null set $X\subset[a,b]$, does there always exist a Riemann integrable function f on $[a,b]$ such that X is the set of points of discontinuity of f? Show that the answer is yes if X is a null F_σ-set (see Exercise 7.G).

4

DIFFERENTIATION

13. Differentiable functions

We first present the results proved in calculus courses about the space $\mathscr{D}$ of all differentiable functions on an interval I. We leave the proof to the reader.

THEOREM 13.1. *Let I be an interval.*

(i) *If f, $g \in \mathscr{D}$ and λ, $\mu \in \mathbb{R}$, then $\lambda f + \mu g \in \mathscr{D}$ and $fg \in \mathscr{D}$. ($\mathscr{D}$ is an algebra of functions.) If $f \in \mathscr{D}$, then f is continuous.*

(ii) *If $f \in \mathscr{D}$ and $f(x) \neq 0$ for all $x \in I$, then $1/f \in \mathscr{D}$.*

(iii) *If $f \in \mathscr{D}$, if J is an interval such that $f(I) \subset J$ and if $g : J \to R$ is differentiable, then $g \circ f \in \mathscr{D}$.*

(iv) *If f is a differentiable bijection of I onto an interval J and if $f'(x) \neq 0$ for all $x \in I$, then the inverse map $f^{-1} : J \to I$ is differentiable.*

(v) *If p, $q \in I$, $p < q$ and $f \in \mathscr{D}$, then there is a $\xi \in (p, q)$ such that $f(q) - f(p) = (q-p)f'(\xi)$* (mean value theorem).

(vi) *If $f \in \mathscr{D}$ and $f'(\xi) > 0$ for some $\xi \in I$, then f is increasing at ξ.*

(vii) *If $f \in \mathscr{D}$ and $f' \geqslant 0$, then f is increasing on I.*

If $f \in \mathscr{D}$ and $f' = 0$, then f is constant.

If $f \in \mathscr{D}$ and $f'(x) > 0$ for all $x \in I$, then f is strictly increasing.

Exercise 13.A. Show that $\mathscr{D}$ is not closed with respect to sup, inf and uniform limits.

Exercise 13.B. Let $f \in \mathscr{D}$. Show that f' is continuous if and only if the function $\Phi_1 f$, defined via

$$\Phi_1 f(x, y) := \frac{f(x) - f(y)}{x - y} \quad (x, y \in I, x \neq y)$$

can be extended to a continuous function on $I \times I$.

Exercise 13.C. Let X be a nonempty subset of $\mathbb{R}$ without isolated points. For an $f : X \to \mathbb{R}$ we define the notions 'continuous', 'differentiable', 'increasing at $x \in X$', 'monotone' in the natural way.

Now let $f : \mathbb{Q} \to \mathbb{R}$.

(i) Let f be increasing at every point of $\mathbb{Q}$. Is f increasing?
(ii) Let f be differentiable. Is f' continuous?
(iii) Let f be increasing and differentiable. Is $f' \geqslant 0$?
(iv) Let f be differentiable and $f' \geqslant 0$. Is f increasing?
(v) Let f be differentiable. Is there a ξ between 0 and 1 such that $f(1)-f(0)=f'(\xi)$?
(vi) Let f be monotone. Is f somewhere differentiable?

Try to answer similar questions for $f : \mathbb{D} \to \mathbb{R}$.

Exercise 13.D. By Exercise 4.D, $\mathbb{D}$ is the set of all numbers $\sum_{n=1}^{\infty} a_n 3^{-n}$ where $a_n \in \{0, 2\}$. Define a function $f : \mathbb{D} \to \mathbb{R}$ as follows.

$$f\left(\sum_{n=1}^{\infty} a_n 3^{-n}\right) := \sum_{n=1}^{\infty} a_n 3^{-n!} \qquad (a_1, a_2, \ldots \in \{0, 2\})$$

(i) Show that f is injective by proving that, if $x, y \in \mathbb{D}$, $N \in \mathbb{N}$, then

$$\text{if } y-x > 3^{-N}, \text{ then } f(y)-f(x) \geqslant 3^{-N!}$$

(ii) For each $m \in \mathbb{N}$ there exists a $c_m \in \mathbb{R}$ with

$$|f(x)-f(y)| \leqslant c_m |x-y|^m \qquad (x, y \in \mathbb{D})$$

(Thus, f is differentiable and $f'=0$. f satisfies Lipschitz conditions of all positive orders.)

(iii) For every $m \in \mathbb{N}$ we have

$$\lim_{\substack{x \in \mathbb{D} \\ x \to y}} \frac{f(x)-f(y)}{(x-y)^m} = 0 \qquad (y \in \mathbb{D})$$

Exercise 13.E (A weird function on a weird set) Let $X := \{\sum_{n=1}^{\infty} a_n 3^{-n!} : a_1, a_2, \ldots \in \{0, 1\}\}$, and define $f : X \to \mathbb{R}$ by

$$f\left(\sum_{n=1}^{\infty} a_n 3^{-n!}\right) := \sum_{n=1}^{\infty} a_n 27^{-n!} \qquad (a_1, a_2, \ldots \in \{0, 1\})$$

(i) Show that X is a closed subset of $\mathbb{R}$ without isolated points.
(ii) Show that for every $y \in X$ we have

$$\lim_{\substack{x \in X \\ x \to y}} \frac{f(x)-f(y)}{(x-y)^3} = 1$$

Exercise 13.F. Prove the following extension of the second part of (vii) of Theorem 13.1.

Let f be a differentiable function on $[0, 1]$ such that $f'(x)=0$ for every $x \in [0, 1] \setminus \mathbb{D}$. Then f is constant. (Hint. For simplicity, assume that $f(0)=0$, $f(1)=1$. Find a sequence $[0, 1]=[a_0, b_0] \supset [a_1, b_1] \supset \ldots$ such that for each n, $b_n - a_n = 3^{-n}$ and $f(b_n)-f(a_n) \geqslant 2^{-n}$. Apply Exercise 7.O.)

It is interesting to observe that in Exercise 7.P(vi) we introduced 'the Cantor function' on $[0, 1]$ which is not constant although its derivative vanishes everywhere on $[0, 1] \setminus \mathbb{D}$. It follows that at some points of $\mathbb{D}$ the Cantor function cannot be differentiable.

Exercise 13.G. (Another extension of Theorem 13.1(vii)) Let $f: I \to \mathbb{R}$ be differentiable and suppose that $f'(x)=0$ for all but countably many points x of I. Show that f is constant. (Hint. f' is Darboux continuous.)

Looking back at the program we outlined in the Introduction (and knowing that a differentiable function is continuous) we see that the remaining question is: How monotone is a differentiable function?

Let us work with a differentiable function $f: \mathbb{R} \to \mathbb{R}$. Clearly, if the sign of f' changes at most finitely many times, then f is piecewise monotone. Even if the sign of f' changes only at the points $1, \frac{1}{2}, \frac{1}{3}, \frac{1}{4}, \ldots$ one may very well say sensible things about monotony properties of f and we have a clear idea of how to sketch a graph of such a function. The situation becomes misty if we want to sketch the graph of a differentiable function f, constant on no interval, but such that $f'(q)=0$ for all rational q. As we have been taught in calculus courses that graphs of differentiable functions are 'smooth' and as we cannot imagine f to have a smooth graph we may jump to the conjecture that such f do not exist and express the hope that, say, a differentiable function must be monotone on some subinterval. The following exercise illustrates the situation.

Exercise 13.H. Let $f: \mathbb{R} \to \mathbb{R}$ be differentiable and such that f is monotone on no interval. Let $N := \{x : f'(x)=0\}$, $S := \{x : f'$ is continuous at $x\}$. Then $S \subset N$, N is dense, $\mathbb{R} \setminus N$ is dense and meagre.

Nevertheless, we shall proceed to construct an example of a differentiable function that is monotone on no interval. The example is due to Y. Katznelson & K. Stromberg (*Am. Math. Monthly* **81** (1974), 349–53).

EXAMPLE 13.2. *There exists a function* $L: \mathbb{R} \to \mathbb{R}$ *with the following properties.*

(i) L *is differentiable.*

(ii) $|L'(x)| \leqslant 1$ *for all* $x \in \mathbb{R}$.

(iii) *Both* $\{x : L'(x) > 0\}$ *and* $\{x : L'(x) < 0\}$ *are dense in* $\mathbb{R}$, *so* L *is monotone on no interval.*

(iv) L *is the difference of two monotone functions.*

(v) $\{x : L'(x) = 0\}$ *is dense in* $\mathbb{R}$.

(vi) L' *is Riemann integrable over no interval.*

Proof. We construct an L satisfying (i)–(iv) (see Exercise 13.I). Let Ω be the collection of all continuous functions $\omega: \mathbb{R} \to (0, \infty)$ having the following property. For all $a, b \in \mathbb{R}$ with $a \neq b$,

$$\left| \frac{1}{b-a} \int_a^b \omega(x)\, \mathrm{d}x \right| \leqslant 4\omega(a)$$

(A) The function $x \mapsto (1+|x|)^{-1/2}$ is an element of Ω.
(B) If $\omega \in \Omega$ and $c \in \mathbb{R}$, then $x \mapsto \omega(x+c)$ is an element of Ω.
If $\omega \in \Omega$ and $r>0$, then $x \mapsto \omega(rx)$ is an element of Ω.
If $\omega_1, \ldots, \omega_n \in \Omega$ and $\lambda_1, \ldots, \lambda_n > 0$, then $\lambda_1\omega_1 + \ldots + \lambda_n\omega_n \in \Omega$.
(C) Let $\varepsilon>0$, let $g : \mathbb{R} \to \mathbb{R}$ be continuous and $g \geqslant \varepsilon$. Let $\alpha_1, \ldots, \alpha_n, \beta_1, \ldots, \beta_n$ be pairwise distinct real numbers. Let $s_1, \ldots, s_n, t_1, \ldots, t_n \in \mathbb{R}$ be such that $0 \leqslant s_i < g(\alpha_i)$ for all i and $t_j > 0$ for all j. Then there exists an $\omega \in \Omega$ such that $\omega \leqslant g$, $\omega(\alpha_i) \geqslant s_i$ for each i and $\omega(\beta_j) < t_j$ for each j. (See Fig. 11.)

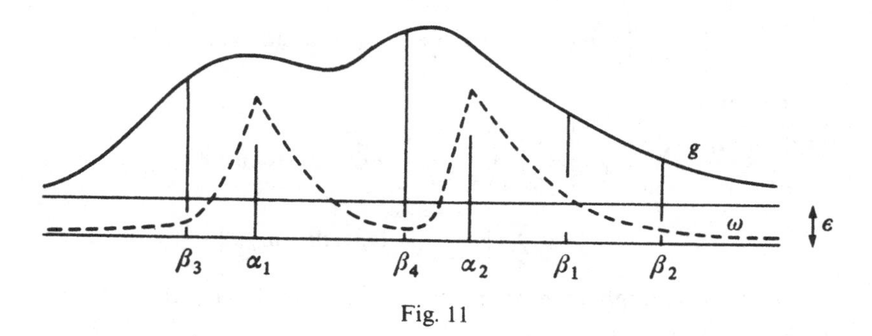

Fig. 11

Proof. We may assume that $\varepsilon < t_j$ for all j and $\varepsilon < g(\alpha_i) - s_i$ for all i. Choose $\delta > 0$ such that $|\alpha_i - \alpha_{i'}| > 2\delta$ whenever $i \neq i'$; $|\alpha_i - \beta_j| > 2\delta$ for all i, j; $g(x) > s_i + \varepsilon$ whenever $|x - \alpha_i| \leqslant \delta$ $(i = 1, \ldots, n)$. Choose $r > 0$ such that $s_i + \ldots + s_n < \varepsilon(1 + r\delta)^{1/2}$. Now let

$$\omega(x) := \sum_{i=1}^{n} s_i(1 + r|x - \alpha_i|)^{-1/2} \qquad (x \in \mathbb{R})$$

By (A) and (B) we have $\omega \in \Omega$. The inequalities $\omega(\alpha_i) \geqslant s_i$ (all i) are clear. Since $1 + r|\beta_j - \alpha_i| \geqslant 1 + 2r\delta$ for each i and j, we have for each j

$$\omega(\beta_j) \leqslant (1 + 2r\delta)^{-1/2} \sum_{i=1}^{n} s_i \leqslant \varepsilon(1 + 2r\delta)^{-1/2}(1 + r\delta)^{1/2} \leqslant \varepsilon < t_j$$

Finally, to prove $\omega \leqslant g$, let $x \in \mathbb{R}$. If for each i $|x - \alpha_i| > \delta$, then $\omega(x) \leqslant (1 + r\delta)^{-1/2}(s_1 + \ldots + s_n) < \varepsilon \leqslant g(x)$. If not, then there is exactly one i for which $|x - \alpha_i| \leqslant \delta$. For this i we obtain

$$s_i(1 + r|x - \alpha_i|)^{-1/2} \leqslant s_i < g(x) - \varepsilon$$

whereas

$$\sum_{k \neq i} s_k(1 + r|x - \alpha_k|)^{-1/2} \leqslant \sum_{k \neq i} s_k(1 + r\delta)^{-1/2} \leqslant (1 + r\delta)^{-1/2} \sum_k s_k \leqslant \varepsilon$$

so that $\omega(x) \leqslant g(x) - \varepsilon + \varepsilon = g(x)$.

(D) Let $\omega_1, \omega_2, \ldots \in \Omega$ be such that the series $\Sigma\omega_n$ converges pointwise. Then the function $x \mapsto \Sigma_{n=1}^{\infty} \int_0^x \omega_n(t)\,dt$ is an antiderivative of $\Sigma_{n=1}^{\infty} \omega_n$.

Proof. For every $x \in \mathbb{R}$ and $n \in \mathbb{N}$,

$$\left|\int_0^x \omega_n(t)\,dt\right| \leqslant 4|x|\omega_n(0)$$

So $F(x) := \Sigma_{n=1}^{\infty} \int_0^x \omega_n(t)\,dt$ exists for each x. Let $a \in \mathbb{R}$. We prove that $F'(a) = \Sigma_{n=1}^{\infty} \omega_n(a)$. Let $\varepsilon > 0$. Choose a positive integer N such that $\Sigma_{n>N} \omega_n(a) < \frac{1}{10}\varepsilon$. For $x \neq a$ and $n \in \mathbb{N}$ we obtain

$$\left|\frac{1}{x-a}\int_a^x \omega_n(t)\,dt - \omega_n(a)\right| \leqslant 4\omega_n(a) + \omega_n(a) = 5\omega_n(a)$$

Hence, for all $x \neq a$:

$$\left|\frac{F(x)-F(a)}{x-a} - \sum_{n=1}^{\infty} \omega_n(a)\right| = \left|\sum_{n=1}^{\infty}\left(\frac{1}{x-a}\int_a^x \omega_n(t)\,dt - \omega_n(a)\right)\right|$$

$$\leqslant \sum_{n \leqslant N} \left|\frac{1}{x-a}\int_a^x \omega_n(t)\,dt - \omega_n(a)\right| + \frac{\varepsilon}{2}$$

If x is close enough to a, then $|(x-a)^{-1}\int_a^x \omega_n(t)\,dt - \omega_n(a)| < \varepsilon/2N$ for $n = 1, \ldots, N$, so that

$$\left|\frac{F(x)-F(a)}{x-a} - \sum_{n=1}^{\infty} \omega_n(a)\right| < \varepsilon.$$

We see that $F'(a) = \Sigma_{n=1}^{\infty} \omega_n(a)$.

(E) Let $\alpha_1, \beta_1, \alpha_2, \beta_2, \ldots$ be pairwise distinct real numbers. Then inductively one can construct $\omega_1, \omega_2, \ldots \in \Omega$ such that for each $n \in \mathbb{N}$ we have, writing $f_n := \omega_1 + \ldots + \omega_n$:

$$(*) \qquad \begin{cases} f_n \leqslant \left(\dfrac{n}{n+1}\right)^2 \\[2ex] f_n(\alpha_i) \geqslant \left(\dfrac{n-1}{n}\right)^2 \text{ for } i = 1, \ldots, n \\[2ex] f_n(\beta_i) < \dfrac{n}{n+1} \cdot \dfrac{i}{i+1} \text{ for } i = 1, \ldots, n \end{cases}$$

Proof. Choose $f_1 := \omega_1 := \frac{1}{8}$. Then (*) holds for $n = 1$. Suppose we have $N \in \{2, 3, \ldots\}$ and $\omega_1, \ldots, \omega_{N-1}$ such that $f_n := \omega_1 + \ldots + \omega_n$ satisfies (*) for $n = 1, \ldots, N-1$. Define

$$\varepsilon := \left(\frac{N}{N+1}\right)^2 - \left(\frac{N-1}{N}\right)^2$$

$$g := \left(\frac{N}{N+1}\right)^2 - f_{N-1}$$

$$s_i := \left(\frac{N-1}{N}\right)^2 - f_{N-1}(\alpha_i) \qquad (i=1, \ldots, N)$$

$$t_i := \frac{N-1}{N} \cdot \frac{i}{i+1} - f_{N-1}(\beta_i) \qquad (i=1, \ldots, N)$$

By (C) there is an $\omega_N \in \Omega$ with $\omega_N \leqslant g$, $\omega_N(\alpha_i) \geqslant s_i$, $\omega_N(\beta_i) < t_i$ for all $i \in \{1, \ldots, N\}$. Then (*) holds for $n=N$, where $f_N = f_{N-1} + \omega_N$.

(F) Let α_i, β_i, ω_i, f_n be as in (E). For each n we have

$$\omega_1 + \omega_2 + \ldots + \omega_n = f_n \leqslant \left(\frac{n}{n+1}\right)^2 \leqslant 1$$

so $\Sigma_{n=1}^{\infty} \omega_n$ converges everywhere. Then by (D), $F(x) := \Sigma_{n=1}^{\infty} \int_0^x \omega_n(t)\,dt$ exists for all $x \in \mathbb{R}$ and $F' = \Sigma_{n=1}^{\infty} \omega_n$. Thus we have constructed *a differentiable function* $F: \mathbb{R} \to \mathbb{R}$ *with* $0 \leqslant F' \leqslant 1$, $F'(\alpha_i) = 1$, $F'(\beta_i) < 1$ *for each* $i \in \mathbb{N}$.

(G) Let $(\alpha_1, \alpha_2, \ldots)$ be an enumeration of $\mathbb{Q}$ and let $\beta_i := \alpha_i + \sqrt{2}$ for all $i \in \mathbb{N}$. Construct F as in (F). By interchanging the roles of the α_i and the β_j we obtain a differentiable $G: \mathbb{R} \to \mathbb{R}$ such that $0 \leqslant G' \leqslant 1$, $G'(\beta_i) = 1$ and $G'(\alpha_i) < 1$ for all i. Now let $L := F - G$. Then L is differentiable, L is the difference of two increasing functions, $-1 \leqslant L' \leqslant 1$, and the sets $\{x : L'(x) > 0\}$ and $\{x : L'(x) < 0\}$ are dense since they contain $\mathbb{Q}$ and $\mathbb{Q} + \sqrt{2}$, respectively.

**Exercise* 13.I. Show that the properties (v) and (vi), mentioned in Example 13.2, follow from (i)–(iv). (Use Exercise 12.F.)

**Exercise* 13.J

(i) Choose α_i, β_i, F as in (G) of the proof of 13.2. Set $H(x) := x - F(x)$ $(x \in \mathbb{R})$. Show that *H is differentiable and strictly increasing while $H'(x) = 0$ for all $x \in \mathbb{Q}$.*

(ii) It follows that in 13.2 we may replace (v) by

$$L'(x) = 0 \text{ for all } x \in \mathbb{Q}$$

(Let L be as in the proof of 13.2; consider $L \circ H$.)

EXAMPLE 13.3. Related to Example 13.2 is the following construction, due to Pompeiu. It yields *a strictly increasing function f defined on some interval $[a, b]$ which is differentiable everywhere (with a bounded derivative) and such that f' vanishes on a dense subset of $[a, b]$.* Actually, we have already obtained such a function in Exercise 13.J(i), but Pompeiu's definition is quite elementary and understandable.

Let $(q_1, q_2, \ldots)$ be an enumeration of $\mathbb{Q} \cap [0, 1]$. For $x \in [0, 1]$, set

$$(*) \qquad g(x) := \sum_{n=1}^{\infty} \frac{1}{n!} \sqrt[3]{(x - q_n)}$$

g is a continuous and strictly increasing map of $[0, 1]$ onto some closed interval $[a, b]$. For our f we take the inverse map $[a, b] \to [0, 1]$. This f is continuous, strictly increasing.

For every $n \in \mathbb{N}$ we have

$$\lim_{x \to q_n} \frac{g(x) - g(q_n)}{x - q_n} \geqslant \lim_{x \to q_n} \frac{1}{n!} \frac{\sqrt[3]{(x - q_n)}}{x - q_n} = \infty$$

so that $f'(g(q_n)) = 0$. We see that f' vanishes on a dense set in $[a, b]$.

For $x, y \in [0, 1]$ with $x < y$ we have $g(y) - g(x) \geqslant \sqrt[3]{(y - q_1)} - \sqrt[3]{(x - q_1)} \geqslant \frac{1}{3}(y - x)$. It follows that all difference quotients of f lie in $[0, 3]$.

Thus, we are done if f is differentiable at every point of the set $[a, b] \setminus \{g(q_n) : n \in \mathbb{N}\}$. This, in turn, will be the case if for every $x \in [0, 1] \setminus \mathbb{Q}$, $\lim_{y \to x}(g(y) - g(x))/(y - x)$ exists in $\mathbb{R} \cup \{\infty\}$.

For $n \in \mathbb{N}$ and $x \in [0, 1]$, put $g_n(x) := (1/n!)\sqrt[3]{(x - q_n)}$: then $g = \Sigma_{n=1}^{\infty} g_n$. Every g_n is differentiable at every point of $[0, 1] \setminus \mathbb{Q}$. Let X be the set $\{x \in [0, 1] : \Sigma_{n=1}^{\infty} g_n'(x) \text{ is finite}\}$. Take $x \in [0, 1] \setminus \mathbb{Q}$.

If $x \notin X$, then for every $\varepsilon > 0$ there exists an $N \in \mathbb{N}$ with $\Sigma_{n=1}^{N} g_n'(x) > \varepsilon^{-1}$; then there is a $\delta > 0$ such that $(x - \delta, x + \delta) \subset [0, 1]$ and $\Sigma_{n=1}^{N} g_n'(y) > \varepsilon^{-1}$ for all $y \in (x - \delta, x + \delta)$; but then for all $y \in (x - \delta, x + \delta) \setminus \{x\}$ the mean value theorem yields

$$\frac{g(y) - g(x)}{y - x} \geqslant (y - x)^{-1} \left(\sum_{n=1}^{N} g_n(y) - \sum_{n=1}^{N} g_n(x) \right) \geqslant \varepsilon^{-1}$$

Thus, $\lim_{y \to x}(g(y) - g(x))/(y - x) = \infty$ if $x \notin X$.

Now suppose $x \in X$. Observing that $|s^{1/3} - t^{1/3}| \leqslant \frac{4}{3} s^{-2/3} |s - t|$ for all $s, t \in \mathbb{R}$, $s \neq 0$, we see that $|g_n(y) - g_n(x)| \leqslant 4 g_n'(x)|y - x|$ for all $y \in \mathbb{R}$ and all $n \in \mathbb{N}$. Hence, for $y \neq x$ and $N \in \mathbb{N}$,

$$\left| \frac{g(y) - g(x)}{y - x} - \sum_{n=1}^{\infty} g_n'(x) \right| = \left| \sum_{n=1}^{\infty} \left(\frac{g_n(y) - g_n(x)}{y - x} - g_n'(x) \right) \right|$$
$$\leqslant \sum_{n=1}^{N} \left| \frac{g_n(y) - g_n(x)}{y - x} - g_n'(x) \right| + \sum_{n > N} 5 g_n'(x)$$

It follows easily that $g'(x) = \Sigma_{n=1}^{\infty} g_x'(x)$.

Notes to Section 13

The existence of nowhere monotone functions that are differentiable and have bounded derivatives is proved by C. E. Weil (*Proc. Am. Math. Soc.* **56** (1976), 388–9) as an application of the Baire category theorem. (See the notes to Section 7.)

The question, for what functions $f : [0, 1] \to \mathbb{R}$ there exists a homeomorphism ϕ of $[0, 1]$ such that $f \circ \phi$ is differentiable (or, more generally, what curves in $\mathbb{R}^n$ admit a differentiable parametrization) has occupied several authors; see A. M. Bruckner's book *Differentiation of Real Functions* (Lecture Notes in Mathematics 659, Springer-Verlag 1978).

G. Choquet in his thesis (*J. Math. pures appl.* **26** (1947), 115–226) has proved that, if $f : [0, 1] \to \mathbb{R}$ is continuous and of bounded variation, then there exists a homeomorphism ϕ of $[0, 1]$ such that $f \circ \phi$ is differentiable and has bounded derivative.

14. Derivatives

Let I be an interval. We denote by $\mathscr{D}'(I)$ or $\mathscr{D}'$ the set of all derivatives of differentiable functions on I, i.e. the set of all functions $I \to \mathbb{R}$ that have antiderivatives. Then, of course, $\mathscr{D}'$ is a vector space of functions. We know that $\mathscr{C} \subset \mathscr{D}'$, $\mathscr{D}' \subset \mathscr{D}\mathscr{C}$, $\mathscr{D}' \subset \mathscr{B}^1$ and that $\mathscr{D}'$ contains discontinuous and unbounded functions.

Exercise 14.A. Find an example of an $f \in \mathscr{D}\mathscr{C} \cap \mathscr{B}^1$ without an antiderivative.

In the Introduction we constructed a noncontinuous $f \in \mathscr{D}'[0, 1]$:

$$f(x) := \begin{cases} \sin x^{-1} & \text{if } 0 < x \leqslant 1 \\ 0 & \text{if } x = 0 \end{cases}$$

The role that the sine function plays in this construction can be taken over by any periodic element of $\mathscr{D}'$:

EXAMPLE 14.1. Let $j \in \mathscr{D}'[0, \infty)$ be such that $j(x+1) = j(x)$ for all $x \geqslant 0$. Let J be an antiderivative of j, and $A := J(1) - J(0)$. Define a function h on $[0, 1]$ by

$$h(x) := \begin{cases} j(x^{-1}) & \text{if } 0 < x \leqslant 1 \\ A & \text{if } x = 0 \end{cases}$$

We prove that $h \in \mathscr{D}'[0, 1]$. (Note the following consequence. If $k \in \mathscr{D}'[0, 1]$ and if $k(x) = j(x^{-1})$ for all $x \in (0, 1]$, then necessarily $k(0) = A$.)

First, observe that we may assume $A = 0$. (Otherwise, replace j by $j - A$, h by $h - A$ and J by $x \mapsto J(x) - Ax$.) The function $x \mapsto J(x+1)$ is an antiderivative of j, so $x \mapsto J(x+1) - J(x)$ is constant. As $J(1) = J(0)$, we see that J is periodic. Then J is bounded. Choose a number C such that $|J(x)| \leqslant C$ for all $x \geqslant 0$.

Define $H : [0, 1] \to \mathbb{R}$ by

$$H(x) := \begin{cases} -x^2 J(x^{-1}) + 2\int_{x^{-1}}^{\infty} J(s) s^{-3} \, ds & \text{if } 0 < x \leqslant 1 \\ 0 & \text{if } x = 0 \end{cases}$$

It is elementary to verify that $H'(x)=h(x)$ for all $x>0$. For $x>0$ we have $|H(x)|\leqslant x^2C+2\int_{x^{-1}}^{\infty} Cs^{-3}\,ds=2x^2C$, so $|(H(x)-H(0))/x|\leqslant 2xC$. Hence $H'(0)=0=A=h(0)$. Consequently, $h=H'\in\mathscr{D}'[0,\infty)$.

Exercise 14.B. By applying the foregoing to the functions j_1, j_2, j_3 with

$$j_1(x):=\sin 2\pi x \quad (x\geqslant 0)$$
$$j_2(x):=|\sin 2\pi x| \quad (x\geqslant 0)$$
$$j_3(x):=\sin^2 2\pi x \quad (x\geqslant 0)$$

show that the following statements are *false*.
(i) If $f\in\mathscr{D}'[0,1]$, then $|f|\in\mathscr{D}'[0,1]$.
(ii) If $f\in\mathscr{D}'[0,1]$, then $f^2\in\mathscr{D}'[0,1]$.

Exercise 14.C. Define $j:[0,\infty)\to\mathbb{R}$ by

$$j(x):=\begin{cases} x & \text{if } 0\leqslant x<\frac{1}{2} \\ 1-x & \text{if } \frac{1}{2}\leqslant x<1\end{cases}$$
$$j(x)=j(x+1) \text{ for all } x\geqslant 0$$

By applying the method of Example 14.1 to the functions $j+1$ and $1/(j+1)$ show that there exists an $f\in\mathscr{D}'[0,1]$ with $1\leqslant f\leqslant 2$ and $1/f\notin\mathscr{D}'[0,1]$.

Exercise 14.D. (We extend Exercises 14.B and 14.C.) Let $\phi:\mathbb{R}\to\mathbb{R}$. Prove the equivalence of the following two statements.
(α) If $f\in\mathscr{D}'[0,1]$, then $\phi\circ f\in\mathscr{D}'[0,1]$.
(β) There exist $\alpha,\beta\in\mathbb{R}$ such that ϕ is the function $x\mapsto\alpha x+\beta$ ($x\in\mathbb{R}$).
(Hint for the implication (α)$\Rightarrow$(β). Prove that $\phi\in\mathscr{D}'(\mathbb{R})$. Let Φ be an antiderivative of ϕ. Let j be as in the previous exercise. Let $a,b\in\mathbb{R}$, $a\neq 0$. Apply the results of Example 14.1 to the functions $j_1:=2aj+b$ and $j_2:=\phi\circ j_1$, and show that $a^{-1}\Phi(a+b)-a^{-1}\Phi(b)=\phi(\frac{1}{2}a+b)$. Thus,

$$\Phi(x+y)-\Phi(x-y)=2y\phi(x) \qquad (x,y\in\mathbb{R})$$

From this formula (and the relation $\Phi'=\phi$) property (β) can be deduced in many ways.)

THEOREM 14.2. *$\mathscr{D}'$ is uniformly closed, that is, if* $\lim_{n\to\infty} f_n=f$ *uniformly and if each f_n has an antiderivative, then so has f.*
Proof. Let $f_1, f_2,\ldots\in\mathscr{D}'$ with $\lim_{n\to\infty} f_n=f$ uniformly. To show that $f\in\mathscr{D}'$ we may assume $|f(x)-f_n(x)|\leqslant 2^{-n-1}$ for all n and x. (Take a suitable subsequence of $f_1, f_2,\ldots$) Then $f=f_1+\Sigma_{n=1}^{\infty}(f_{n+1}-f_n)$, where $|f_{n+1}-f_n|\leqslant 2^{-n}$ for each n. So we are done if we can prove the following: *if $g_1, g_2,\ldots\in\mathscr{D}'$ and $|g_n|\leqslant 2^{-n}$ for all n, then $g:=\Sigma_{n=1}^{\infty} g_n\in\mathscr{D}'$.*

Let I be the interval on which our functions are defined. Choose $x_0\in I$ and for each $n\in\mathbb{N}$ let G_n be an antiderivative of g_n with $G_n(x_0)=0$.

For $x, y\in I$ and $n\in\mathbb{N}$, by the mean value theorem there exists a ξ

between x and y such that $G_n(x)-G_n(y)=(x-y)g_n(\xi)$. Hence,

$$(*) \qquad |G_n(x)-G_n(y)|\leqslant 2^{-n}|x-y| \qquad (x, y \in I, n \in \mathbb{N})$$

In particular, $|G_n(x)|\leqslant 2^{-n}|x-x_0|$ for all x and n. Therefore, $G(x) := \sum_{n=1}^{\infty} G_n(x)$ exists for all $x \in I$ and we have obtained a function G on I. We prove that $G'=g$.

Let $x \in I$, $\varepsilon>0$. Take $N \in \mathbb{N}$ such that $2 \cdot 2^{-N}\leqslant \frac{1}{2}\varepsilon$. For all $y \in I$, $y \neq x$, we have, from (*),

$$\begin{aligned}
\left|\frac{G(y)-G(x)}{y-x}-g(x)\right| &= \left|\sum_{n=1}^{\infty}\left(\frac{G_n(y)-G_n(x)}{y-x}-g_n(x)\right)\right| \\
&\leqslant \left|\sum_{n\leqslant N}\left(\frac{G_n(y)-G_n(x)}{y-x}-g_n(x)\right)\right| + \left|\sum_{n>N}\left(\frac{G_n(y)-G_n(x)}{y-x}-g_n(x)\right)\right| \\
&\leqslant \left|\sum_{n\leqslant N}\left(\frac{G_n(y)-G_n(x)}{y-x}-g_n(x)\right)\right| + \sum_{n>N}(2^{-n}+2^{-n}) \\
&\leqslant \left|\sum_{n\leqslant N}\left(\frac{G_n(y)-G_n(x)}{y-x}-g_n(x)\right)\right| + \frac{\varepsilon}{2}
\end{aligned}$$

We see that $|(G(y)-G(x))/(y-x)-g(x)|\leqslant\varepsilon$ if y is sufficiently close to x.

Exercise 14.E. Prove the following extension of Theorem 14.2. Let I be an interval. Let $f_1, f_2, \ldots$ be a sequence of elements of $\mathscr{D}'(I)$ that converges uniformly to a function $f : I \to \mathbb{R}$. Let $x_0 \in I$. For each n, let F_n be an antiderivative of f_n for which $F_n(x_0)=0$. Then the sequence $F_1, F_2, \ldots$ converges to an antiderivative of f. The convergence is uniform if I is bounded.

Exercise 14.F

(i) If $f \in \mathscr{D}'[a, b]$ is bounded and $g : [a, b] \to \mathbb{R}$ is continuous, then $fg \in \mathscr{D}'[a, b]$. (Hint. First show that $x \mapsto xf(x)$ is in $\mathscr{D}'$. Then use Weierstrass' approximation theorem (8.1(vii)) and Theorem 14.2.)

(ii) Let $f, g : [0, 1] \to \mathbb{R}$ be defined as follows. $f(0):=g(0):=0$. For $x>0$, $f(x):= x^{-1/2}\sin x^{-1}$ and $g(x) := x^{1/2}\sin x^{-1}$. Then $f \in \mathscr{D}'$, g is continuous, but $fg \notin \mathscr{D}'$.

Exercise 14.G. Let $f \in \mathscr{D}'[a, b]$ be bounded and let $g : [0, 1] \to [a, b]$ have a continuous derivative with $g'(x)>0$ for all $x \in [0, 1]$. Then $f \circ g \in \mathscr{D}'[0, 1]$ (apply Exercise 14.F).

Exercise 14.H. (An extension of Exercise 14.F(i))

(i) Let $h \in \mathscr{D}'[0, 1]$ be bounded and such that $h \geqslant 0$, $h(0)=0$ while h is continuous at every point of $(0, 1]$. Then $fh \in \mathscr{D}'[0, 1]$ for every bounded $f \in \mathscr{D}'[0, 1]$. (Hint. Let $f \in \mathscr{D}'$, $0 \leqslant f \leqslant 1$. Use Exercise 14.F to construct a function G on $(0, 1]$ for which $G'=fh$. Show that

$$0 \leqslant G(y)-G(x) \leqslant H(y)-H(x) \qquad (0 \leqslant x \leqslant y \leqslant 1)$$

where $H : [0, 1] \to \mathbb{R}$ is an antiderivative of h. Deduce that G can be extended to a continuous function G_1 on $[0, 1]$ for which $G_1'(0)=0=f(0)h(0)$.)

(ii) Find a function h having the properties mentioned in (i) without being continuous at 0. (For example, first find a continuous function h on $(0, 1]$ with $h(1/n)=1$ for every $n \in \mathbb{N}$ and $\int_0^x h(t)\,dt \leqslant x^2$ for all $x \in (0, 1]$.)

Exercise 14.I. If f has a continuous derivative and if $g \in \mathscr{D}'$, then $fg \in \mathscr{D}'$. (Observe that here we have no boundedness condition. Compare Exercises 14.F and 14.H.)

EXAMPLE 14.3. (A bounded derivative that is not Riemann integrable) Such an example has already been given in 13.2. We now present a construction that is somewhat more direct.

Let $I_1, I_2, \ldots$ be pairwise disjoint open intervals in $[0, 1]$. For each i, let $I_i=(p_i, q_i)$ and choose $s_i, t_i \in I_i$ with $p_i<s_i<t_i<q_i$ and such that $s_i-p_i=q_i-t_i$. For each i, choose a differentiable function $f_i : I_i \to \mathbb{R}$ for which (see Fig. 12)
(i) $f_i(x)=0$ if $x \leqslant s_i$ or $x \geqslant t_i$ $(x \in I_i)$
(ii) $0 \leqslant f_i(x) \leqslant (s_i-p_i)^2$ $(x \in I_i)$
(iii) $\max\{|f_i'(x)| : x \in I_i\}=1$
(It is clear that there exist such functions f_i.)

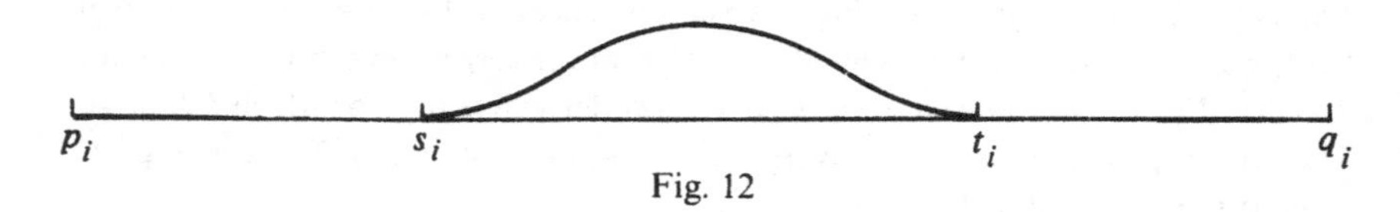

Fig. 12

Now define $f : [0, 1] \to \mathbb{R}$ as follows.

$$f(x) := \begin{cases} f_i(x) & \text{if } i \in \mathbb{N},\ x \in I_i \\ 0 & \text{if } x \in [0, 1] \setminus \bigcup_i I_i \end{cases}$$

We prove that f is differentiable. It is obvious that f is differentiable at every point of $\bigcup_i I_i$, so let $a \in [0, 1] \setminus \bigcup_i I_i$. Then $f(a)=0$. If for any $x \in [0, 1]$, $x \neq a$, the difference quotient $(f(x)-f(a))/(x-a)$ is not 0, then there is an i with $x \in (s_i, t_i)$, so that

$$\left|\frac{f(x)-f(a)}{x-a}\right| = \frac{f_i(x)}{|x-a|} \leqslant \frac{(s_i-p_i)^2}{|x-a|} \leqslant |x-a|$$

Thus, for each $x \in [0, 1]$ we have $|f(x)-f(a)| \leqslant |x-a|^2$, whence $f'(a)=0$. We conclude that f is, indeed, differentiable.

By the construction preceding Theorem 5.5 we can arrange that the set $A := [0, 1] \setminus \bigcup_i I_i$ is (closed and) nowhere dense but not null. Let a be an accumulation point of A. We show that f' is discontinuous at a. In fact, there exist $a_1, a_2, \ldots \in A$ such that $\lim_{n\to\infty} a_n=a$ while $a_n \neq a$ for each n. As A does not contain any interval, between a and each a_n there must lie an I_i and thereby a point x_n with $|f'(x_n)|=1$ (by property (iii) of f_i). Now $\lim_{n\to\infty} x_n=a$ but $\lim_{n\to\infty}|f'(x_n)|=1\neq 0=|f'(a)|$ and f' is discontinuous at a.

Thus, f' is discontinuous at every accumulation point of A. Now A has only countably many isolated points. (For each isolated point a of A there exist $s, t \in \mathbb{Q}$ such that $(s, t)\cap A=\{a\}$; further, $\mathbb{Q}$ is countable.) Hence, f' is discontinuous at almost every point of A. Consequently, the set $\{x : f'$ is discontinuous at $x\}$ is not a null set, so f' is not Riemann integrable. We have found: *f is differentiable, $|f'|\leqslant 1$, and f' is not Riemann integrable.*

14.4. We can choose f in such a way that, in addition to the above, *f' is lower semicontinuous.* In fact, let A, I_i, p_i, q_i be as in 14.3. Define $s_i, t_i \in I_i$ by $s_i-p_i=q_i-t_i=\frac{1}{4}(q_i-p_i)$. For each $i \in \mathbb{N}$, let f_i be the function on I_i such that $f_i=0$ on (p_i, s_i) while f_i' is the function whose graph is sketched in Fig. 13. (Here $L=\frac{1}{4}(q_i-p_i)$.) Then f_i has the properties (i), (ii) and (iii) mentioned in Example 14.3. Moreover, we see that

(iv) $f_i'\geqslant -\frac{1}{4}(q_i-p_i)$

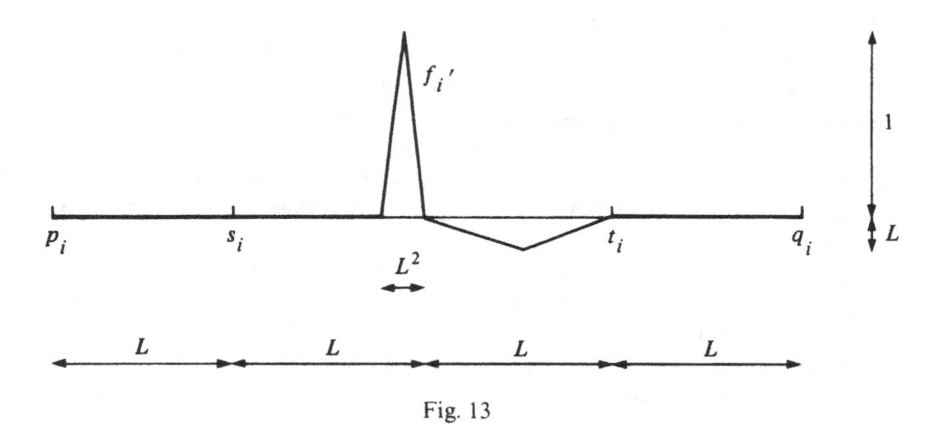

Fig. 13

From these $f_1, f_2, \ldots$ we construct f as before. We claim that now f' is lower semicontinuous. Again, f' is continuous on every I_i. Now take a point a of A. Our claim is proved if we show that

$$f'(x)-f'(a)\geqslant -|x-a| \quad \text{for all } x \in [0, 1]$$

i.e. that

$$f'(x)\geqslant -|x-a| \quad \text{for all } x \in [0, 1]$$

It suffices to consider the points x for which $f'(x)<0$. For such an x there is an i with $x \in (s_i, t_i)$. As $a \notin (p_i, q_i)$, we have, using (iv), $|x-a| \geqslant s_i - p_i = \frac{1}{4}(q_i - p_i) \geqslant -f'(x)$, and we are done.

14.5. We can do still better (or worse). There exists a sequence $A_1, A_2, \ldots$ of closed nowhere dense subsets of $[0, 1]$ such that $[0, 1] \setminus \bigcup_n A_n$ is a null set (see Theorem 5.5). For each n the procedure given above yields a lower semicontinuous function $g_n : [0, 1] \to [-1, 1]$ that is discontinuous at almost every point of A_n but has an antiderivative. Set $g := \sum_{n=1}^{\infty} 2^{-n} g_n$. By Theorem 14.2, g has an antiderivative, f, say. Furthermore, g is bounded and lower semicontinuous. It follows from Exercise 10.W that g is discontinuous at every point where one of the g_n is discontinuous, so g is discontinuous at almost every point of $\bigcup_n A_n$, hence at almost every point of $[0, 1]$. We have now made a function f on $[0, 1]$ such that *f is differentiable, while its derivative is bounded, lower semicontinuous and discontinuous at almost every point of* $[0, 1]$. (Thus, f' is not Riemann integrable over any subinterval of $[0, 1]$.)

Exercise 14.J. A curious property of derivatives is that they all have connected graphs. More generally, if $f : \mathbb{R} \to \mathbb{R}$ is Darboux continuous and of the first class of Baire, then its graph Γ is connected.

We outline a proof. Suppose there exist open subsets U_1, U_2 of $\mathbb{R}^2$ such that $U_1 \cup U_2 \supset \Gamma$, $U_1 \cap U_2 = \varnothing$, $U_1 \cap \Gamma \neq \varnothing$, $U_2 \cap \Gamma \neq \varnothing$. For $i=1, 2$, let $A_i := \{x \in \mathbb{R} : (x, f(x)) \in U_i\}$. Then $A_1 \cup A_2 = \mathbb{R}$, $A_1 \cap A_2 = \varnothing$. Put $X := \bar{A}_1 \cap \bar{A}_2$.
(i) If $p \in A_1$, $q \in A_2$ and $p<q$, then $[p, q]$ intersects X.
(ii) If $a \in A_1$, there exist $a_1, a_2, \ldots \in A_1$ such that $a_1 < a_2 < \ldots$ and $\lim_{n\to\infty} a_n = a$.
(iii) If J is a component of $\mathbb{R} \setminus X$, then $\bar{J} \subset A_1$ or $\bar{J} \subset A_2$. (This follows from obvious variations of (i) and (ii).)
(iv) Let g be the restriction of f to X. There exists a $c \in X$ such that g is continuous at c. Without loss of generality, assume $c \in A_1$. Then there is an $\varepsilon>0$ with $X \cap (c-\varepsilon, c+\varepsilon) \subset A_1$.
(v) All components of $(c-\varepsilon, c+\varepsilon) \setminus X$ are contained in A_1.
(vi) $c \notin \bar{A}_2$; a contradiction.

Notes to Section 14

Although much work has been done on characterizing the class of all derivatives, the results are not satisfactory. See Bruckner (1978).

Exercise 14.D raises the question of which maps $\phi : [0, 1] \to [0, 1]$ have the property that for all $f \in \mathscr{D}'[0, 1]$ it follows that $f \circ \phi \in \mathscr{D}'[0, 1]$. This problem seems to be much harder than the one we attacked in Exercise 14.D. An answer is given by M. Laczkovich & G. Petruska in *Fund. Math.* **100** (1978), 179–99. A sufficient condition on ϕ is that ϕ have a continuous derivative such that $1/\phi'$ is of bounded variation (R. J. Fleissner). On the other hand, it is not enough that there exist numbers m, $M>0$ such that $m(x-y) \leqslant \phi(x)-\phi(y) \leqslant M(x-y)$ $(0 \leqslant x \leqslant y \leqslant 1)$ (Bruckner).

Related to the above is the question for what functions $f: [0, 1] \to \mathbb{R}$ one has $f \circ \phi \in \mathscr{D}'[0, 1]$ for a suitable (or for every) homeomorphism ϕ of $[0, 1]$. The answers are as follows. (For details, see Bruckner (1978).)
(i) $f \circ \phi \in \mathscr{D}'[0, 1]$ for *all* homeomorphisms ϕ, if and only if f is continuous (J. Lipiński).
(ii) $f \circ \phi \in \mathscr{D}'[0, 1]$ for *some* homeomorphism ϕ, if and only if f is Darboux continuous and of the first class of Baire (I. Maximoff).

15. The fundamental theorem of calculus

The processes of differentiation and integration are well known to be, in some sense, each other's inverses. Slightly more precisely, if D is the operation that assigns to every $f \in \mathscr{D}[a, b]$ its derivative $Df := f'$ and if J is the operation that assigns to every $g \in \mathscr{R}[a, b]$ its indefinite integral

$$Jg : x \mapsto \int_a^x g(t)\,dt \qquad (a \leqslant x \leqslant b)$$

then $J(Df)=f$ and $D(Jg)=g$ for many functions f and g. Several theorems to this effect exist, such as the fundamental theorem of calculus which says that $D(Jg)=g$ for every continuous function g. (This is our Theorem 8.1(viii).)

In this section we try to obtain more general results. Let us first consider the identity $D(Jg)=g$. It implies, firstly, that g can be reconstructed from Jg, and, secondly, that this can be done by differentiation. However, an integrable function is not determined by its indefinite integral. ($J(\xi_{\{a\}}) = J(0)$.) Thus, we first have to deal with the question: if $Jh=Jg$, then in what respect can h differ from g? But there is another problem, since Jg may not be differentiable for some g (find an example). Hence, in order to 'reconstruct' a function from its indefinite integral differentiation will not always be adequate. We may surmount this difficulty by extending the operation D to a $\tilde{D}$ defined on a class of functions containing all indefinite integrals (e.g. D_r^+: see Definition 4.4).

The formula $J(Df)=f$ yields similar problems. First, we know that f is not completely determined by Df. This obstacle is not serious, as we also know that f is determined by Df up to a constant. The second problem is deeper: can f (up to a constant) be obtained from Df by integration? The Riemann integral will not suffice, since we have seen in Examples 13.2 and 14.3 that Df need not be Riemann integrable. Again, the problem may be solved by extending the operation J to a $\tilde{J}$ which is defined on a larger class of functions.

Unfortunately, with the introduction of $\tilde{D}$ and $\tilde{J}$ we run up against the same difficulties which we were trying to avoid, but which now concern the identities $\tilde{D}(\tilde{J}g)=g$ and $\tilde{J}(\tilde{D}f)=f$. In sections 21 and 22 we shall see that our problem can more or less be solved by making suitable choices for $\tilde{D}$ and $\tilde{J}$. The extensions of J, meant here, (the Lebesgue and the Perron integrals) involve a lot of machinery. In this section we shall see what we can do if we admit generalizations of the differentiation D, whereas for J we stick to the Riemann integral. Thus, we set ourselves the following tasks. ($\tilde{D}$ will be some reasonable extension of D.)
1*a*. If $Jf=Jg$, then compare f and g.
1*b*. If $\tilde{D}f=\tilde{D}g$, then compare f and g.
2*a*. If $\tilde{D}f$ is Riemann integrable, compare $J\tilde{D}f$ and f.
2*b*. If $\tilde{D}Jf$ is defined, compare $\tilde{D}Jf$ and f.
In the following exercise we discuss 1*a*.

**Exercise* 15.A. For $f, g \in \mathscr{R}[a, b]$ the following conditions are equivalent.
(α) $Jf=Jg$, i.e. $\int_a^x f(t)\,dt=\int_a^x g(t)\,dt$ for all $x \in [a, b]$.
(β) If $c, d \in [a, b]$, then $\int_c^d f(t)dt=\int_c^d g(t)dt$.
(γ) $f(x)=g(x)$ for every point x of $[a, b]$ where both f and g are continuous.
(δ) $f=g$ a.e. on $[a, b]$.
(ε) $\{x : f(x)=g(x)\}$ is dense in $[a, b]$.
(ζ) $\int_a^b |f(x)-g(x)|dx=0$.
(Prove $(\alpha)\Leftrightarrow(\beta)\Rightarrow(\gamma)\Rightarrow(\delta)\Rightarrow(\varepsilon)\Rightarrow(\zeta)\Rightarrow(\beta)$, using the results of Section 12.)

Next, we turn to 1*b*. We first show that $Df=Dg$ implies that $f-g$ is constant. Clearly, it suffices to prove that, if $Df=0$, then f is constant. It will be convenient to show also that, if $Df\geqslant 0$, then f is increasing. (This theorem is, of course, extremely well known. Our proof differs from the one generally found in textbooks.)

THEOREM 15.1. *Let* $f \in \mathscr{D}[a, b]$.
(i) *If* $f'(x)>0$ *for all* $x \in [a, b]$, *then* f *is strictly increasing.*
(ii) *If* $f'(x)\geqslant 0$ *for all* $x \in [a, b]$, *then* f *is increasing.*
(iii) *If* $f'(x)=0$ *for all* $x \in [a, b]$, *then* f *is constant.*

Usually, one deduces this theorem from the mean value theorem, which is quite hard to prove and which is much overrated in importance. We shall present a more elementary proof of Theorem 15.1 which has the advantage of working also for D_r^+f instead of f'. (See Exercise 15.B.) Further, as a corollary of Theorem 15.1 we obtain the following weak form of the mean value theorem (which for most practical purposes is sufficient).

COROLLARY 15.2. *Let $f \in \mathscr{D}[a, b]$. Let $A, B \in \mathbb{R}$ be such that $A \leqslant f' \leqslant B$. Then $A(b-a) \leqslant f(b)-f(a) \leqslant B(b-a)$.*
Proof. By Theorem 15.1(ii), the functions $x \mapsto f(x)-Ax$ and $x \mapsto Bx-f(x)$ are increasing.

It will turn out that the conclusion of Corollary 15.2 holds already under the assumption that f is continuous and that $A \leqslant D_r^+f \leqslant B$ (Exercise 15.E). Compare this with the fact that, for $f(x) := |x|$ $(x \in [-2, 1])$, we have $f(1)-f(-2)=(1-(-2))D_r^+f(x)$ for no x.

Proof of Theorem 15.1. (i) It suffices to show that f is increasing. (If f were increasing but not strictly increasing, then f would be constant on a subinterval of $[a, b]$, contradicting the assumption.) Without loss of generality we assume $f(a)>f(b)$. Choose any t with $f(b)<t<f(a)$. As f is continuous, the set $\{x \in [a, b] : f(x)=t\}$ has a largest element, c, say. Clearly, $c<b$. On $(c, b]$ the function f does not take the value t, while $f(b)<t$, so for all $x \in (c, b]$ we have $f(x)<t=f(c)$, whence $(f(x)-f(c))/(x-c)<0$. It follows that $f'(c) \leqslant 0$, a contradiction.
(ii) By (i), for each $\varepsilon>0$ the function $x \mapsto f(x)+\varepsilon x$ is increasing. Then so is f.
(iii) By (ii), both f and $-f$ are increasing.

Theorem 15.1 (or, properly speaking, part (iii) of it) takes care of 1b for the case $\tilde{D}=D$. Now let us consider $\tilde{D}=D_r^+$: if f and g are *continuous* functions on $[a, b]$ with $D_r^+f=D_r^+g$, does it follow that $f-g$ is constant? This time, it is not clear that we lose nothing by assuming $g=0$, since the operation D_r^+ is not linear. Still, it will be useful first to consider the case $g=0$.

**Exercise* 15.B. Let $f : [a, b] \to \mathbb{R}$ be continuous.
(i) If $D_r^+f(x)>0$ for all $x \in (a, b)$, then f is strictly increasing.
(ii) If $D_r^+f(x) \geqslant 0$ for all $x \in (a, b)$, then f is increasing.
(iii) If $D_r^+f(x)=0$ for all $x \in (a, b)$, then f is constant.
(Warning. The proof of (iii) is different from the one of Theorem 15.1(iii). Infer from (ii) that f is increasing. Then $D_r^-f=0$. Consider $-f$.)

One may ask what conclusions can be drawn from $D_r^+f<0$ and $D_r^+f\leqslant 0$, respectively. Since $D_r^+f=-D_r^-(-f)$, the answer can be found in

**Exercise* 15.C. Let $f:[a,b]\to\mathbb{R}$ be continuous.
(i) If $D_r^-f(x)>0$ for all $x\in(a,b)$, then f is strictly increasing.
(ii) If $D_r^-f(x)\geqslant 0$ for all $x\in(a,b)$, then f is increasing.
(iii) If $D_r^-f(x)=0$ for all $x\in(a,b)$, then f is constant.
(To prove (i) and (ii), simple adaptations of the proofs of 15.1(i) and (ii) suffice. (Another way is to use Exercise 15.B.) For (iii), note that $D_r^+(-f)=-D_r^-f$.)

Now we consider the general question: if $D_r^+f\geqslant D_r^+g$, is $f-g$ necessarily increasing? Since D_r^+ is not a linear operation the answer does not follow at once from Exercises 15.B and 15.C (and, in fact, the answer turns out to be negative).

**Exercise* 15.D
(i) Let $f,g:[a,b]\to\mathbb{R}$, $x\in[a,b]$. If $D_r^+f(x)+D_r^+g(x)$ is defined (e.g. if $D_r^+g(x)$ is finite), then

$$D_r^+(f+g)(x)\leqslant D_r^+f(x)+D_r^+g(x)$$

(ii) Let $f,g:[a,b]\to\mathbb{R}$ be continuous. Let $D_r^+f(x)\geqslant D_r^+g(x)$ for all $x\in(a,b)$ and suppose $D_r^+g(x)\in\mathbb{R}$ for all $x\in(a,b)$. Then $f-g$ is increasing. (Hint. $D_r^+f\leqslant D_r^+(f-g)+D_r^+g$.) (In Exercise 15.P we shall discover continuous functions f,g (with $D_r^+g(x)=\infty$ for some x) such that $D_r^+f=D_r^+g$ and $f-g$ is decreasing, not constant.)
(iii) Let $f,g:[a,b]\to\mathbb{R}$ be continuous. If $D_r^+f(x)=D_r^+g(x)\in\mathbb{R}$ for all $x\in(a,b)$, then $f-g$ is constant. If $D_r^-f(x)=D_r^-g(x)\in\mathbb{R}$ for all $x\in(a,b)$, then $f-g$ is constant.

In the next exercise we formulate the weak form of the mean value theorem which we can use for 2a (the comparison of $J\tilde{D}f$ and f).

**Exercise* 15.E. Let $f:[a,b]\to\mathbb{R}$ be continuous. Let $A,B\in\mathbb{R}$ such that $A\leqslant D_r^+f(x)\leqslant B$ for all $x\in(a,b)$. Then $A(b-a)\leqslant f(b)-f(a)\leqslant B(b-a)$. Let $C,E\in\mathbb{R}$ such that $C\leqslant D_r^-f(x)\leqslant E$ for all $x\in(a,b)$. Then $C(b-a)\leqslant f(b)-f(a)\leqslant E(b-a)$.

We are now going to study the connection between $f(b)-f(a)$ and the Riemann integrals of the Dini derivatives of f. These Dini derivatives are not defined at both end points of $[a,b]$, so that formally they cannot be Riemann integrable. Therefore, we define

$$D_r^+f(b):=0,\qquad D_r^-f(b):=0$$
$$D_l^+f(a):=0,\qquad D_l^-f(a):=0$$

Let $f:[a,b]\to\mathbb{R}$ be continuous. Let $a\leqslant p<q\leqslant b$. If $A,B\in\mathbb{R}$ and

$$A\leqslant D_r^+f(x)\leqslant B\qquad (x\in(p,q))$$

then by Exercise 15.E we have

$$A \leqslant \frac{f(y)-f(x)}{y-x} \leqslant B \qquad (p \leqslant x < y \leqslant q)$$

It follows that on (p, q) the functions D_r^-f, D_l^+f, D_l^-f are all between A and B. Thus, if one of the Dini derivatives of f is bounded, then so are all the others and for each open subinterval I of $[a, b]$ we have

$$\sup_I D_r^+f = \sup_I D_r^-f = \sup_I D_l^+f = \sup_I D_l^-f = \sup_{\substack{x,y \in I \\ x \neq y}} \frac{f(y)-f(x)}{y-x}$$

while similar equalities hold for the infima. Consequently, if one of the Dini derivatives of f is Riemann integrable, then so are the others and their Riemann integrals over any subinterval of $[a, b]$ are equal.

This leads to the following extension of Exercise 8.C(i).

THEOREM 15.3. *Let $f : [a, b] \to \mathbb{R}$ be continuous. Let one of the Dini derivatives of f be Riemann integrable. Then so are the others, and all Dini derivatives are equal a.e. on $[a, b]$. If $\tilde{D}f$ denotes any one of the four Dini derivatives of f, then*

$$\int_x^y \tilde{D}f(t)\,dt = f(y) - f(x) \qquad (x, y \in [a, b])$$

Proof. By the preceding remarks and by Exercise 15.A it suffices to prove, for example, that $\int_x^y D_r^+f(t)\,dt = f(y) - f(x)$ $(x, y \in [a, b])$. Without restriction, let $x = a$, $y = b$. Let $a = x_0 < x_1 < \ldots < x_n = b$ be a partition of $[a, b]$. Then

$$f(b) - f(a) = \sum_{k=1}^{n} (f(x_k) - f(x_{k-1}))$$

and, by Exercise 15.E,

$$j_k(x_k - x_{k-1}) \leqslant f(x_k) - f(x_{k-1}) \leqslant h_k(x_k - x_{k-1}) \qquad (k = 1, 2, \ldots, n)$$

where j_k and h_k are the infimum and the supremum of D_r^+f on (x_{k-1}, x_k). Hence,

$$\sum_{k=1}^{n} j_k(x_k - x_{k-1}) \leqslant f(b) - f(a) \leqslant \sum_{k=1}^{n} h_k(x_k - x_{k-1})$$

Now we can choose our partition such that both $\Sigma_{k=1}^n j_k(x_k - x_{k-1})$ and $\Sigma_{k=1}^n h_k(x_k - x_{k-1})$ are close to $\int_a^b D_r^+f(t)\,dt$. The desired formula follows.

Exercise 15.F. Let $f : [a, b] \to \mathbb{R}$ be continuous. Let $c \in (a, b)$. If D_r^+f is continuous at c, then so are the other Dini derivatives of f, and $f'(c)$ exists. (Hint. Assume $D_r^+f(c) = 0$. Use Exercise 15.E to show that for every $\varepsilon > 0$ there exists a $\delta > 0$ such that $|f(x) - f(y)| \leqslant \varepsilon|x - y|$ for all $x, y \in [a, b] \cap (c - \delta, c + \delta)$.)

Exercise 15.G. Let $f : [a, b] \to \mathbb{R}$ be continuous. Let $c \in [a, b)$. If $\lim_{x \downarrow c} D_r^+f(x)$ exists, then $\lim_{x \downarrow c} D_r^+f(x) = D_r f(c)$. (Compare the previous exercise.)

Exercises 15.B–15.E and Theorem 15.3 show that the Dini derivatives are generalizations of the ordinary derivative that suit our purposes. Now we look at other ones.

DEFINITION 15.4. Let I be an interval and let $f: I \to \mathbb{R}$. For $x \in I$ we define elements $D^+f(x)$ and $D^-f(x)$ of $\mathbb{R} \cup \{\infty\} \cup \{-\infty\}$ by

$$D^+f(x) := \overline{\lim_{y \to x}} \frac{f(y)-f(x)}{y-x}, \qquad D^-f(x) := \underline{\lim_{y \to x}} \frac{f(y)-f(x)}{y-x}$$

Trivially, $D^-f \leqslant D^+f$. If x is an interior point of I, then we have $D^+f(x) = \max(D_r^+f(x), D_l^+f(x))$ and $D^-f(x) = \min(D_r^-f(x), D_l^-f(x))$. f is differentiable at x if and only if $D^+f(x) = D^-f(x) \in \mathbb{R}$.

The next exercise shows that D^+f and D^-f are not so harmless as they look.

**Exercise* 15.H. Let $f: [a, b] \to \mathbb{R}$ be continuous. If $D^-f(x) > 0$ for all $x \in [a, b]$, then f is strictly increasing. If $D^-f(x) \geqslant 0$ for all $x \in [a, b]$, then f is increasing. But there exists a continuous function $g: [0, 1] \to \mathbb{R}$ which is not decreasing although $D^-g(x) \leqslant 0$ for all $x \in [0, 1]$. (Try Weierstrass' example (7.16).)

(The reader will wonder if a continuous function f with $D^-f = 0$ is necessarily constant. It is, but for the proof we need deep results from Lebesgue integration theory. See Exercise 21.N.)

Exercise 15.I. (The symmetric derivative) Let $f: \mathbb{R} \to \mathbb{R}$. If for every $x \in \mathbb{R}$ the limit

$$(*) \qquad f^s(x) := \lim_{h \downarrow 0} \frac{f(x+h) - f(x-h)}{2h}$$

exists, then the formula (*) defines a function f^s on $\mathbb{R}$, which is called the *symmetric derivative* of f. The characteristic function of the set $\{n^{-1} : n \in \mathbb{Z}, n \neq 0\}$ is a nontrivial example of a function that has a symmetric derivative. Show that the following are true.

(i) If f is differentiable, then f^s exists and $f^s = f'$.

(ii) Let $f(x) = 2|x| + x$ $(x \in \mathbb{R})$. Then f is continuous, f has a symmetric derivative but is not differentiable; f attains an absolute minimum at 0, but $f^s(0) \neq 0$.

(iii) Let f be continuous. Let $a < b$ and suppose that $f^s(x)$ exists for all $x \in [a, b]$. If $f^s(x) > 0$ (or $f^s(x) \geqslant 0$, $f^s(x) = 0$) for all $x \in [a, b]$, then f is strictly increasing (or increasing, constant) on $[a, b]$. Hint. Follow the proof of 15.1, but with $c := \sup\{x \in [a, b] : f(x) > t\}$.

(iv) The continuity condition in (iii) is not redundant.

(v) If f and f^s are both continuous, then f is differentiable. (If g is an antiderivative of f^s, then $(g - f)^s = 0$.)

Exercise 15.J. Let $f: \mathbb{R} \to \mathbb{R}$ be continuous and suppose that $f^s(x)$ exists for all $x \in [a, b]$. Let $A, B \in \mathbb{R}$. If $A \leqslant f^s \leqslant B$ on $[a, b]$, then $A(b-a) \leqslant f(b) - f(a) \leqslant B(b-a)$.

If f^s is Riemann integrable over $[a, b]$, then $\int_x^y f^s(t)\,dt = f(y) - f(x)$ $(x, y \in [a, b])$.

We now turn to our last task (2*b*), the reconstruction of g from Jg via a differentiation process. By Exercise 15.A we can only expect to find $DJg = g$ a.e. We already have the result of Exercise 12.O:

THEOREM 15.5. *Let $g : [a, b] \to \mathbb{R}$ be Riemann integrable. Then the function $Jg : x \mapsto \int_a^x g(t)\,dt$ $(x \in [a, b])$ is differentiable a.e. on $[a, b]$ and $(Jg)' = g$ a.e. on $[a, b]$.*

Observe, however, that Theorem 15.5 gives us, starting from Jg, only an a.e. defined function and not a Riemann integrable function that is equal a.e. to g. We can do better by using our generalizations of D. In fact, let $a < x < y < b$. Then

$$\frac{Jg(y) - Jg(x)}{y - x} = \frac{1}{y - x} \int_x^y g(t)\,dt$$

For any p, q for which $x < p < q < y$ we therefore have

$$\inf_{(x,y)} g \leqslant \frac{Jg(q) - Jg(p)}{q - p} \leqslant \sup_{(x,y)} g$$

Now let $\tilde{D}$ be any one of the operations D_r^+, D_r^-, D_l^+, D_l^-, D^+, D^-. We obtain

$$\inf_{(x,y)} g \leqslant \inf_{(x,y)} \tilde{D}Jg \leqslant \sup_{(x,y)} \tilde{D}Jg \leqslant \sup_{(x,y)} g$$

Since g is Riemann integrable, we have also that $\tilde{D}Jg$ is Riemann integrable and that

$$\int_a^b g(t)\,dt = \int_a^b (\tilde{D}Jg)(t)\,dt$$

We have proved:

THEOREM 15.6. *Let $g : [a, b] \to \mathbb{R}$ be Riemann integrable, let Jg be its indefinite integral $x \mapsto \int_a^x g(t)\,dt$ $(x \in [a, b])$ and let $\tilde{D}$ be any of the operations D_r^+, D_r^-, D_l^+, D_l^-, D^+, D^-. Then $\tilde{D}Jg$ is Riemann integrable and $\tilde{D}Jg = g$ almost everywhere. In particular, if Jg is differentiable, then $(Jg)'$ is Riemann integrable and $(Jg)' = g$ almost everywhere.*

**Exercise* 15.K. Let $g : [a, b] \to \mathbb{R}$ be monotone and let $x \mapsto \int_a^x g(t)\,dt$ be differentiable. Then its derivative equals g everywhere on (a, b) and g is continuous on (a, b). (Compare Exercise 2.F.)

**Exercise* 15.L. Prove or disprove the following statements.
(i) If $g \in \mathscr{R}[a, b]$ and $g \geqslant 0$ a.e., then Jg is increasing.

(ii) If $g \in \mathscr{R}[a, b]$ and $g=0$ a.e., then $Jg=0$.
(iii) If $g \in \mathscr{R}[a, b]$ and $g>0$ a.e., then Jg is strictly increasing.
(i)′ If $g \in \mathscr{R}[a, b]$ and Jg is increasing, then $g \geqslant 0$ a.e.
(ii)′ If $g \in \mathscr{R}[a, b]$ and $Jg=0$, then $g=0$ a.e.
(iii)′ If $g \in \mathscr{R}[a, b]$ and Jg is strictly increasing, then $g>0$ a.e.

We now come back to other generalizations of D that follow a different direction. First of all we observe that for the conclusion of Theorem 15.1 we do not need to know that $f'(x)>0$ ($\geqslant 0$, $=0$) for *all* $x \in [a, b]$. (Compare Exercises 7.N and 13.G.)

THEOREM 15.7. *Let $f:[a, b]\rightarrow\mathbb{R}$ be continuous. Assume that $f'(x)>0$ (or $f'(x)\geqslant 0$, $f'(x)=0$) for all but countably many points x of $[a, b]$. Then f is strictly increasing (or increasing, constant).*
Proof. We prove: if $f'(x)>0$ for all $x \in [a, b]\setminus A$ where A is a countable set, then f is increasing. As in the proof of Theorem 15.1 we may assume that $f(a)>f(b)$. Now $f(A)$ is countable, so we can choose $t \in \mathbb{R}$ with $f(b)<t<f(a)$, $t \notin f(A)$. From here on we may simply copy the proof of 15.1, the point being that $c \notin A$.

The function f' of Theorem 15.7 is defined on $[a, b]\setminus B$ where B is countable. In order to reconstruct f from f' via an integration process we would need a definition of an integral for functions that are not everywhere defined. We shall leave this for Section 19.

15.8. If we proceed in the spirit of 15.7 and try to admit larger sets of 'exceptional' points x, we run into subtleties.

There is a differentiable function which is not constant although its derivative vanishes everywhere outside a meagre set. In fact, let L be as in Example 13.2. Then set $\{x : L'(x)=0\}$ is a dense G_δ (Example 13.2(v) and Theorem 11.3), so its complement is meagre. But L is not constant on any interval.

The situation changes drastically if in the above we replace 'meagre' by 'null'. A differentiable function whose derivative vanishes almost everywhere has to be constant. A proof of this fact requires much preparative work: we give it in Section 21. However, in Theorem 15.7 the function f is not necessarily differentiable everywhere. Let us consider the following problem: *if $f:[a, b]\rightarrow\mathbb{R}$ is continuous and if $f'(x)=0$ for almost every $x \in [a, b]$, must f be constant?* We show that the answer is no.

15.9. The Cantor function. We shall construct a function Φ on $[0, 1]$ with the properties
(i) Φ is increasing and continuous.
(ii) Φ is not constant.
(iii) $\Phi'(x)=0$ for almost all $x \in [0, 1]$.
Actually, instead of (ii) and (iii) we shall have the stronger propositions:
(ii)′ Φ maps $\mathbb{D}$ onto $[0, 1]$.
(iii)′ Φ is constant on every component of $[0, 1]\setminus\mathbb{D}$.
Fig. 14 may give the reader some idea of the graph of Φ, the so-called 'Devil's staircase'. (The intervals I_{in} are the components of $[0, 1]\setminus\mathbb{D}$: see Example 4.2.) It is possible to define Φ first on the intervals I_{in} and then extend it continuously to an increasing function on $[0, 1]$. The following approach may be more convenient.

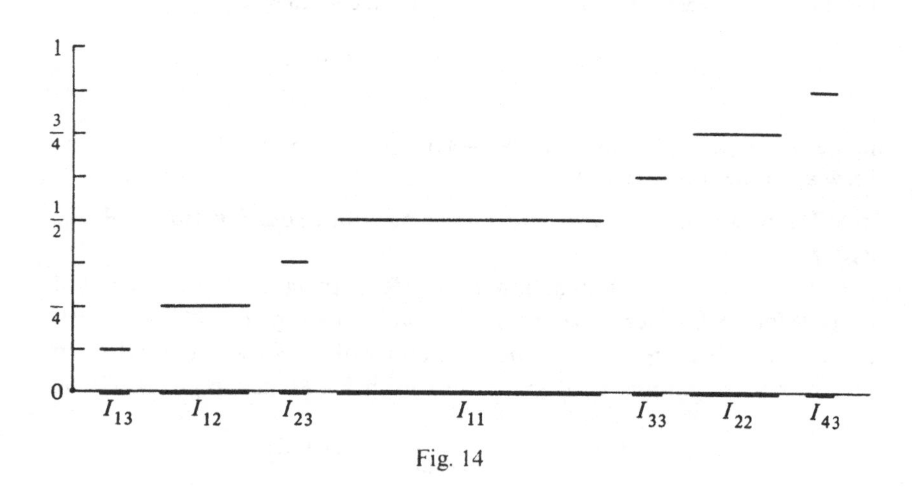

Fig. 14

By Exercise 4.D, the formula

$$\phi\left(2\sum_{i=1}^{\infty} \alpha_i 3^{-i}\right):=\sum_{i=1}^{\infty} \alpha_i 2^{-i} \qquad (\alpha_1, \alpha_2, \ldots \in \{0, 1\})$$

defines a function ϕ on $\mathbb{D}$. Note that $\phi(\mathbb{D})=[0, 1]$. We prove ϕ to be increasing. Let $\alpha_1, \alpha_2, \ldots, \beta_1, \beta_2, \ldots \in \{0, 1\}$ be such that $2\Sigma_{i=1}^{\infty} \alpha_i 3^{-i} < 2\Sigma_{i=1}^{\infty} \beta_i 3^{-i}$. Take $n:=\min\ \{i : \alpha_i \neq \beta_i\}$. Then $0<\Sigma_{i=1}^{\infty} (\beta_i-\alpha_i)3^{-i} =(\beta_n-\alpha_n)3^{-n}+\Sigma_{i>n}(\beta_i-\alpha_i)3^{-i} \leqslant (\beta_n-\alpha_n)3^{-n}+\Sigma_{i>n}2\cdot 3^{-i}=(\beta_n-\alpha_n+1)3^{-n}$, so $\beta_n \geqslant \alpha_n$. As $\alpha_n, \beta_n \in \{0, 1\}$ and $\alpha_n \neq \beta_n$, it follows that $\alpha_n=0, \beta_n=1$. But then $\phi(2\Sigma_{i=1}^{\infty}\alpha_i 3^{-i})=\Sigma_{i=1}^{\infty}\alpha_i 2^{-i} \leqslant \Sigma_{i>n}2^{-i}=2^{-n} \leqslant \Sigma_{i=1}^{\infty}\beta_i 2^{-i}=\phi(2\Sigma_{i=1}^{\infty}\beta_i 3^{-i})$.

Thus, ϕ is, indeed, increasing. We extend ϕ to an increasing function Φ

on $[0, 1]$ by

$$\Phi(x) := \sup \{\phi(y) : y \in \mathbb{D}, y \leqslant x\}$$

Φ maps $[0, 1]$ *onto* $[0, 1]$, hence is continuous (Exercise 1.I). It is evident that Φ is constant on each component of $[0, 1] \setminus \mathbb{D}$, so (i), (ii)′ and (iii)′ hold.

Φ is *the Cantor function.*

**Exercise* 15.M

(i) If $x, y \in \mathbb{D}$, $x < y$ and $\phi(x) = \phi(y)$, then the interval (x, y) is a component of $[0, 1] \setminus \mathbb{D}$.
(ii) Show that our definition of the Cantor function is in accordance with Exercise 7.P(vi).

Exercise 15.N

(i) Let $x, y \in \mathbb{D}$, $x < y$. Write $x = 2\Sigma_{i=1}^{\infty} \alpha_i 3^{-i}$, $y = 2\Sigma_{i=1}^{\infty} \beta_i 3^{-i}$ where $\alpha_i, \beta_i \in \{0, 1\}$. Set $n := \min \{i : \alpha_i \neq \beta_i\}$. Then $3^{-n} \leqslant y - x \leqslant 3^{-n+1}$.
(ii) Show that for each $n \in \mathbb{N}$ the function $\mathbb{D} \to \{0, 1\}$, defined by

$$2 \sum_{i=1}^{\infty} \alpha_i 3^{-i} \mapsto \alpha_n \qquad (\alpha_1, \alpha_2, \ldots \in \{0, 1\})$$

is continuous.
(iii) Let $\alpha := \log 2/\log 3$. Show that $|\Phi(x) - \Phi(y)| \leqslant 2|x - y|^{\alpha}$ $(x, y \in [0, 1])$ (i.e. $\Phi \in \mathrm{Lip}_{\alpha}$; see Exercise 8.G).

15.10. We now construct a *strictly* increasing continuous function $f : \mathbb{R} \to \mathbb{R}$ with $f' = 0$ a.e.

First, extend Φ to a function $g : \mathbb{R} \to \mathbb{R}$ with $g(x) = 1$ for $x \geqslant 1$ and $g(x) = 0$ for $x \leqslant 0$. Then g is continuous and increasing, $0 \leqslant g \leqslant 1$ and $g' = 0$ a.e. Let $((a_1, b_1), (a_2, b_2), \ldots)$ be an enumeration of the set of all open intervals with rational end points. For each $k \in \mathbb{N}$, define $g_k : \mathbb{R} \to \mathbb{R}$ by

$$g_k(x) := 2^{-k} g\left(\frac{x - a_k}{b_k - a_k}\right) \qquad (x \in \mathbb{R})$$

For each k, g_k is continuous and increasing, $0 \leqslant g_k \leqslant 2^{-k}$, $g_k(b_k) = 2^{-k}$, $g_k(a_k) = 0$ and $g_k' = 0$ a.e. Now put $f := \Sigma_{k=1}^{\infty} g_k$. Then f is continuous and increasing. By the theorem of Fubini (4.14), $f' = 0$ a.e. Finally, if $x, y \in \mathbb{R}$ and $x < y$, there is a $k \in \mathbb{N}$ with $x < a_k < b_k < y$: then we have $f(y) - f(x) \geqslant f(b_k) - f(a_k) \geqslant g_k(b_k) - g_k(a_k) = 2^{-k} > 0$. Thus, f is strictly increasing.

Exercise 15.O. A function $f : [a, b] \to \mathbb{R}$ with $f' = 0$ a.e. is sometimes called *singular.* The Cantor function is an example of a continuous singular function that is not constant. Prove that a singular function that satisfies a Lipschitz condition (see Corollary 4.13) must be constant. (Hint. It suffices to prove that, if $g : [a, b] \to \mathbb{R}$ satisfies a Lipschitz condition and if $g'(x) > 0$ for all x outside a null set A, then $g(b) \geqslant g(a)$. Show that for such g and A, the set $g(A)$ does not contain an interval. Now conclude the proof along the lines of the proof of Theorem 15.7. Later on, in Theorem 21.10, we shall apply the same technique to obtain stronger results.)

**Exercise* 15.P. Consider the following slight generalization of differentiation. For $f: [a, b] \to \mathbb{R}$ and $x \in [a, b]$, if D^+f and D^-f have the same value at x, we denote this value by $f^\nabla(x)$. (The difference from ordinary differentiation lies in the fact that we now admit the values ∞ and $-\infty$.)

Obviously, if $f^\nabla(x)=0$ for all $x \in [a, b]$, then $f'=0$ and f is constant on $[a, b]$. However, we shall give *functions g and h on an interval $[a, b]$ such that $g^\nabla(x)=h^\nabla(x)$ (and therefore $D_r^+g(x)=D_r^+h(x)$) for all $x \in [a, b]$ while $g-h$ is not constant.*

By Exercise 1.P there exists a strictly increasing function f on $[0, 1]$, with a continuous derivative and such that $\{x : f'(x)=0\} = \mathbb{D}$. Such an f maps $[0, 1]$ bijectively onto an interval $[a, b]$. Let $g : [a, b] \to [0, 1]$ be the inverse map and put $h := g + \Phi \circ g$ where Φ is the Cantor function. Show that $h'(x)=g'(x)$ for all $x \in [a, b] \setminus f(\mathbb{D})$ and that $g^\nabla(x)=\infty$ for $x \in f(\mathbb{D})$. Since $\Phi \circ g$ is increasing, we see that $h^\nabla(x)=\infty$ for $x \in f(\mathbb{D})$. Thus, $g^\nabla=h^\nabla$. But $g-h$ is not constant.

(This example is an adaptation of the one given by S. Ruziewicz in *Fund. Math.* **1** (1920), 148–51.)

Notes to Section 15

We have presented several variants of the theorem that says that a function on an interval must be increasing if it has a nonnegative derivative. Further elaborations on this theme, together with historical notes, are given in Saks (pp. 203–7, 225–6, 272, 275). See also H. W. Pu (*Coll. Math.* **31** (1974), 289–92).

A few references for the symmetric derivative are H. Auerbach (*Fund. Math.* **8** (1926), 49–55), W. Sierpiński (*Fund. Math.* **11** (1928), 148–50), E. M. Stein & A. Zygmund (*Studia Math.* **23** 1963), 247–83) *and* L. Filipczak (*Coll. Math.* **20** (1969), 249–53).

5

BOREL MEASURABILITY

16. The classes of Baire

For the time being we deal with functions that have $\mathbb{R}$ as their domain of definition. By $\mathscr{C}$ we denote the set of all continuous functions on $\mathbb{R}$.

In Section 11 we have introduced $\mathscr{B}^1$, the set of all limits of sequences of continuous functions. We have seen that limits of elements of $\mathscr{B}^1$ may themselves fail to be elements of $\mathscr{B}^1$ (e.g. $\xi_{\mathbb{Q}}$). Let $\mathscr{B}^2$ (the second class of Baire) be the set of all limits of elements of $\mathscr{B}^1$, $\mathscr{B}^3$ the set of all limits of elements of $\mathscr{B}^2$, etc. Trivially, $\mathscr{C} \subset \mathscr{B}^1 \subset \mathscr{B}^2 \subset \mathscr{B}^3 \subset \ldots$; we know that the inclusions $\mathscr{C} \subset \mathscr{B}^1$ and $\mathscr{B}^1 \subset \mathscr{B}^2$ are strict. Before investigating this sequence more thoroughly we first start with a few exercises on the higher classes of Baire.

Exercise 16.A. Show that the $\phi : \mathbb{R} \to \mathbb{R}$ constructed in Exercise 9.M is in $\mathscr{B}^3$.

Exercise 16.B. Let $f : \mathbb{R} \to \mathbb{R}$ be a function such that

$$D_r f(x) := \lim_{y \downarrow x} \frac{f(y) - f(x)}{y - x}$$

exists for every $x \in \mathbb{R}$. Show that $D_r f \in \mathscr{B}^2$. (Hint. Use Theorem 7.7.)

Exercise 16.C. Let $f_1, f_2, \ldots$ be a sequence of continuous functions such that $f := \limsup_{n \to \infty} f_n$ is finite everywhere. Then $f \in \mathscr{B}^2$.

Exercise 16.D

(i) $f \in \mathscr{C}$, $g \in \mathscr{B}^m$ implies $f \circ g \in \mathscr{B}^m$ ($m \in \mathbb{N}$).

(ii) $f \in \mathscr{B}^n$, $g \in \mathscr{B}^m$ implies $f \circ g \in \mathscr{B}^{n+m}$ ($n, m \in \mathbb{N}$).

Exercise 16.E

(i) Let $X \subset \mathbb{R}$ be an F_σ or a G_δ. Then $\xi_X \in \mathscr{B}^2$.

(ii) If $\xi_X \in \mathscr{B}^2$ then X is both an $F_{\sigma\delta}$ and a $G_{\delta\sigma}$.

(iii) If X is both an $F_{\sigma\delta}$ and a $G_{\delta\sigma}$, then $\xi_X \in \mathscr{B}^2$.

(Hint. Use Lemma 11.11.)

Exercise 16.F. Let $f : [a, b] \to \mathbb{R}$ be continuous. Suppose that for every $y \in \mathbb{R}$ the set $\{x : f(x) = y\}$ is finite. For $y \in \mathbb{R}$, let $N_f(y)$ be the number of elements of $\{x : f(x) = y\}$. Thus we obtain a function N_f on $\mathbb{R}$ with values in $\mathbb{N} \cup \{0\}$ (see Exercise 9.N). Let Y_0 be the set of all local extrema of f. Recall that Y_0 is countable (Theorem 7.2).
(i) Show that $f(a) \in Y_0$ and $f(b) \in Y_0$.
(ii) Let $c \in \mathbb{R}$. Prove that there exists an $\varepsilon > 0$ such that $N_f(y) \geqslant N_f(c)$ for all $y \in (c - \varepsilon, c + \varepsilon) \setminus Y_0$. (Hint. Let $x_1, \dots, x_n$ be the elements of $\{x : f(x) = y\}$, $x_1 < x_2 < \dots < x_n$. Choose $a_0, \dots, a_n \in [a, b]$ in such a way that $a_0 < x_1 < a_1 < x_2 < \dots < a_{n-1} < x_n < a_n$. Take $\varepsilon := \min_i |f(a_i) - c|$.)
(iii) Prove now that the function N_f is of the second class. (Define

$$g(y) := \begin{cases} \arctan N_f(y) & \text{if } y \in \mathbb{R} \setminus Y_0 \\ \tfrac{1}{2}\pi & \text{if } y \in Y_0 \end{cases}$$

Then $g(y) = g^{\uparrow}(y)$ for every $y \in \mathbb{R} \setminus Y_0$. Infer that the function $y \mapsto \arctan N_f(y)$ is of the second class and that, consequently, so is N_f.)

16.1. The fact that in the sequence $\mathscr{C} \subset \mathscr{B}^1 \subset \mathscr{B}^2 \subset \mathscr{B}^3 \subset \dots$ the first two inclusions are strict leads naturally to the following two considerations.

(1) Do we have a smallest class of functions containing $\mathscr{C}$ and closed for formation of limits? The (affirmative) answer can be obtained by the following standard reasoning. Call (momentarily) a set $\mathscr{F}$ of functions $\mathbb{R} \to \mathbb{R}$ an L-set if $\mathscr{C} \subset \mathscr{F}$ and $\mathscr{F}$ is closed with respect to formation of limits. The set of all functions is clearly an L-set. Intersections of L-sets are themselves L-sets. Thus, the intersection $\mathscr{B}$ of *all* L-sets is the smallest L-set.

(2) Will the sequence $\mathscr{C} \subset \mathscr{B}^1 \subset \mathscr{B}^2 \subset \mathscr{B}^3 \subset \dots$ ever become stationary? If, for some k, $\mathscr{B}^k$ were equal to $\mathscr{B}^{k+1}$, then $\mathscr{B}^k = \mathscr{B}^{k+1} = \mathscr{B}^{k+2} = \dots = \mathscr{B}$, and we would have a simple construction of $\mathscr{B}$ out of $\mathscr{C}$ in k steps. But, as we shall see, it turns out that $\mathscr{B}^k \neq \mathscr{B}^{k+1}$ for all $k \in \mathbb{N}$. So if we want to construct $\mathscr{B}$ starting off with $\mathscr{C}$, we are led to consider

$$\mathscr{B}^{\omega} := \bigcup_{k \in \mathbb{N}} \mathscr{B}^k$$

If even $\mathscr{B}^{\omega}$ is not closed for formation of limits we define $\mathscr{B}^{\omega+1}, \mathscr{B}^{\omega+2}, \dots$ in an obvious way. If we want to, we can put

$$\mathscr{B}^{2\omega} := \bigcup_{k \in \mathbb{N}} \mathscr{B}^{\omega+k}$$

and continue.

In the second part of this section we shall see that in the 'sequence'

$$\mathscr{C} \subset \mathscr{B}^1 \subset \mathscr{B}^2 \subset \dots \mathscr{B}^{\omega} \subset \mathscr{B}^{\omega+1} \subset \dots \mathscr{B}^{2\omega} \subset \dots \mathscr{B}^{3\omega} \subset \dots$$

the inclusions are strict as far as the eye can see. On the other hand, in

Section 17 we shall 'construct' $\mathscr{B}$ out of $\mathscr{C}$ by making use of some theory of well-ordered sets.

But first we study the class $\mathscr{B}$.

DEFINITION 16.2. A set $\mathscr{F}$ of functions $\mathbb{R} \to \mathbb{R}$ is called an *L-set* if
(i) $\mathscr{C} \subset \mathscr{F}$;
(ii) if $f_1, f_2, \ldots \in \mathscr{F}$ and if $f = \lim_{n\to\infty} f_n$, then $f \in \mathscr{F}$.
The collection of all functions $\mathbb{R} \to \mathbb{R}$ is an L-set. By $\mathscr{B}$ we denote the intersection of all L-sets. This $\mathscr{B}$ is itself an L-set and it is contained in every L-set. The elements of $\mathscr{B}$ are called *Borel functions* or *Borel measurable functions.*

16.3. If $f \in \mathscr{C}$, then the set of all functions $g : \mathbb{R} \to \mathbb{R}$ for which $f+g \in \mathscr{B}$ is an L-set, hence contains $\mathscr{B}$. In other words, if $f \in \mathscr{C}$ and $g \in \mathscr{B}$, then $f+g \in \mathscr{B}$.

Consequently, if $g \in \mathscr{B}$, then the set of all functions $f : \mathbb{R} \to \mathbb{R}$ for which $f+g \in \mathscr{B}$ is an L-set, hence contains $\mathscr{B}$.

Thus, if $f \in \mathscr{B}$ and $g \in \mathscr{B}$, then $f+g \in \mathscr{B}$. In formally the same way one proves that $fg \in \mathscr{B}$, $f \vee g \in \mathscr{B}$, $f \wedge g \in \mathscr{B}$ and even $f \circ g \in \mathscr{B}$. (Take the trouble to verify this!) We have:

THEOREM 16.4. *$\mathscr{B}$ is a vector space and a ring containing $\mathscr{C}(\mathbb{R})$, $\mathscr{B}^1$, $\mathscr{B}^2, \ldots$ If $f \in \mathscr{B}$ and $g \in \mathscr{B}$, then $f \vee g \in \mathscr{B}$, $f \wedge g \in \mathscr{B}$, $f \circ g \in \mathscr{B}$, $|f| \in \mathscr{B}$. If $f_1, f_2, \ldots \in \mathscr{B}$ and if $\lim_{n\to\infty} f_n(x)$ exists for all x, then $\lim_{n\to\infty} f_n \in \mathscr{B}$. If $f_1, f_2, \ldots \in \mathscr{B}$ and if $\sup_{n\in\mathbb{N}} f_n(x)$ is finite for every $x \in \mathbb{R}$, then $\sup_{n\in\mathbb{N}} f_n \in \mathscr{B}$.* (In fact, $\sup_{n\in\mathbb{N}} f_n = \lim_{n\to\infty} f_1 \vee f_2 \vee \ldots f_n$.)

DEFINITION 16.5. A subset E of $\mathbb{R}$ is a *Borel set* (is *Borel measurable*) if $\xi_E \in \mathscr{B}$. We denote the collection of all Borel sets by Ω.

From Theorem 16.4 one obtains easily:

THEOREM 16.6
(i) $\varnothing \in \Omega$, $\mathbb{R} \in \Omega$.
(ii) *If $A \in \Omega$ and $B \in \Omega$, then $A \cup B \in \Omega$, $\mathbb{R} \setminus A \in \Omega$, $B \setminus A \in \Omega$.*
(iii) *If $A_1, A_2, \ldots \in \Omega$, then $\bigcup_n A_n \in \Omega$ and $\bigcap_n A_n \in \Omega$.*
(iv) *All open sets and all closed sets are elements of Ω* (see Theorem 10.2(ii)).

It takes a very sophisticated argument to show that not all functions and sets are Borel measurable. See 18.20 and Example 20.10.

Ω determines $\mathscr{B}$ in the following way. (Compare Theorem 11.12.)

THEOREM 16.7. *Let $f:\mathbb{R}\to\mathbb{R}$. The conditions* ($\alpha$)–($\varepsilon$) *are equivalent.*
(α) $f\in\mathscr{B}$.
(β) *For every* $a\in\mathbb{R}$, $\{x:f(x)\geq a\}\in\Omega$.
(γ) *For every* $a\in\mathbb{R}$, $\{x:f(x)>a\}\in\Omega$.
(δ) *For every open set* U, $f^{-1}(U)\in\Omega$.
(ε) *For every* $E\in\Omega$, $f^{-1}(E)\in\Omega$.
Proof. Let $f\in\mathscr{B}$, $E\in\Omega$. Then $\xi_{f^{-1}(E)}=\xi_E\circ f\in\mathscr{B}$, so $f^{-1}(E)\in\Omega$. Hence, (α)$\Rightarrow$(ε). The implications (ε)$\Rightarrow$(δ)$\Rightarrow$(γ) are obvious in view of the facts that every open set belongs to Ω and that (a,∞) is open for every $a\in\mathbb{R}$. (γ)$\Rightarrow$(β) since $\{x:f(x)\geq a\}=\bigcup_{n\in\mathbb{N}}\{x:f(x)>a-n^{-1}\}$. Now assume ($\beta$) : we prove ($\alpha$). For $a\in\mathbb{R}$, let $A(a):=\{x:f(x)\geq a\}$. Then $\xi_{A(a)}\in\mathscr{B}$ and $\xi_{\mathbb{R}\setminus A(a)}=1-\xi_{A(a)}\in\mathscr{B}$. Now

$$f\vee 0=\sup_{\substack{a\in\mathbb{Q}\\ a\geq 0}}a\xi_{A(a)},\qquad (-f)\vee 0=\sup_{\substack{a\in\mathbb{Q}\\ a\geq 0}}a\xi_{\mathbb{R}\setminus A(-a)}$$

As $f=f\vee 0-(-f)\vee 0$, we have $f\in\mathscr{B}$.

**Exercise* 16.G. Let $\mathscr{A}$ be a collection of subsets of $\mathbb{R}$ with the following two properties.
(i) If $A_1, A_2,\ldots\in\mathscr{A}$, then $\bigcup_n A_n\in\mathscr{A}$ and $\bigcap_n A_n\in\mathscr{A}$.
(ii) Every closed subset of $\mathbb{R}$ is an element of $\mathscr{A}$.
Then every Borel set is an element of $\mathscr{A}$. (To prove this, show that the set of all functions $f:\mathbb{R}\to\mathbb{R}$ with

$$\{x:f(x)\geq a\}\in\mathscr{A}\text{ for all }a\in\mathbb{R}$$

is an L-set.)

We see that we could also have defined Ω as the smallest $\mathscr{A}$ having the properties (i) and (ii).

Exercise 16.H. Consider all sets $\mathscr{F}$ of functions $\mathbb{R}\to\mathbb{R}$ for which
(i) $\mathscr{C}\subset\mathscr{F}$;
(ii) if $f_1, f_2,\ldots\in\mathscr{F}$ and the series Σf_n converges, then $\Sigma_{n=1}^{\infty}f_n\in\mathscr{F}$.
Show that among these sets there is a smallest one and that this smallest set is, in fact, $\mathscr{B}$. (Hint. First, show that $\mathscr{B}$ is such an $\mathscr{F}$. Each $\mathscr{F}$ is closed under addition and the intersection, $\mathscr{B}_1$, of all sets $\mathscr{F}$ is a group under addition. Now show that $\mathscr{B}_1$ is closed under pointwise limits. Use the fact that, if $f=\lim_{n\to\infty}f_n$, then $f=f_1+\Sigma_{n=1}^{\infty}(f_{n+1}-f_n)$.)

Exercise 16.I. Consider all sets $\mathscr{F}$ of functions $\mathbb{R}\to\mathbb{R}$ for which
(i) $\mathscr{C}\subset\mathscr{F}$;
(ii) if $f_1, f_2,\ldots\in\mathscr{F}$ is a monotone sequence converging to f, then $f\in\mathscr{F}$.
Show that the smallest among these sets is $\mathscr{B}$. (Hint. Let $\mathscr{B}_2$ be the intersection of all these sets $\mathscr{F}$. Show that, if $f,g\in\mathscr{B}_2$, then $f\vee g\in\mathscr{B}_2$. Now use the formula $\lim_{n\to\infty}f_n=\lim_{p\to\infty}\lim_{q\to\infty}(f_p\vee f_{p+1}\vee\ldots\vee f_{p+q})$ to prove that $\mathscr{B}_2$ is closed under limits.)

Exercise 16.J. Consider all sets $\mathscr{F}$ of functions $\mathbb{R}\to\mathbb{R}$ for which
(i) $\mathscr{C}\subset\mathscr{F}$;

(ii) if $f_1, f_2, \ldots \in \mathscr{F}$ and the series $\Sigma|f_n|$ converges, then $\Sigma_{n=1}^{\infty} f_n \in \mathscr{F}$.
Show that the smallest of these sets is $\mathscr{B}$. (Hint. Let $\mathscr{B}_3$ be the intersection of all these sets $\mathscr{F}$. Show that $\mathscr{B}_3$ is a group under addition, and has the properties (i) and (ii) mentioned in the preceding exercise.)

Exercise 16.K. Let $f_1, f_2, \ldots$ be Borel measurable functions on $\mathbb{R}$. Show that the following sets are Borel measurable.
(i) The set of all points x for which the sequence $f_1(x), f_2(x), \ldots$ is increasing.
(ii) The set of all points x for which the sequence $f_1(x), f_2(x), \ldots$ is bounded.
(iii) The set of all points x for which $\Sigma_{n=1}^{\infty} |f_n(x)| < \infty$.
(iv) The set of all points x for which $\lim_{n\to\infty} f_n(x) = 0$.
(v) The set of all points x for which there exist infinitely many values of n with $f_n(x) > 0$.

Exercise 16.L. Let $X_1, X_2, \ldots$ be Borel sets whose union is $\mathbb{R}$. Suppose that $f_1, f_2, \ldots \in \mathscr{B}$ and that $f: \mathbb{R} \to \mathbb{R}$ is such that for each n,

$$f(x) = f_n(x) \qquad (x \in X_n)$$

Then $f \in \mathscr{B}$.

**Exercise* 16.M. (A curious property of Borel sets) For subsets A and B of $\mathbb{R}$, define $A \sim B$ if both $A \setminus B$ and $B \setminus A$ are meagre.
(i) $\sim$ is an equivalence relation. If $A \sim B$, then $\mathbb{R} \setminus A \sim \mathbb{R} \setminus B$. If $A_1, A_2, \ldots$ and $B_1, B_2, \ldots$ are subsets of $\mathbb{R}$ with $A_n \sim B_n$ for all n, then we have $\bigcup_n A_n \sim \bigcup_n B_n$ and $\bigcap_n A_n \sim \bigcap_n B_n$.
(ii) Let $V \subset \mathbb{R}$ and let U be the interior of $\bar{V}$. Show that $\bar{V} \setminus U$ is meagre and that $U \setminus W$ is meagre for every open set W that contains V. Deduce that $V \sim U$ if V is a G_δ.
(iii) Now let $\mathscr{W}$ be the collection of all subsets A of $\mathbb{R}$ for which there exists an open set U with $A \sim U$. Prove that $\mathscr{W}$ has the following properties.
(*a*) $\mathscr{W}$ contains all open sets and all meagre sets.
(*b*) If $A \in \mathscr{W}$, then $\mathbb{R} \setminus A \in \mathscr{W}$,
(*c*) If $A_1, A_2, \ldots \in \mathscr{W}$, then $\bigcup_n A_n \in \mathscr{W}$ and $\bigcap_n A_n \in \mathscr{W}$.
(iv) Use Exercise 16.G to show that $\mathscr{W} \supset \Omega$.
Thus, a Borel set is, except for a meagre set, equal to an open set.

**Exercise* 16.N. For a sequence $a_1, a_2, \ldots$ of real numbers we define a real number $\mathrm{Lim}_{n\to\infty} a_n$ by

$$\operatorname*{Lim}_{n\to\infty} a_n := \begin{cases} \lim_{n\to\infty} a_n & \text{if the sequence converges} \\ 0 & \text{otherwise} \end{cases}$$

Let $f_1, f_2, \ldots \in \mathscr{B}$. Then the set $E := \{x \in \mathbb{R} : \lim_{n\to\infty} f_n(x) \text{ exists}\}$ is a Borel set. (Hint. $x \in E$ if and only if the sequence $f_1(x), f_2(x), \ldots$ is Cauchy.)
The function $x \mapsto \mathrm{Lim}_{n\to\infty} f_n(x)$ is a Borel function. (Hint. Consider the sequence $f_1\xi_E, f_2\xi_E, \ldots$)

**Exercise* 16.O. Let $X \subset \mathbb{R}$ be a Borel set and let $f: X \to \mathbb{R}$ be continuous. Define $f_X: \mathbb{R} \to \mathbb{R}$ by

$$f_X(x) := \begin{cases} f(x) & \text{if } x \in X \\ 0 & \text{if } x \in \mathbb{R} \setminus X \end{cases}$$

Then f_X is a Borel function.

16.8. We shall need a slight generalization of the preceding theory. Let $m \in \mathbb{N}$. We call a set $\mathscr{F}$ of functions $\mathbb{R}^m \to \mathbb{R}$ an *L-set* if it has the following two properties.
(i) $\mathscr{F}$ contains all continuous functions $\mathbb{R}^m \to \mathbb{R}$.
(ii) If $f_1, f_2, \ldots \in \mathscr{F}$ and if $f = \lim_{n\to\infty} f_n$, then $f \in \mathscr{F}$.
The intersection of all L-sets of functions on $\mathbb{R}^m$ is denoted $\mathscr{B}(\mathbb{R}^m)$. Its elements are called *Borel functions.* A subset of $\mathbb{R}^m$ is said to be a *Borel set* if its characteristic function is a Borel function. Theorems 16.4, 16.6 and 16.7 and Exercises 16.G and 16.N have immediate analogues.

**Exercise* 16.P. Let $m, k \in \mathbb{N}$. Let Φ be a continuous map $\mathbb{R}^m \to \mathbb{R}^k$. Then $f \circ \Phi \in \mathscr{B}(\mathbb{R}^m)$ for every $f \in \mathscr{B}(\mathbb{R}^k)$.

**Exercise* 16.Q. Let $m \in \mathbb{N}$.
(i) If $f_1, \ldots, f_m \in \mathscr{B}$, then the function

$$(x_1, \ldots, x_m) \mapsto f_1(x_1) f_2(x_2) \ldots f_m(x_m) \qquad ((x_1, \ldots, x_m) \in \mathbb{R}^m)$$

is a Borel function on $\mathbb{R}^m$.
(ii) If $A_1, \ldots, A_m$ are Borel subsets of $\mathbb{R}$, then $A_1 \times \ldots \times A_m$ is a Borel subset of $\mathbb{R}^m$.

**Exercise* 16.R. Let I be an interval. Define the set $\mathscr{B}(I)$ of Borel functions $I \to \mathbb{R}$ by a simple modification of Definition 16.2 and 16.8.
(i) If $f \in \mathscr{B}$, then the restriction of f to I is a Borel function on I.
(ii) Conversely, if $f \in \mathscr{B}(I)$, then the function $f_I : \mathbb{R} \to \mathbb{R}$, defined by

$$f_I(x) := \begin{cases} f(x) & \text{if } x \in I \\ 0 & \text{if } x \in \mathbb{R} \setminus I \end{cases}$$

is an element of $\mathscr{B}$ (see Exercises 11.G and 16.O).
(iii) If $A \subset I$, then $\xi_A \in \mathscr{B}$ if and only if the restriction of ξ_A to I is an element of $\mathscr{B}(I)$. Thus, a subset of I is Borel if and only if it is 'Borel relative to I'.

We now return to the sequence $\mathscr{C} \subset \mathscr{B}^1 \subset \mathscr{B}^2 \subset \ldots$ and the question whether it is strictly increasing. First, however, two exercises: 16.S is an illustration of what we may expect if $\mathscr{B}^k \neq \mathscr{B}^{k+1}$ for all $k \in \mathbb{N}$, while 16.T is needed for a later proof, but is also interesting in its own right.

**Exercise* 16.S. Assuming that $\mathscr{B}^k \neq \mathscr{B}^{k+1}$ for all $k \in \mathbb{N}$, prove that $\mathscr{B}^\omega := \bigcup_k \mathscr{B}^k$ is not closed under limits, and not even under uniform limits. (Define $\mathscr{B}^k(I)$ for intervals I and show that $\mathscr{B}^k(I) \neq \mathscr{B}^{k+1}(I)$ for all k and I (see Exercise 11.G). Choose pairwise disjoint intervals $I_1, I_2, \ldots$ and make $f_k \in \mathscr{B}^{k+1}(I_k) \setminus \mathscr{B}^k(I_k)$, $|f_k| \leqslant 1/k$.)

**Exercise* 16.T. Let ϕ be the function $\mathbb{R}\to[0, 1]$ that is periodic with period 1 and whose graph is sketched in Fig. 15.
(i) Let $\alpha_1, \alpha_2, \ldots \in \{0, 1\}$ and $x:=\sum_{k=1}^{\infty} \alpha_k 3^{-k}$. Prove $\phi(3^{m-1}x)=\alpha_m$ $(m \in \mathbb{N})$
(ii) For $x \in \mathbb{R}$, define

$$\psi_1(x):=\sum_{n=1}^{\infty} 2^{-n}\phi(3^{2n}x)$$

$$\psi_2(x):=\sum_{n=1}^{\infty} 2^{-n}\phi(3^{2n-1}x)$$

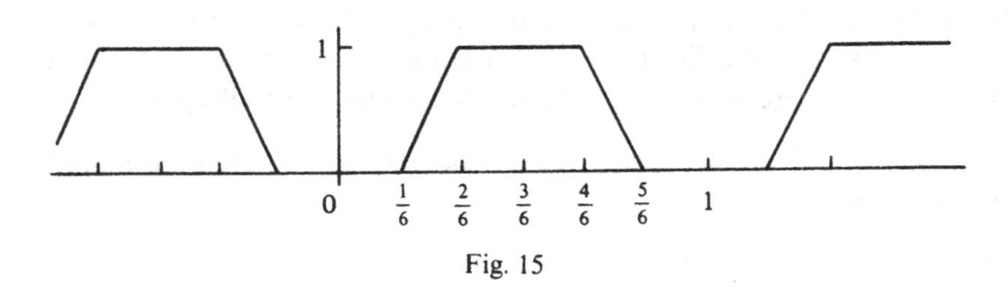

Fig. 15

Both ψ_1 and ψ_2 are continuous functions $\mathbb{R}\to[0, 1]$. For all $y, z \in [0, 1]$ there exists an $x \in [0, 1]$ such that $\psi_1(x)=y$, $\psi_2(x)=z$.

Thus, the formula

$$F(x):=(\psi_1(x), \psi_2(x)) \qquad (0 \leq x \leq 1)$$

yields *a continuous surjection F of* $[0, 1]$ *onto the square* $[0, 1]\times[0, 1]$. (Such an F is called a *Peano curve*.)
(iii) (An extension of (ii)) Let λ be a bijection $\mathbb{N}^2\to\mathbb{N}$, for example

$$\lambda(m, n):=2^m(2n-1) \qquad (m, n \in \mathbb{N})$$

Define functions $\tau_1, \tau_2, \ldots$ on $\mathbb{R}$ by

$$\tau_m(x):=\sum_{n=1}^{\infty} 2^{-n}\phi(3^{\lambda(m,n)}x) \qquad (x \in \mathbb{R})$$

Then each τ_m is a continuous map of $\mathbb{R}$ into $[0, 1]$. If $y_1, y_2, \ldots \in [0, 1]$, then there exists an $x \in [0, 1]$ such that $\tau_1(x)=y_1$, $\tau_2(x)=y_2$, etc.

For a set $\mathscr{A}$ of functions on $\mathbb{R}$ we denote by $\mathscr{A}^*$ the set of all functions that can be written as $\lim_{n\to\infty} f_n$ for certain $f_1, f_2, \ldots \in \mathscr{A}$. Thus, $\mathscr{C}^*=\mathscr{B}^1$, $\mathscr{B}^{1*}=\mathscr{B}^2, \ldots$

Let $\mathscr{A}$ be a set of functions on $\mathbb{R}$. A function $F:\mathbb{R}^2\to\mathbb{R}$ is said to be a *catalogue* of $\mathscr{A}$ if
(i) F is Borel measurable,
(ii) for every $f \in \mathscr{A}$ there exists an $s \in [0, 1]$ such that $f(x)=F(x, s)$ for all $x \in \mathbb{R}$.

LEMMA 16.9

(i) *If* $\mathscr{A}_1, \mathscr{A}_2, \ldots$ *have catalogues, then so does* $\bigcup_n \mathscr{A}_n$.
(ii) *If* $\mathscr{A}$ *has a catalogue, then so does* $\mathscr{A}^*$.
(iii) $\mathscr{C}$ *has a catalogue.*
(iv) $\mathscr{B}$ *does not have a catalogue.*
(v) *If* $\mathscr{A} \supset \mathscr{C}$ *and* $\mathscr{A}$ *has a catalogue, then* $\mathscr{A}^* \neq \mathscr{A}$.

Proof. (i) For each $n \in \mathbb{N}$, let F_n be a catalogue of $\mathscr{A}_n$. By Exercise 16.P the functions $(x, y, z) \mapsto F_n(x, y)$ and $(x, y, z) \mapsto \xi_{\{1/n\}}(z)$ are Borel functions on $\mathbb{R}^3$. Then so is the function H, defined by

$$H(x, y, z) := \sum_{n=1}^{\infty} F_n(x, y)\xi_{\{1/n\}}(z) \qquad (x, y, z \in \mathbb{R})$$

Let ψ_1, ψ_2 be as in Exercise 16.T. Define

$$F(x, s) := H(x, \psi_1(s), \psi_2(s)) \qquad (x, s \in \mathbb{R})$$

According to Exercise 16.P, F is Borel measurable. We show that F is a catalogue for $\bigcup_n \mathscr{A}_n$. Take $n \in \mathbb{N}$ and $f \in \mathscr{A}_n$. There is a $y \in [0, 1]$ such that $f(x) = F_n(x, y)$ for all x. There is an $s \in [0, 1]$ with $\psi_1(s) = y$, $\psi_2(s) = 1/n$. Then for all x we have $f(x) = F_n(x, y) = H(x, y, 1/n) = H(x, \psi_1(s), \psi_2(s)) = F(x, s)$.

(ii) Let F be a catalogue for $\mathscr{A}$. With $\tau_1, \tau_2, \ldots$ as in Exercise 16.T(iii), set (see Exercise 16.N for the definition of 'Lim').

$$F^*(x, s) := \operatorname*{Lim}_{n \to \infty} F(x, \tau_n(s)) \qquad (x, s \in \mathbb{R})$$

By a straightforward generalization of Exercise 16.N, F^* is Borel measurable. If $f \in \mathscr{A}^*$ there exist $f_1, f_2, \ldots \in \mathscr{A}$ with $f = \lim_{n\to\infty} f_n$. For every n there is an $s_n \in [0, 1]$ with $f_n(x) = F(x, s_n)$ for all x. By the construction of the τ_n there exists an $s \in [0, 1]$ such that $s_n = \tau_n(s)$ for every n. Now for all $x \in \mathbb{R}$ we obtain $f(x) = \lim_{n\to\infty} f_n(x) = \lim_{n\to\infty} F(x, \tau_n(s)) = F^*(x, s)$. We see that F^* is a catalogue for $\mathscr{A}^*$.

(iii) Obviously, if $f \in \mathscr{C}$, then $\{f\}$ has a catalogue. By (i), every countable subset of $\mathscr{C}$ has a catalogue. Let $\mathscr{P}$ be the set of all polynomial functions with rational coefficients. Being countable, $\mathscr{P}$ has a catalogue: then so has $\mathscr{P}^*$. We are done if $\mathscr{C} \subset \mathscr{P}^*$. Take $g \in \mathscr{C}$. By the Weierstrass approximation theorem, for every $n \in \mathbb{N}$ there exists a $g_n \in \mathscr{P}$ with

$$|f(x) - g_n(x)| \leqslant 1/n \qquad \text{for all } x \in [-n, n]$$

Then $f = \lim g_n \in \mathscr{P}^*$.

(iv) Suppose F is a catalogue for $\mathscr{B}$. Define $g : \mathbb{R} \to \mathbb{R}$ by

$$g(x) := F(x, x) \qquad (x \in \mathbb{R})$$

By Exercise 16.P, $g \in \mathscr{B}$, so $g+1 \in \mathscr{B}$ and there is an $s \in [0, 1]$ with

$$g(x)+1=F(x, s) \qquad \text{for all } x \in \mathbb{R}$$

In particular, $g(s)+1=F(s, s)=g(s)$: a contradiction.
(v) If $\mathscr{A} \supset \mathscr{C}$ and $\mathscr{A}^*=\mathscr{A}$, then $\mathscr{A} \supset \mathscr{B}$. As we have seen, $\mathscr{B}$ does not have a catalogue: then neither does $\mathscr{A}$.

16.10. Now we return to the subject matter of the beginning of this section. By (iii) and (ii), each $\mathscr{B}^n$ ($n \in \mathbb{N}$) has a catalogue, so $\mathscr{B}^n \neq \mathscr{B}^{n+1}$ by (v). Apparently, the sequence

$$\mathscr{C} \subset \mathscr{B}^1 \subset \mathscr{B}^2 \subset \mathscr{B}^3 \subset \dots \tag{1}$$

is strictly increasing. By (i), $\mathscr{B}^\omega := \bigcup_n \mathscr{B}^n$ has a catalogue, so $\mathscr{B}^\omega \neq (\mathscr{B}^\omega)^*$. Defining $\mathscr{B}^{\omega+1} := (\mathscr{B}^\omega)^*$, $\mathscr{B}^{\omega+n+1} := (\mathscr{B}^{\omega+n})^*$ ($n \in \mathbb{N}$) we obtain a strictly increasing 'sequence'

$$\mathscr{C} \subset \mathscr{B}^1 \subset \mathscr{B}^2 \subset \mathscr{B}^3 \subset \dots \subset \mathscr{B}^\omega \subset \mathscr{B}^{\omega+1} \subset \mathscr{B}^{\omega+2} \subset \dots \tag{2}$$

There is no reason to stop here. Setting $\mathscr{B}^{2\omega} := \mathscr{B}^{\omega+\omega} := \bigcup_n \mathscr{B}^{\omega+n}$, $\mathscr{B}^{2\omega+1} := (\mathscr{B}^{2\omega})^*, \dots, \mathscr{B}^{3\omega} := \bigcup_n \mathscr{B}^{2\omega+n}, \dots, \mathscr{B}^{(m+1)\omega} := \bigcup_n \mathscr{B}^{m\omega+n}, \dots$ we can construct an even longer 'sequence'

$$\mathscr{C} \subset \mathscr{B}^1 \subset \mathscr{B}^2 \subset \dots \subset \mathscr{B}^\omega \subset \mathscr{B}^{\omega+1} \subset \dots \subset \mathscr{B}^{2\omega} \subset \mathscr{B}^{2\omega+1} \subset \dots \subset \mathscr{B}^{3\omega} \subset \dots \subset \mathscr{B}^{4\omega} \subset \dots \tag{3}$$

which is still strictly increasing. One may further introduce sets like $\mathscr{B}^{\omega^2} := \mathscr{B}^{\omega\omega} := \bigcup_n \mathscr{B}^{n\omega}$, $\mathscr{B}^{\omega^2+\omega} := \bigcup_n \mathscr{B}^{\omega^2+n}$, $\mathscr{B}^{2\omega^2} := \bigcup_n \mathscr{B}^{\omega^2+n\omega}$, $\mathscr{B}^{\omega^3} := \bigcup_n \mathscr{B}^{n\omega^2}$, $\mathscr{B}^{\omega^\omega} := \bigcup_n \mathscr{B}^{\omega^n}$, etc., making longer and longer increasing 'sequences'. Observe that all entries of such 'sequences' are proper subsets of $\mathscr{B}$.

One can go on as long as one's patience endures. In the next section we shall see what happens after that.

Exercise 16.U. The reader is invited to investigate what happens to the theory in this section if one considers functions whose domain of definition is $[0, 1]$ instead of $\mathbb{R}$.

Exercise 16.V. A function $f: \mathbb{R} \to [0, 1]$ that maps every subinterval of $\mathbb{R}$ onto all of $[0, 1]$ cannot be of the first class of Baire, because functions of the first class have continuity points. However, we can now construct such a function that belongs to the second class. (Compare Exercise 16.A.)

Let ϕ be as in Exercise 16.T(i). Define

$$f(x) := \limsup_{n\to\infty} \frac{1}{n} \sum_{i=0}^{n-1} \phi(3^i x) \qquad (x \in \mathbb{R})$$

(i) Show that $f \in \mathscr{B}^2$.
(ii) Clearly, f maps $\mathbb{R}$ into $[0, 1]$. We want to prove that f maps any subinterval of $\mathbb{R}$ onto $[0, 1]$. First, observe that $f(3x)=f(x)$ and that $f(x+1)=f(x)$ for all $x \in \mathbb{R}$.

Deduce that $f(x+3^{-p})=f(x)$ for all $x \in \mathbb{R}$ and $p \in \{0, 1, 2, \ldots\}$. Therefore, it is enough to prove that f maps $[0, 1]$ onto $[0, 1]$. Take $t \in [0, 1]$. Find $\alpha_1, \alpha_2, \ldots \in \{0, 1\}$ such that

$$t = \limsup_{n\to\infty} \frac{1}{n}(\alpha_1 + \alpha_2 + \ldots + \alpha_n)$$

(see Exercise 9.M). Show that $t=f(x)$, where $x := \sum_{k=1}^{\infty} \alpha_k 3^{-k}$.

Notes to Section 16

We have defined $\mathscr{B}$ to be the smallest L-set; in the next section we shall see that $\mathscr{B}$ can also be described as the set of all functions obtainable by a certain transfinite (but countable) construction. The historical development was different. Originally, a function was called Borel measurable (or rather analytically representable) if it could be built in countably many steps, starting from continuous functions. It was Sierpiński (*Fund.Math.***1** (1920), 159–65) who pointed out the short-cut that by-passes the transfinite induction procedure but requires quantification over all sets of functions. Nowadays his point of view is generally accepted and in 'mainstream' mathematics the constructive definition is no longer used.

Exercise 16.D(ii) raises the question: if $n, m \in \mathbb{N}$, can every element of $\mathscr{B}^{n+m}$ be written as $f \circ g$ for suitable $f \in \mathscr{B}^n$ and $g \in \mathscr{B}^m$? A. Lindenbaum (*Fund. Math.* **23** (1934), 15–37 and 304) proved that this is indeed the case. His paper gives a good survey of related results, with useful references. It also deals with the following.

For $n \in \mathbb{N}$ let $\mathscr{F}_n$ be a set of functions $\mathbb{R} \to \mathbb{R}$. Define $\ldots \circ \mathscr{F}_2 \circ \mathscr{F}_1$ to be the set of all functions of the form $\lim_{n\to\infty} f_n \circ \ldots \circ f_2 \circ f_1$ where $f_n \in \mathscr{F}_n$ for all n. Lindenbaum proves a very general formula implying that $\ldots \circ \mathscr{B}^1 \circ \mathscr{B}^1 = \mathscr{B}^{\omega+1}$. (It had been shown by Sierpiński that $\ldots \circ \mathscr{C} \circ \mathscr{C}$ is a proper subset of $\mathscr{B}^1$.)

In a similar way, if $\mathscr{F}_1, \mathscr{F}_2, \ldots$ are sets of functions $\mathbb{R} \to \mathbb{R}$ one can define $\mathscr{F}_1 \circ \mathscr{F}_2 \circ \ldots$ Then $\mathscr{C} \circ \mathscr{C} \circ \ldots = \mathscr{B}^1$ (W. Sierpiński, *Fund. Math.* **24** (1935), 1–7) and $\mathscr{B}^1 \circ \mathscr{B}^1 \circ \ldots = \mathscr{B}^{\omega+1}$ (H. Fried, *Fund. Math.* **26** (1936), 196–201).

Clearly, a function $f: \mathbb{R} \to \mathbb{R}$ belongs to $\mathscr{B}^1$ if and only if f is the sum of a pointwise summable sequence of continuous functions. We define $\mathscr{S}^1$ to be the class of all functions on $\mathbb{R}$ that are the sum of a pointwise *absolutely* summable sequence of continuous functions, i.e. $f \in \mathscr{S}^1$ if and only if there exist $f_1, f_2, \ldots \in \mathscr{C}$ such that for all $x \in \mathbb{R}$ one has $\sum_{i=1}^{\infty} |f_i(x)| < \infty$ and $\sum_{i=1}^{\infty} f_i(x) = f(x)$. It is not hard to see that $\mathscr{S}^1 = \{g - h : g, h \in \mathscr{C}^+\}$.

The class $\mathscr{S}^1$ was considered by W. Sierpiński and S. Kempisty in several papers in *Fund. Math.* **2** (1921). $\mathscr{S}^1$ turns out to be a proper subset of $\mathscr{B}^1$: there exists a bounded function that has only countably many discontinuities (hence belongs to $\mathscr{B}^1$) but is not an element of $\mathscr{S}^1$. However, all elements of $\mathscr{B}^1$ are uniform limits of elements of $\mathscr{S}^1$.

The definition of $\mathscr{S}^1$ leads in a natural way to the classes $\mathscr{S}^2, \mathscr{S}^3, \ldots$: a function f belongs to $\mathscr{S}^{n+1}$ if there exist $f_1, f_2, \ldots \in \mathscr{S}^n$ such that $\Sigma_{i=1}^{\infty} |f_i|$ is everywhere finite and $\Sigma_{i=1}^{\infty} f_i = f$ pointwise. The relations between the sequences $\mathscr{S}^1, \mathscr{S}^2, \ldots$ and $\mathscr{B}^1, \mathscr{B}^2, \ldots$ are studied by, among others, A. Lindenbaum (*Fund. Math.* **23** (1934), 15–37 and 304). See also R. D. Mauldin (*Adv. Math.* **12** (1974), 418–50).

17. Transfinite construction of the Borel functions

In the preceding section we have seen how, by repeating the operation indicated by * one generates strictly increasing 'sequences'

$$\mathscr{C}, \mathscr{B}^1, \mathscr{B}^2, \ldots, \mathscr{B}^{\omega}, \ldots, \mathscr{B}^{2\omega}, \ldots, \mathscr{B}^{3\omega}, \ldots, \mathscr{B}^{\omega^2}, \ldots, \mathscr{B}^{\omega^2+\omega},$$
$$\ldots, \mathscr{B}^{\omega^3}, \ldots, \mathscr{B}^{\omega^{\omega}}, \ldots, \mathscr{B}^{\omega^{\omega^{\omega}}}, \ldots$$

consisting of proper subsets of $\mathscr{B}$. We are now going to construct the whole set $\mathscr{B}$ by combining all such 'sequences'. The idea is the following.

The (ordinary) sequence $\mathscr{C}, \mathscr{B}^1, \mathscr{B}^2, \ldots$ may be considered as a set which is indexed by $\{0, 1, 2, \ldots\}$. In order to facilitate discussion of extensions of such an infinite sequence we prefer to admit other indexing sets, such as $\{0, \frac{1}{2}, \frac{2}{3}, \frac{3}{4}, \ldots\}$. For simplicity of notation we put $\mathscr{B}^0 := \mathscr{C}$. Thus we obtain a correspondence

$$n(n+1)^{-1} \mapsto \mathscr{B}^n \qquad (n=0, 1, 2, \ldots)$$

This correspondence can be extended by setting

$$1+n(n+1)^{-1} \mapsto \mathscr{B}^{\omega+n} \qquad (n=0, 1, 2, \ldots)$$

to obtain the increasing 'sequence' (2) of page 110. In a similar way, the formula

$$m+n(n+1)^{-1} \mapsto \mathscr{B}^{m\omega+n} \qquad (m, n=0, 1, 2, \ldots)$$

yields the increasing 'sequence' (3). We now have $\{m+n(n+1)^{-1} : m, n=0, 1, \ldots\}$ as our indexing set. For a further extension this indexing is not helpful, but replacing $m+n(n+1)^{-1}$ by, say, arctan $(m+n(n+1)^{-1})$ will enable us to continue.

It turns out that the most convenient indexing sets are the so-called 'well-ordered' subsets of $\mathbb{Q}$. (For a formal definition, see 17.1.)

We shall proceed roughly along the following lines.

(1) For every well-ordered set $D \subset \mathbb{Q}$, in a standard way we build a 'sequence' $x \mapsto \mathscr{A}_D(x)$ $(x \in D)$ of subsets of $\mathscr{B}$ (e.g. if D is the set $\{m+n(n+1)^{-1} : m, n=0, 1, 2, \ldots\}$ and if $x=m+n(n+1)^{-1} \in D$, then $\mathscr{A}_D(x) = \mathscr{B}^{m\omega+n}$.) These 'sequences' are strictly increasing in the sense that $\mathscr{A}_D(x)$ is a proper subset of $\mathscr{A}_D(y)$ as soon as $x, y \in D$ and $x<y$.

(2) If $D, E \subset \mathbb{Q}$ are well-ordered sets, then one of the 'sequences' $x \mapsto \mathscr{A}_D(x)$ and $y \mapsto \mathscr{A}_E(y)$ is an extension of the other.

(3) By forming the union of all the sets $\mathscr{A}_D(x)$ (over all $x \in D$ and all well-ordered sets $D \subset \mathbb{Q}$) we obtain $\mathscr{B}$.

There is an extensive theory of abstract well-ordered sets. As we want to keep our theory as elementary as possible, we shall consider only well-ordered subsets of $\mathbb{Q}$. The interested reader may extend Definition 17.1 to arbitrary ordered sets and verify the validity of 17.5–17.9 in this general setting. In the remainder of this section the countability of our well-ordered sets is essential.

DEFINITION 17.1. A subset D of $\mathbb{Q}$ is said to be *well-ordered* if every nonempty subset of D has a smallest element. Obvious examples are $\mathbb{N}$ and the finite subsets of $\mathbb{Q}$, while $\mathbb{Z}$ and $\mathbb{Q}$ itself are not well-ordered. Every subset of a well-ordered set is well-ordered.

Exercise 17.A. The following subsets of $\mathbb{Q}$ are well-ordered.
(i) $\{0, \frac{1}{2}, \frac{2}{3}, \frac{3}{4}, \ldots\}$
(ii) $\{m+n(n+1)^{-1} : m, n=0, 1, 2, \ldots\}$
(iii) $\{-3^{-m}-3^{-m-n} : m, n=0, 1, 2, \ldots\}$
(iv) $\{-3^{-k}(1+3^{-m}(1+3^{-n})) : k, m, n=0, 1, 2, \ldots\}$

**Exercise* 17.B
(i) A union of finitely many well-ordered sets is well-ordered.
(ii) Unions of countably many well-ordered sets need not be well-ordered.
(iii) If for every $n \in \mathbb{N}$ D_n is a well-ordered subset of $\mathbb{Q} \cap [n, \infty)$, then $\bigcup_n D_n$ is well-ordered.

Exercise 17.C. A subset D of $\mathbb{Q}$ is well-ordered if and only if there do *not* exist $x_1, x_2, \ldots \in D$ with $x_1 > x_2 > x_3 \ldots$

17.2. Let $D \subset \mathbb{Q}$ be well-ordered. *If $D \neq \varnothing$, then D has a smallest element.* If $x \in D$ and if x is not the largest element of D, then $\{y \in D : y > x\}$ has a smallest element: this is called the *successor* of x.

DEFINITION 17.3. Let D_1, D_2 be well-ordered sets. We say that they are *isomorphic* $(D_1 \sim D_2)$ if there exists a strictly increasing bijection of D_1

onto D_2. Such a strictly increasing bijection (whose inverse is, of course, again strictly increasing) is said to be an *isomorphism*.

Exercise 17.D. The sets mentioned in (ii) and (iii) of Exercise 17.A are isomorphic. The sets of (i) and (ii) are not isomorphic.

DEFINITION 17.4. Let D be well-ordered. An *initial interval* of D is a subset X of D with the property: if $x \in X$, $y \in D$, $y \leqslant x$, then $y \in X$. If X is such an initial interval of D, then either $X = D$ or $D \setminus X$ has a smallest element, a, say. In the latter case, $X = \{x \in D : x < a\}$. We denote this set by D^a.

LEMMA 17.5. *Let D be a well-ordered set. Let X be an initial interval of D and let $\phi : X \to D$ be strictly increasing. Then $\phi(x) \geqslant x$ for all $x \in X$.*
Proof. Assume that the conclusion of the lemma is false. The set $\{x \in X : \phi(x) < x\}$ has a smallest element, a. For all $x \in D^a$ we have $x \in X$, $x < a$, and therefore $x \leqslant \phi(x) < \phi(a)$. Thus, $\phi(a)$, being larger than every element of D^a, is not in D^a, i.e. $\phi(a) \geqslant a$. But this is a contradiction.

COROLLARY 17.6. *Let X and Y be initial intervals of a well-ordered set D and let ϕ be an isomorphism of X onto Y. Then $X = Y$ and ϕ is the identity map of X.*
Proof. By the above, $\phi(x) \geqslant x$ for all $x \in X$. But, similarly, $\phi^{-1}(y) \geqslant y$ for all $y \in Y$, i.e. $\phi(x) = x$ for all $x \in X$.

Let D_1, D_2 be well-ordered sets. Define

$$\Gamma := \{(a, b) \in \mathbb{R}^2 : a \in D_1, b \in D_2, D_1{}^a \sim D_2{}^b\},$$
$$E_1 := \{a \in D_1 : \text{there exists a } b \in D_2 \text{ with } D_1{}^a \sim D_2{}^b\},$$
$$E_2 := \{b \in D_2 : \text{there exists an } a \in D_1 \text{ with } D_1{}^a \sim D_2{}^b\}.$$

If $(a, b) \in \Gamma$ and $(a, b') \in \Gamma$, then $D_1{}^a \sim D_2{}^b$ and $D_1{}^a \sim D_2{}^{b'}$, so that $D_2{}^b \sim D_2{}^{b'}$; by Corollary 17.6, $b = b'$. Similarly, if $(a, b) \in \Gamma$ and also $(a', b) \in \Gamma$, then $a = a'$. Thus, Γ is the graph of a bijection $f : E_1 \to E_2$.

If $a \in E_1$, then there exists an isomorphism ϕ of $D_1{}^a$ onto $D_2{}^{f(a)}$. Then for every $x \in D_1{}^a$ we have $D_1{}^x \sim D_2{}^{\phi(x)}$. In particular, $x \in E_1$. We see that *E_1 is an initial interval of D_1*. In the same way, E_2 turns out to be *an initial interval of D_2*. Furthermore, with a, ϕ, x as above, the fact that $D_1{}^x \sim D_2{}^{\phi(x)}$ implies that $(x, \phi(x)) \in \Gamma$, so $f(x) = \phi(x) \in \phi(D_1{}^a) = D_2{}^{f(a)}$ and $f(x) < f(a)$. Thus, *f is an isomorphism of E_1 onto E_2*. Observe that, consequently, f^{-1} is an isomorphism of E_2 onto E_1.

Assume $E_1 \neq D_1$ and $E_2 \neq D_2$. Then there exist $a \in D_1$ and $b \in D_2$ with

$E_1 = D_1{}^a$ and $E_2 = D_2{}^b$. But now $D_1{}^a \sim D_2{}^b$, so $a \in E_1 = D_1{}^a$: a contradiction. It follows that either $E_1 = D_1$ or $E_2 = D_2$. We have proved:

THEOREM 17.7. *Of any two well-ordered sets, one is isomorphic to an initial interval of the other.*

LEMMA 17.8. *Let D be a well-ordered set and $X \subset D$. Then, as a well-ordered set, X is isomorphic to an initial interval of D.*

Proof. Otherwise there would exist an isomorphism ϕ of D onto an initial interval of X with $\phi(D) \neq X$. Let a be the smallest element of $X \setminus \phi(D)$. By Lemma 17.5, $\phi(x) \geqslant x$ for all $x \in D$. In particular, $\phi(a) \geqslant a$, so $\phi(a) \notin X^a$. But $X^a = \phi(D)$.

**Exercise* 17.E

(i) For $x \in \mathbb{Q}$, define

$$\phi(x) := \begin{cases} 1 - (2+2x)^{-1} & \text{if } x \geqslant 0, \\ (2-2x)^{-1} & \text{if } x < 0 \end{cases}$$

Then ϕ defines a strictly increasing bijection $\mathbb{Q} \to \mathbb{Q} \cap (0, 1)$. Thus, if D is a well-ordered subset of $\mathbb{Q}$, then $\phi(D)$ is a well-ordered subset of $\mathbb{Q} \cap (0, 1)$.

(ii) If D is any well-ordered subset of $\mathbb{Q}$, then there exists a well-ordered set $D' \subset \mathbb{Q}$, having a largest element, a, say, and such that $D \sim D' \setminus \{a\}$.

(iii) If $D_1, D_2, \ldots$ are well-ordered subsets of $\mathbb{Q}$, then there exists a well-ordered set $D \subset \mathbb{Q}$ such that each D_n is isomorphic to an initial interval of D. (Choose $D := \bigcup_{n \in \mathbb{N}} \{\phi(x) + n : x \in D_n\}$.)

Let D be a well-ordered set. By a *hypersequence* of *length* D we shall mean a map $\mathscr{A}$, assigning to every element x of D a subset $\mathscr{A}(x)$ of $\mathscr{B}$ such that (using * as in Lemma 16.9)

(*a*) if x_0 is the smallest element of D, then $\mathscr{A}(x_0) = \mathscr{C}$.

(*b*) if $x, x' \in D$ and x' is the successor of x, then $\mathscr{A}(x') = \mathscr{A}(x)^*$.

(*c*) if $x \in D$ and if x is neither the smallest element of D nor the successor of any element of D, then $\mathscr{A}(x) = \bigcup_{y<x} \mathscr{A}(y)$.

THEOREM 17.9. *For every well-ordered set D there exists exactly one hypersequence of length D.*

Proof. First, let $\mathscr{A}_1$ and $\mathscr{A}_2$ be hypersequences of length D. It is easy to see that the set $\{x \in D : \mathscr{A}_1(x) \neq \mathscr{A}_2(x)\}$ does not have a smallest element and therefore has to be empty. Then $\mathscr{A}_1 = \mathscr{A}_2$. This proves the uniqueness part of the theorem.

To prove the existence, take a well-ordered set D. For every $t \in D$, define $\bar{D}^t := \{x \in D : x \leqslant t\} = D^t \cup \{t\}$. Let $D_0 := \{t \in D :$ there exists a hypersequence of length $\bar{D}^t\}$. If $t \in D_0$, $s \in D$ and $s<t$, and if $\mathscr{A}_t$ is a hypersequence of length $\bar{D}^t$, then the restriction of $\mathscr{A}_t$ to $\bar{D}^s$ is a hypersequence of length $\bar{D}^s$. Hence, D_0 is an initial interval of D. Moreover, by using the uniqueness of hypersequences, we obtain a map $\mathscr{A}$, defined on D_0, such that for every $t \in D_0$ the restriction of $\mathscr{A}$ to $\bar{D}^t$ is *the* hypersequence of length $\bar{D}^t$. Clearly, this map $\mathscr{A}$ is a hypersequence of length D_0 and we are done if $D_0=D$.

Suppose $D_0 \neq D$. Let a be the smallest element of $D \setminus D_0$. Then in an obvious way one can construct a hypersequence of length $D_0 \cup \{a\}$. But $D_0 \cup \{a\} = \bar{D}^a$, so we find $a \in D_0$, which is a contradiction.

For a well-ordered set D we indicate by $\mathscr{A}_D$ *the* hypersequence of length D. The results of the preceding section may now be formalized as follows.

THEOREM 17.10. *Let D be a well-ordered subset of $\mathbb{Q}$. Then for every $x \in D$ $\mathscr{A}_D(x)$ has a catalogue. If $x, y \in D$ and $x<y$, then $\mathscr{A}_D(x)$ is a proper subset of $\mathscr{A}_D(y)$.*

Proof. It follows from Lemma 16.9 that the set $\{x \in D : \mathscr{A}_D(x)$ does not have a catalogue$\}$ does not have a smallest element, hence must be empty. Thus, every $\mathscr{A}_D(x)$ has a catalogue.

Let $x, y \in D$, $x<y$. Let x' be the successor of x in D. As $\mathscr{A}_D(x)$ has a catalogue, $\mathscr{A}_D(x')$ (which is $\mathscr{A}_D(x)^*$) properly contains $\mathscr{A}_D(x)$, and it suffices to prove that $\mathscr{A}_D(y) \supset \mathscr{A}_D(x')$. But one shows without difficulty that $\{z \in D : z \geqslant x', \mathscr{A}_D(z) \not\supset \mathscr{A}_D(x')\}$ has no smallest element and therefore is empty.

In order to construct $\mathscr{B}$ we have defined the notion of hypersequence as a generalization of 'sequence'. But now we see that one hypersequence will not suffice. In fact, if $D \subset \mathbb{Q}$ is well ordered, then $\bigcup_{x \in D} \mathscr{A}_D(x)$ (by the countability of D) has a catalogue, so cannot be equal to $\mathscr{B}$ (see Lemma 16.9). Thus, in order to reach $\mathscr{B}$, we have once more to carry out a new construction.

There are two ways open to us. In the first lines of this section we have hinted at the possibility of making one extremely long 'sequence' that has all hypersequences as initial intervals. The union of the entries of this 'sequence' ought to be $\mathscr{B}$. This program puts a nice finishing touch to a theory of hypersequences but it requires more abstract reasoning than is really necessary for the construction of the Borel functions, which is our goal. Therefore we shelve it for the moment. The interested reader may turn to Appendix C.

Here, we take a different track. Let $D \subset \mathbb{Q}$ be well-ordered. We view D as a recipe for constructing sets of functions. We start at the smallest element of D, having in our possession the set $\mathscr{C}$ of all continuous functions. Now we make a walk from the left to the right, successively passing all points of D and enlarging our set of functions: after one step we have $\mathscr{B}^1$, after the next $\mathscr{B}^2$; at the first 'limit point' of D we have $\mathscr{B}^\omega$, one step later $\mathscr{B}^{\omega+1}$, and so on. By $\mathscr{B}_D$ we denote the collection of all functions that we have at our disposal at the end of our trip. Formally:

$$\mathscr{B}_D := \bigcup_{x \in D} \mathscr{A}_D(x)$$

Thus, $\mathscr{B}_{\{0\}} = \mathscr{C}$, $\mathscr{B}_{\{0,1\}} = \mathscr{B}^1$, $\mathscr{B}_{\{2,5,9\}} = \mathscr{B}^2$, $\mathscr{B}_{\mathbb{N}} = \mathscr{B}^\omega$.

Always, $\mathscr{B}_D$ is a subset of $\mathscr{B}$. We now claim that every Borel function can be constructed in this way by choosing a suitable recipe D. This is the content of Theorem 17.11 and the main result of Section 17:

THEOREM 17.11. $\mathscr{B} = \bigcup \{\mathscr{B}_D : D \text{ is a well-ordered subset of } \mathbb{Q}\}$.

For the proof we first formalize two statements about the sets $\mathscr{B}_D$.

**Exercise* 17.F

(i) Let D_1, D_2 be well-ordered sets. Let ϕ be an isomorphism of D_1 onto an initial interval of D_2. Then $\mathscr{A}_{D_1} = \mathscr{A}_{D_2} \circ \phi$, and, consequently, $\mathscr{B}_{D_1} \subset \mathscr{B}_{D_2}$.

(ii) Let D be a well-ordered subset of $\mathbb{Q}$ having a largest element, a. Then $D \cup \{a+1\}$ is well ordered and $\mathscr{B}_D^* = \mathscr{B}_{D \cup \{a+1\}}$.

Proof of Theorem 17.11. Let $\mathscr{B}^w$ be the union of all sets $\mathscr{B}_D$ where D runs through the well-ordered subsets of $\mathbb{Q}$. We want to show that $\mathscr{B}^w = \mathscr{B}$. Since, trivially, $\mathscr{C} \subset \mathscr{B}^w \subset \mathscr{B}$, it suffices to prove that $\mathscr{B}^w$ is closed for limits.

Let $f_1, f_2, \ldots \in \mathscr{B}^w$ and $f = \lim_{n \to \infty} f_n$. For each n there exists a well-ordered $D_n \subset \mathbb{Q}$ such that $f_n \in \mathscr{B}_{D_n}$. Let D be as in Exercise 17.E(iii); according to Exercise 17.E(ii) we can choose D such that it has a largest element, a, say. By Exercise 17.F(i), for all n we have $f_n \in \mathscr{B}_{D_n} \subset \mathscr{B}_D$; therefore (using Exercise 17.F(ii)), $f \in \mathscr{B}_D^* = \mathscr{B}_{D \cup \{a+1\}}$, so that $f \in \mathscr{B}^w$ and we are done.

Theorem 17.11 enables us to show that, among all functions $\mathbb{R} \to \mathbb{R}$, the Borel functions are quite rare. Indeed, it is well known that the set of all functions does not have cardinality $\leqslant c$ (see Appendix B, Example 8(viii)). On the other hand, we have:

COROLLARY 17.12. *$\mathscr{B}$ has the cardinality of the continuum.*

Proof (in the notation of Appendix B). $\mathbb{R} < \mathscr{B}$ since all constant functions are Borel. It remains to prove that $\mathscr{B}$ has cardinality $\leqslant c$.

Let $D \subset \mathbb{Q}$ be well-ordered. For every $x \in D$, by Theorem 17.10 $\mathscr{A}_D(x)$ has a catalogue, which yields us an injection $\mathscr{A}_D(x) \to \mathbb{R}$; thus, $\mathscr{A}_D(x) < \mathbb{R}$ for every $x \in D$. As $D < \mathbb{R}$ it follows from Theorem B.9(ii) that $\mathscr{B}_D < \mathbb{R}$.

Thus, each $\mathscr{B}_D$ has cardinality $\leqslant c$. The set of all well-ordered subsets D of $\mathbb{Q}$ also has cardinality $\leqslant c$ (Corollary B.6). Therefore, by another application of Theorem B.9(ii) and by Theorem 17.11, $\mathscr{B}$ has cardinality $\leqslant c$.

18. Analytic sets

In Theorem 16.7 (the implication $(\alpha) \Rightarrow (\varepsilon)$) we have seen that, if E is a Borel subset of $\mathbb{R}$ and if $f: \mathbb{R} \to \mathbb{R}$ is Borel measurable, then the set $f^{-1}(E)$ is Borel. A natural question now is the following. *If $E \subset \mathbb{R}$ is a Borel set and if $f: \mathbb{R} \to \mathbb{R}$ is Borel measurable, does $f(E)$ have to be Borel?*

We are now going to study the subsets A of $\mathbb{R}$ that have the property

(α) A is the image of a Borel set E under a Borel measurable function $f: \mathbb{R} \to \mathbb{R}$.

We shall prove that (α) is equivalent to the following property (β) which seems to be much more restrictive:

(β) A is the image of a G_δ-set E under a continuous function $f: \mathbb{R} \to \mathbb{R}$.

Now compare (α) and (β) with

(γ) A is the image of a G_δ-set E under a continuous function $f: E \to \mathbb{R}$.

Obviously, $(\beta) \Rightarrow (\gamma)$. By Exercise 16.O, $(\gamma) \Rightarrow (\alpha)$. Thus, if the equivalence of (α) and (β) is granted, we have $(\alpha) \Leftrightarrow (\beta) \Leftrightarrow (\gamma)$. Actually, we shall see that, without loss of generality, in (γ) we can restrict ourselves to one special G_δ-set, namely $\mathbb{R} \setminus \mathbb{Q}$. It will turn out that (α), (β) and (γ) are equivalent to

(δ) There exists a continuous map of $\mathbb{R} \setminus \mathbb{Q}$ onto A.

(There are other sets we could have taken instead of $\mathbb{R} \setminus \mathbb{Q}$, such as $[0, 1] \setminus \mathbb{Q}$. See Exercise 18.C.)

The question raised at the beginning of this section may now be formulated as: *do (α), (β), (γ), (δ) imply that A is a Borel set*? We shall obtain a negative answer to this question. The techniques needed to arrive at that answer will also supply a (positive) answer to another question: *if $f: \mathbb{R} \to \mathbb{R}$ is Borel measurable and bijective, is its inverse map necessarily Borel measurable*?

18.1. By $\mathscr{N}$ we denote the collection of all sequences $a = (a_1, a_2, \ldots)$ of positive integers. It will be convenient to take none of the properties

$(\alpha)-(\delta)$ as a starting point, but (see Definition 18.3) take instead

(ε) There exists a continuous map of $\mathcal{N}$ onto A.

Of course we shall have to say here what is meant by 'continuous', since $\mathcal{N}$ is not a subset of $\mathbb{R}$. We shall explain this below. (The reader who is acquainted with topology may keep in mind throughout Section 18 that we impose on $\mathcal{N}=\mathbb{N}^{\mathbb{N}}$ the product topology, giving each factor the discrete topology.)

We shall also prove that (ε) is equivalent to (α), (β), (γ), (δ) (see Exercise 18.J).

Let $a, a^1, a^2, \ldots$ be elements of $\mathcal{N}$. (Thus, a^3 is a sequence $(a_1{}^3, a_2{}^3, \ldots)$ where $a_k{}^3 \in \mathbb{N}$ for all $k \in \mathbb{N}$.) We say that $\lim_{n\to\infty} a^n = a$ *coordinatewise* if $\lim_{n\to\infty} a_1{}^n = a_1$, $\lim_{n\to\infty} a_2{}^n = a_2$, etc.; equivalently: for all $k \in \mathbb{N}$ there is an $N \in \mathbb{N}$ with the property that $a_k{}^N = a_k{}^{N+1} = a_k{}^{N+2} = \ldots = a_k$.

A subset S of $\mathcal{N}$ is said to be *closed* if it has the following property. If $a^1, a^2, \ldots \in S$ and $\lim_{n\to\infty} a^n = a \in \mathcal{N}$ coordinatewise, then $a \in S$.

A function $f\colon \mathcal{N} \to \mathbb{R}$ is said to be *continuous* if $\lim_{n\to\infty} f(a^n) = f(a)$ for all $a, a^1, a^2, \ldots \in \mathcal{N}$ for which $\lim_{n\to\infty} a^n = a$ coordinatewise.

The reader will be able to formulate definitions for continuity of maps $\mathcal{N} \to \mathcal{N}$ and $\mathbb{R} \to \mathcal{N}$. All compositions of continuous maps between $\mathcal{N}$ and $\mathbb{R}$ are continuous.

18.2. As an example, for $a = (a_1, a_2, \ldots) \in \mathcal{N}$, set

$$f(a) := 2^{-a_1} + 2^{-a_1-a_2} + 2^{-a_1-a_2-a_3} + \ldots$$

(Note that the sum converges!) If $a, b \in \mathcal{N}$, $n \in \mathbb{N}$ and $a_k = b_k$ for $k = 1, 2, \ldots, n$, then

$$\begin{aligned} |f(a)-f(b)| &\leqslant \sum_{k>n} |2^{-a_1-\ldots-a_k} - 2^{-b_1-\ldots-b_k}| \\ &\leqslant \sum_{k>n} 2^{-k} = 2^{-n} \end{aligned}$$

It follows easily that f is continuous. Trivially, $f(\mathcal{N}) \subset (0, 1]$. Conversely, if $x \in (0, 1]$, then by considering the dyadic development(s) of x one sees that there is exactly one $a \in \mathcal{N}$ with $f(a) = x$. Thus, f is a continuous bijection $\mathcal{N} \to (0, 1]$. Observe, however, that its inverse map $(0, 1] \to \mathcal{N}$ is not continuous since

$$\begin{aligned} &\tfrac{1}{2} = f(2, 1, 1, 1, \ldots) \\ &\tfrac{1}{2} + (\tfrac{1}{2})^n = f(1, n, 1, 1, \ldots) \qquad (n \in \mathbb{N}) \end{aligned}$$

In Exercise 18.A we shall see that the inverse of f (which is going to play a role later on) is not completely wild.

**Exercise* 18.A. Let $g:(0,1]\to\mathcal{N}$ be the inverse of f. We show that g is continuous at 'most' points of $(0, 1]$. Define $p_1, p_2, \ldots : (0,1]\to\mathcal{N}$ by

$$p_n(x) := g(x)_1 + g(x)_2 + \ldots + g(x)_n \qquad (x\in(0,1],\ n\in\mathbb{N})$$

(i) Show that for all x and n, $p_n(x) = [1 - {}^2\log(x - \Sigma_{j<n} 2^{-p_j(x)})]$.
(ii) Prove inductively that for every $x\in(0,1]$ and $n\in\mathbb{N}$ there exists a $\delta>0$ such that $p_1, \ldots, p_n$ are constant on $(x-\delta, x]$. Deduce that g *is left continuous* in the following sense. If $x, x_1, x_2, \ldots \in (0,1]$, $\lim_{i\to\infty} x_i = x$ and $x_i \leqslant x$ for all $i\in\mathbb{N}$, then $\lim_{i\to\infty} g(x_i) = g(x)$ coordinatewise.
(iii) Show in a similar way that g is continuous at every point of $(0, 1]$ that does not lie in the countable set $Q_2 := \{m2^{-n} : m\in\mathbb{Z}, n\in\mathbb{N}\}$.

DEFINITION 18.3. We call a subset A of $\mathbb{R}$ *analytic* if either $A=\varnothing$ or there exists a continuous surjection $\mathcal{N}\to A$.

18.4. Obviously, *if* $A\subset\mathbb{R}$ *is analytic and if* $f: A\to\mathbb{R}$ *is continuous, then* $f(A)$ *is analytic.*

18.5. Examples of analytic sets

(i) In 18.2 we have obtained a continuous surjection $f:\mathcal{N}\to(0,1]$, so $(0, 1]$ is analytic. By 18.4 every bounded closed interval $[a, b]$ is analytic. So is $\mathbb{R}$. ($x\mapsto x^{-1}\sin x^{-1}$ is continuous on $(0, 1]$!)
(ii) $a\mapsto a_1$ is a continuous surjection $\mathcal{N}\to\mathbb{N}$, so $\mathbb{N}$ is analytic. It follows from 18.4 that all countable subsets of $\mathbb{R}$ are analytic.

Exercise 18.B. Prove that the collection of all analytic sets has the cardinality of the continuum, so that there must exist nonanalytic sets. (Let $\mathcal{N}_1$ be the set of all elements of $\mathcal{N}$ that are eventually constant. $\mathcal{N}_1$ is countable, and a continuous function $\mathcal{N}\to\mathbb{R}$ is completely determined by its restriction to $\mathcal{N}_1$.)

THEOREM 18.6. *Let* $A_1, A_2, \ldots$ *be analytic subsets of* $\mathbb{R}$. *Then their union and intersection are analytic.*

Proof. Without restriction, let each A_n be nonempty, so that for each n we can choose a continuous surjection $f_n:\mathcal{N}\to A_n$. Define $f:\mathcal{N}\to\mathbb{R}$ by

$$f(a) := f_{a_1}(a_2, a_3, \ldots) \qquad (a=(a_1, a_2, \ldots)\in\mathcal{N})$$

It is perfectly easy to see that f is a continuous surjection $\mathcal{N}\to\bigcup_n A_n$.

The proof of the analyticity of the intersection is much trickier. (At first glance one might be inclined to use complementation, but it turns out that complements of analytic sets may not be analytic.) We assume that the intersection is not empty. Define mappings $F_1, F_2, \ldots$ of $\mathcal{N}$ into $\mathcal{N}$ by

$$F_n(a_1, a_2, \ldots) := (a_{2n-1}, a_{2(2n-1)}, a_{4(2n-1)}, a_{8(2n-1)}, \ldots)$$

Each F_n is continuous and we have the following fact:

(*) if $a^1, a^2, \ldots \in \mathcal{N}$, then there is a unique $a \in \mathcal{N}$ with $a^n = F_n(a)$ for all $n \in \mathbb{N}$

viz. $a = (a_1^1, a_2^1, a_1^2, a_3^1, a_1^3, a_2^2, a_1^4, a_4^1, \ldots)$.

Set

$$S := \{a \in \mathcal{N} : f_1(F_1(a)) = f_2(F_2(a)) = \ldots\}$$

Observe that S is a closed subset of $\mathcal{N}$.

Now $f_1 \circ F_1$ is a continuous function on $\mathcal{N}$ and we claim that it maps S onto $\bigcap_n A_n$. Indeed, if $a \in S$, then for all n we have $(f_1 \circ F_1)(a) = (f_n \circ F_n)(a) \in f_n(\mathcal{N}) = A_n$, so $(f_1 \circ F_1)(S) \subset \bigcap_n A_n$. Conversely, if $x \in \bigcap_n A_n$, then there exist $a^1, a^2, \ldots \in \mathcal{N}$ such that $x = f_n(a^n)$ $(n \in \mathbb{N})$; with a as in (*) we obtain $f_n(F_n(a)) = x$ for all n, so $a \in S$ and $x = (f_1 \circ F_1)(a)$.

It follows that we are done if we can find a continuous $F : \mathcal{N} \to \mathcal{N}$ with $F(\mathcal{N}) = S$: then $f_1 \circ F_1 \circ F$ will be a continuous surjection $\mathcal{N} \to \bigcap_n A_n$. We construct such an F in the following lemma, which reflects a crucial property of $\mathcal{N}$.

LEMMA 18.7. *Let S be a nonempty closed subset of $\mathcal{N}$. Then there exists a continuous surjection $F : \mathcal{N} \to S$ with $F(a) = a$ for all $a \in S$.*

Proof. If $m \in \mathbb{N}$ and $p_1, p_2, \ldots, p_m \in \mathbb{N}$, set

$$\mathcal{N}_{p_1,\ldots,p_m} := \{a \in \mathcal{N} : a_1 = p_1, \ldots, a_m = p_m\}$$

For $a \in \mathcal{N}$ define $(a_1^*, a_2^*, \ldots) \in \mathcal{N}$ inductively:

$$a_1^* := \begin{cases} a_1 & \text{if } S \cap \mathcal{N}_{a_1} \neq \varnothing \\ \min\{s_1 : s \in S\} & \text{otherwise} \end{cases}$$

In any case we can choose an $s^1 \in S \cap \mathcal{N}_{a_1^*}$.

$$a_2^* := \begin{cases} a_2 & \text{if } S \cap \mathcal{N}_{a_1,a_2} \neq \varnothing \\ \min\{s_2 : s \in S \cap \mathcal{N}_{a_1^*}\} & \text{otherwise} \end{cases}$$

Choose $s^2 \in S \cap \mathcal{N}_{a_1^*, a_2^*}$.

$$a_2^* := \begin{cases} a_3 & \text{if } S \cap \mathcal{N}_{a_1,a_2,a_3} \neq \varnothing \\ \min\{s_3 : s \in S \cap \mathcal{N}_{a_1^*,a_2^*}\} & \text{otherwise} \end{cases}$$

Choose $s^3 \in S \cap \mathcal{N}_{a_1^*, a_2^*, a_3^*}$.

Proceeding in this fashion, for every $a \in \mathcal{N}$ we obtain an element $F(a) := (a_1^*, a_2^*, \ldots)$ of $\mathcal{N}$; thus we have a map $F : \mathcal{N} \to \mathcal{N}$. It is clear that $F(a) = a$ if $a \in S$, so $F(\mathcal{N}) \supset S$. On the other hand, in the notation used above, $\lim_{n\to\infty} s^n = F(a)$ coordinatewise, so $F(a) \in S$ for *all* $a \in \mathcal{N}$.

Finally, for each n the numbers $a_1^*, \ldots, a_n^*$ are completely determined by $a_1, \ldots, a_n$. Therefore, F is continuous.

THEOREM 18.8. *All Borel subsets of* $\mathbb{R}$ *are analytic (and have analytic complements).*

(In Theorem 18.16 we shall see that not all analytic sets are Borel.)

Proof. By Exercise 16.G and Theorem 18.6 it is enough to prove that all closed subsets of $\mathbb{R}$ are analytic. We give two proofs, both starting from the observation that all closed intervals $[a, b]$ ($a, b \in \mathbb{R}$, $a<b$) are analytic (18.5).

First, note that every open subset of $\mathbb{R}$ is a union of countably many closed intervals and therefore is analytic by Theorem 18.6. But every closed set is a G_δ, hence is also analytic by 18.6.

The second proof is more sophisticated. Let A be a nonempty closed subset of $\mathbb{R}$. By Theorem 18.5(i), $\mathbb{R}$ is analytic. Choose a continuous surjection $f: \mathcal{N} \to \mathbb{R}$. It is perfectly easy to see that $f^{-1}(A)$ is closed in $\mathcal{N}$. Hence, by Lemma 18.7 we have a continuous surjection $F: \mathcal{N} \to f^{-1}(A)$. Now $f \circ F$ maps $\mathcal{N}$ continuously onto A.

**Exercise* 18.C. Show that for a nonempty $A \subset \mathbb{R}$ the following are equivalent.

(α) A is analytic.

(β) There is a continuous surjection $\mathbb{R} \setminus \mathbb{Q} \to A$.

(γ) There is a continuous surjection $[0, 1] \setminus \mathbb{Q} \to A$.

(For the proof of the implication $(\alpha) \Rightarrow (\beta)$, note that Exercise 18.A provides a continuous map $[0, 1] \setminus \mathbb{Q} \to \mathcal{N}$ whose image contains all but countably many elements of $\mathcal{N}$; it is easy to extend this map to a continuous surjection $\mathbb{R} \setminus \mathbb{Q} \to \mathcal{N}$.)

Exercise 18.D. A nonempty subset A of $\mathbb{R}$ is analytic if and only if A is the set of all values of some left continuous function on $(0, 1]$. (Hint. Exercise 18.A yields a left continuous surjection $(0, 1] \to \mathcal{N}$. On the other hand, if $f: (0, 1] \to \mathbb{R}$ is left continuous, then by Theorem 7.7 the set X of all discontinuity points of f is countable. As a continuous image of a Borel set, $f((0, 1] \setminus X)$ is analytic.)

LEMMA 18.9. (Separation lemma) *If A and B are disjoint analytic sets, then there exists a Borel set E with $A \subset E$, $B \subset \mathbb{R} \setminus E$.*

Proof. Let us call two subsets, S and T, of $\mathbb{R}$ *separated* if there exists a Borel set E with $S \subset E$ and $T \subset \mathbb{R} \setminus E$. Now first observe: *If for all $m, n \in \mathbb{N}$, S_m and T_n are subsets of $\mathbb{R}$ and if $\bigcup_m S_m$ and $\bigcup_n T_n$ are not separated, then there exist m and n such that S_m and T_n are not separated.* (Proof. Otherwise, for all m, n we could select a Borel set $E_{m,n}$ with $S_m \subset E_{m,n}$ and $T_n \subset \mathbb{R} \setminus E_{m,n}$. Let $E := \bigcap_n \bigcup_m E_{m,n}$: then E is Borel, $\bigcup_m S_m \subset E$ and $\bigcup_n T_n \subset \mathbb{R} \setminus E$.)

For $n \in \mathbb{N}$ and $p_1, \ldots, p_n \in \mathbb{N}$ put

$$\mathcal{N}_{p_1,\ldots,p_n} := \{x \in \mathcal{N} : x_1 = p_1, \ldots, x_n = p_n\}$$

Let A and B be analytic sets that are not separated: we prove that they are not disjoint. Take continuous surjections $f: \mathcal{N} \to A$ and $g: \mathcal{N} \to B$. By the above observation there exist $a_1, b_1 \in \mathbb{N}$ such that $f(\mathcal{N}_{a_1})$ and $g(\mathcal{N}_{b_1})$ are not separated; there exist $a_2, b_2 \in \mathbb{N}$ such that $f(\mathcal{N}_{a_1,a_2})$ and $g(\mathcal{N}_{b_1,b_2})$ are not separated; etc. Thus, we obtain $a, b \in \mathcal{N}$ with the property that for all $n \in \mathbb{N}$ the sets $f(\mathcal{N}_{a_1,\ldots,a_n})$ and $g(\mathcal{N}_{b_1,\ldots,b_n})$ are not separated. For each $n \in \mathbb{N}$ the union of all intervals $(f(x)-n^{-1}, f(x)+n^{-1})$, where x runs through $\mathcal{N}_{a_1,\ldots,a_n}$ is an open set (hence a Borel set) that contains $f(\mathcal{N}_{a_1,\ldots,a_n})$; then its complement cannot entirely contain $g(\mathcal{N}_{b_1,\ldots,b_n})$. Consequently, for each n there exist $b^n \in \mathcal{N}_{b_1,\ldots,b_n}$ and $a^n \in \mathcal{N}_{a_1,\ldots,a_n}$ for which $g(b^n) \in (f(a^n)-n^{-1}, f(a^n)+n^{-1})$. Now $\lim_{n\to\infty} a^n = a$ coordinatewise and $\lim_{n\to\infty} b^n = b$ coordinatewise, so $|f(a)-g(b)| = \lim_{n\to\infty} |f(a^n)-g(b^n)| = 0$. Then $f(a) = g(b)$, so $\varnothing \neq f(\mathcal{N}) \cap g(\mathcal{N}) = A \cap B$.

If, in the lemma, A and B are each other's complements, then we must have $E = A$, so A is Borel. Together with Theorem 18.8 this yields:

THEOREM 18.10. *A subset A of $\mathbb{R}$ is Borel if and only if both A and its complement are analytic.*

Exercise 18.E. Let $A_1, A_2, \ldots$ be pairwise disjoint analytic sets. Then there exist pairwise disjoint Borel sets $E_1, E_2, \ldots$ such that $A_n \subset E_n$ for each n.

Exercise 18.F. Let $f: \mathbb{R} \to \mathbb{R}$ be continuous and surjective. Let $A \subset \mathbb{R}$. Then A is Borel if and only if $f^{-1}(A)$ is Borel.

18.11. Let $n \in \mathbb{N}$. We call a subset A of $\mathbb{R}^n$ *analytic* if there exists a continuous surjection $\mathcal{N} \to A$. (For maps $\mathcal{N} \to \mathbb{R}^n$, continuity is defined coordinatewise.)

**Exercise* 18.G. Let $m, n \in \mathbb{N}$. Show that the following are true.
(i) If $A \subset \mathbb{R}^n$ is analytic and $F: A \to \mathbb{R}^m$ is continuous, then $F(A)$ is analytic in $\mathbb{R}^m$.
(ii) If $A \subset \mathbb{R}^n$ and $B \subset \mathbb{R}^m$ are analytic, then $A \times B$ is analytic in $\mathbb{R}^{n+m}$. (Consider the map $F: \mathcal{N} \to \mathcal{N} \times \mathcal{N}$ defined by the formula $F(a) = ((a_1, a_3, a_5, \ldots), (a_2, a_4, a_6, \ldots))$ $(a \in \mathcal{N})$.)
(iii) If $A_1, \ldots, A_n$ are analytic subsets of $\mathbb{R}$, then $A_1 \times \ldots \times A_n$ is analytic in $\mathbb{R}^n$. In particular, $\mathbb{R}^n$ itself is analytic.
(iv) All closed subsets of $\mathbb{R}^n$ are analytic.
(v) Countable unions and intersections of analytic subsets of $\mathbb{R}^n$ are analytic.
(vi) All Borel subsets of $\mathbb{R}^n$ are analytic.

(It would not be overly hard to generalize Theorem 18.10, but our interest lies in analytic subsets of $\mathbb{R}$, not $\mathbb{R}^n$. Exercise 18.G is mainly meant to serve us as a tool.)

THEOREM 18.12. (Sierpiński) *A function $\mathbb{R}\to\mathbb{R}$ is Borel measurable if and only if its graph is a Borel subset of $\mathbb{R}^2$.*

Proof. Let $f:\mathbb{R}\to\mathbb{R}$ and let Γ be its graph. Let $\pi_1:\mathbb{R}^2\to\mathbb{R}$ be the first coordinate function. Assume that Γ is a Borel set. Take $a\in\mathbb{R}$: we prove that the set $\{x\in\mathbb{R}:f(x)\geqslant a\}$ is Borel in $\mathbb{R}$ (see Theorem 16.7). Note that

$$\{x:f(x)\geqslant a\}=\pi_1(\Gamma\cap\mathbb{R}\times[a,\infty))$$

is the image of an analytic subset of $\mathbb{R}^2$ under a continuous map $\mathbb{R}^2\to\mathbb{R}$ and therefore is analytic. But its complement in $\mathbb{R}$ is $\pi_1(\Gamma\cap\mathbb{R}\times(-\infty,a))$, which is also analytic. Thus, $\{x:f(x)\geqslant a\}$ is Borel.

For the converse, assume that f is Borel measurable. We have

$$A:=\{(x,y)\in\mathbb{R}^2:f(x)<y\}=\bigcup_{q\in\mathbb{Q}}\{x\in\mathbb{R}:f(x)<q\}\times[q,\infty)$$

so A is a Borel subset of $\mathbb{R}^2$. Similarly, $B:=\{(x,y)\in\mathbb{R}^2:f(x)>y\}$ is Borel. But then, so is $\mathbb{R}^2\setminus A\cup B$, which is just Γ.

COROLLARY 18.13. *If $f:\mathbb{R}\to\mathbb{R}$ is bijective and Borel measurable, then the inverse map f^{-1} is also Borel measurable.*

Proof. The formula $(x,y)\mapsto(y,x)$ defines a continuous bijection $\mathbb{R}^2\to\mathbb{R}^2$. The graph of f^{-1} is just the inverse image of the graph of f under this map. Hence, if the latter graph is Borel, then so is the former.

Exercise 18.H. Let $f:\mathbb{R}\to\mathbb{R}$ be a function whose graph is an analytic subset of $\mathbb{R}^2$. Then f is Borel measurable and its graph is actually Borel measurable. (Reconsider the proof of Theorem 18.12.)

Exercise 18.I. Let $f:\mathbb{R}\to\mathbb{R}$. Show that f is Borel measurable if and only if for every analytic set $X\subset\mathbb{R}$ the set $f^{-1}(X)$ is analytic. (Using the methods of the proof of Theorem 18.12 and observing that, for all $X\subset\mathbb{R}$, $f^{-1}(X)=\pi_1(\Gamma\cap\mathbb{R}\times X)$, one shows that, if f is Borel and X is analytic, then $f^{-1}(X)$ is analytic. For the converse, use Theorem 18.10 to show that $f^{-1}(X)$ is Borel for every Borel set X.)

THEOREM 18.14. *Let A be an analytic subset of $\mathbb{R}$ and let $f:\mathbb{R}\to\mathbb{R}$ be Borel measurable. Then $f(A)$ is analytic.*

Proof. Let Γ be the graph of f and let $\pi_2:\mathbb{R}^2\to\mathbb{R}$ be the second coordinate function. Then $f(A)=\pi_2(\Gamma\cap A\times\mathbb{R})$. Being a continuous image of an analytic subset of $\mathbb{R}^2$, $f(A)$ is analytic.

**Exercise* 18.J. Prove the equivalence of the statements (α), (γ), (δ), (ε) of the beginning of this section. (That we may add (β) to this list will follow from Theorem 18.15(ii).)

Exercise 18.K. For a continuous function $f:\mathbb{R}\to\mathbb{R}$ let E_f denote the set of points where f is differentiable. Prove the following theorem (due to Poprougénko). A subset X of $\mathbb{R}$ is analytic if and only if there exists an $f\in\mathscr{C}(\mathbb{R})$ with $X=\{f'(x):x\in E_f\}$.

(The 'if' follows easily from Theorems 7.11 and 18.14. The proof of the converse is somewhat more involved. Let X be analytic; for simplicity assume $X \cap [-1, 1] \neq \emptyset$. Use Exercise 18.D to obtain a left continuous $g : (0, \infty) \to \mathbb{R}$ that, for each $n \in \mathbb{N}$, maps $(0, n]$ onto $X \cap [-n, n]$. By Theorem 7.7 there exists an infinite sequence $x_1, x_2, \ldots$ in $(0, \infty)$ such that every discontinuity point of g is an x_k. Extend g to a left continuous $h : \mathbb{R} \to \mathbb{R}$ with $h(x) = g(x_k)$ for all $x \in (-k, -k+1]$ $(k \in \mathbb{N})$. Show that h is Riemann integrable over every interval $[a, b]$: define $f(x) := \int_0^x h(t)\,dt$ $(x \in \mathbb{R})$.)

If we consider Theorem 18.14, it is quite remarkable that we can obtain *all* analytic subsets of $\mathbb{R}$ by means of *one* continuous function. We will do so, however, in two ways.

(i) We prove that any analytic subset of $\mathbb{R}$ can be obtained by applying the ordinary projection map $\mathbb{R}^2 \to \mathbb{R}$ to a suitable G_δ-subset of $\mathbb{R}^2$. (Note that the F_σ-subsets of $\mathbb{R}^2$ behave very differently here: projecting such a set always yields an F_σ-subset of $\mathbb{R}$.)

(ii) It is even possible to construct a continuous $g : \mathbb{R} \to \mathbb{R}$ such that every analytic subset of $\mathbb{R}$ is the image under g of a suitable G_δ-subset of $\mathbb{R}$. This g, however, is a less elementary function than the projection of (i).

THEOREM 18.15

(i) *Let π_2 be the second coordinate map $\mathbb{R}^2 \to \mathbb{R}$. A subset X of $\mathbb{R}$ is analytic if and only if there is a G_δ-subset A of $\mathbb{R}^2$ with $X = \pi_2(A)$.*

(ii) *There exists a continuous function $g : \mathbb{R} \to \mathbb{R}$ such that the analytic subsets of $\mathbb{R}$ are just the sets $g(B)$ where B runs through the G_δ-subsets of $\mathbb{R}$.*

Proof. Let $T := [0, 1] \setminus \mathbb{Q}$. Recall that, by Exercise 18.C, a nonempty set $X \subset \mathbb{R}$ is analytic if and only if there exists a continuous surjection $T \to X$.

(i) If $A \subset \mathbb{R}^2$ is a G_δ, then A is analytic (Exercise 18.G(vi)), so $\pi_2(A)$ is analytic (Exercise 18.G(i)). Conversely, let $X \subset \mathbb{R}$ be nonempty, analytic. Let $f : T \to X$ be a continuous surjection and let Γ be its graph. Clearly, $X = \pi_2(\Gamma)$, so we are done if Γ is a G_δ in $\mathbb{R}^2$. If $\bar{\Gamma}$ is the closure of Γ in $\mathbb{R}^2$, then it is not hard to see that $\Gamma = \bar{\Gamma} \cap T \times \mathbb{R}$. But T is a G_δ-subset of $\mathbb{R}$, so $T \times \mathbb{R}$ is G_δ in $\mathbb{R}^2$. Then so is Γ.

(ii) Whatever continuous $g : \mathbb{R} \to \mathbb{R}$ we take, if $B \subset \mathbb{R}$ is a G_δ, then $g(B)$ will be analytic. It follows from our proof of (i) and the fact that $T \subset [0, 1]$ that $g := \pi_2 \circ G$ will satisfy the requirements of (ii), provided that G is a continuous surjection $\mathbb{R} \to [0, 1] \times \mathbb{R}$. We proceed to construct such a G.

Let ψ_1, ψ_2 be as in Exercise 16.T. ψ_2 is continuous and $\psi_2(n) = 0$ for all integers n. Hence, if we define $h : \mathbb{R} \to \mathbb{R}$ by

$$h(x) := [x]\psi_2(x) \qquad (x \in \mathbb{R})$$

($[x]$ denoting the entire part of x), then h is continuous and

$$h(x) = n\psi_2(x) \qquad (n \in \mathbb{Z};\ x \in [n, n+1])$$

Define

$$G(x) := (\psi_1(x), h(x)) \qquad (x \in \mathbb{R})$$

G is continuous. It follows from Exercise 16.T(ii) that, for every $n \in \mathbb{N}$, G maps $[n, n+1]$ onto the rectangle $[0, 1] \times [0, n]$ and $[-n, -n+1]$ onto the rectangle $[0, 1] \times [-n, 0]$. Hence, $G(\mathbb{R}) = [0, 1] \times \mathbb{R}$.

By a cardinality argument we have already seen (Exercise 18.B) that the class of the analytic subsets of $\mathbb{R}$ is strictly smaller than the class of *all* subsets of $\mathbb{R}$. We proceed to show that it is strictly larger than the class of the Borel sets.

THEOREM 18.16 (Sierpiński) *Not all analytic sets are Borel.*

The proof boils down to a catalogue argument like the one we used in Section 16 and the proof of the uncountability of the continuum (see Appendix A).

First we have to extend our definitions of closedness and analyticity to subsets of $\mathcal{N}^p$ where $p \in \mathbb{N}$.

18.17. Let (a, b), (a^1, b^1), (a^2, b^2), ... be elements of $\mathcal{N}^2$. We say that $\lim_{n\to\infty} (a^n, b^n) = (a, b)$ *coordinatewise* if $\lim_{n\to\infty} a^n = a$ coordinatewise and $\lim_{n\to\infty} b^n = b$ coordinatewise.

The definition extends naturally to coordinatewise convergence in $\mathcal{N}^3$, $\mathcal{N}^4$, ... Also, we leave it to the reader to define *continuity* for maps $\mathcal{N}^3 \to \mathcal{N}$, $\mathcal{N}^2 \to \mathbb{R}$, etc., and *closedness* and *analyticity* for subsets of $\mathcal{N}^p$ ($p \in \mathbb{N}$). Lemma 18.7 implies that all closed subsets of $\mathcal{N}$ are analytic. (The same is true for $\mathcal{N}^2$, $\mathcal{N}^3$, ... instead of $\mathcal{N}$.)

LEMMA 18.18. *There exists a subset W of $\mathcal{N}^3$ such that*
(i) *for every continuous $F : \mathcal{N} \to \mathcal{N}$ there exists an $s \in \mathcal{N}$ such that the graph of F is just $\{(x, y) \in \mathcal{N}^2 : (x, y, s) \in W\}$,*
(ii) *W is closed.*

Proof. We use the notation $\mathcal{N}_{p_1,\dots,p_m}$ as in the proof of Lemma 18.7. Let us call a subset U of $\mathcal{N}^2$ a *square* if there exist $m \in \mathbb{N}$ and $a_1, \dots, a_m, b_1, \dots, b_m \in \mathbb{N}$ such that $U = \mathcal{N}_{a_1,\dots,a_m} \times \mathcal{N}_{b_1,\dots,b_m}$. There are only countably many such squares: let $(U_1, U_2, \dots)$ be an enumeration of them. Now put

$$W := \{(a, b, c) \in \mathcal{N}^3 : \text{for all } n \in \mathbb{N}, (a, b) \notin U_{c_n}\}$$

We prove that this W has the properties (i) and (ii).
(i) Let $F : \mathcal{N} \to \mathcal{N}$ be continuous, let Γ be its graph. For every point

(a, b) of $\mathcal{N}^2 \setminus \Gamma$ there exists an $m \in \mathbb{N}$ with $\mathcal{N}_{a_1,\ldots,a_m} \times \mathcal{N}_{b_1,\ldots,b_m} \subset \mathcal{N}^2 \setminus \Gamma$. (Otherwise, for each m one can choose $a^m \in \mathcal{N}_{a_1,\ldots,a_m}$ and $b^m \in \mathcal{N}_{b_1,\ldots,b_m}$ with $b^m = F(a^m)$; as $\lim_{n\to\infty} a^m = a$, $\lim_{n\to\infty} b^m = b$ coordinatewise one obtains $b = F(a)$, which is false.) Therefore, $\mathcal{N}^2 \setminus \Gamma$ is a union of squares. Then there exists an $s \in \mathcal{N}$ with $\mathcal{N}^2 \setminus \Gamma = U_{s_1} \cup U_{s_2} \cup \ldots$ For $(x, y) \in \mathcal{N}^2$ we now have $(x, y) \in \Gamma$ if and only if $(x, y) \notin U_{s_n}$ for all n, i.e. if and only if $(x, y, s) \in W$.

(ii) Suppose we have $x, x^i, y, y^i, s, s^i \in \mathcal{N}$ $(i \in \mathbb{N})$, $\lim_{i\to\infty} x^i = x$, $\lim_{i\to\infty} y^i = y$, $\lim_{i\to\infty} s^i = s$ and $(x, y, s) \notin W$: we show that, for sufficiently large i, $(x^i, y^i, s^i) \notin W$.

There is an $n \in \mathbb{N}$ with $(x, y) \in U_{s_n}$. Let U_{s_n} be the square $\mathcal{N}_{a_1,\ldots,a_m} \times \mathcal{N}_{b_1,\ldots,b_m}$. Then $x_k = a_k$, $y_k = b_k$ for $k = 1, \ldots, m$. If $i \in \mathbb{N}$ is large enough, we have

$$s_n^i = s_n$$
$$x_k^i = x_k \quad \text{and} \quad y_k^i = y_k \quad \text{for} \quad k = 1, \ldots, m$$

in other words,

$$U_{s_n^i} = \mathcal{N}_{a_1,\ldots,a_m} \times \mathcal{N}_{b_1,\ldots,b_m}$$
$$x_k^i = a_k \quad \text{and} \quad y_k^i = b_k \quad \text{for} \quad k = 1, \ldots, m$$

But then $(x^i, y^i) \in U_{s_n^i}$.

18.19. We are now going to construct a nonanalytic subset B of $\mathcal{N}$ that has an analytic complement. If $A \subset \mathcal{N}$ is analytic, there exists a continuous $F : \mathcal{N} \to \mathcal{N}$ with $F(\mathcal{N}) = A$. By Lemma 18.18 there exists an $s \in \mathcal{N}$ such that

$$y = F(x) \text{ if and only if } (x, y, s) \in W \qquad (x, y \in \mathcal{N})$$

Then $A = \{y \in \mathcal{N} : \text{there is an } x \in \mathcal{N} \text{ with } (x, y, s) \in W\}$.

Thus, if for all $z \in \mathcal{N}$ we define $A_z \subset \mathcal{N}$ by

$$A_z := \{y \in \mathcal{N} : \text{there is an } x \in \mathcal{N} \text{ with } (x, y, z) \in W\}$$

then every analytic set is an A_z. Therefore, if we can find a subset B of $\mathcal{N}$ such that

$$z \in B \text{ if and only if } z \notin A_z \qquad (z \in \mathcal{N})$$

then B will not be analytic. Now trivially such a B exists: take $B := \{z \in \mathcal{N} : z \notin A_z\}$. We proceed to prove that $\mathcal{N} \setminus B$ *is* analytic. That is not hard either:

$$\begin{aligned}\mathcal{N} \setminus B &= \{z \in \mathcal{N} : z \in A_z\} \\ &= \{z \in \mathcal{N} : \text{there is an } x \in \mathcal{N} \text{ with } (x, z, z) \in W\} \\ &= \pi_2(\Delta_{23} \cap W)\end{aligned}$$

where $\Delta_{23} = \{(x, y, z) \in \mathcal{N}^3 : y = z\}$ and π_2 is the second coordinate map $\mathcal{N}^3 \to \mathcal{N}$. As π_2 is continuous, $\pi_2(\Delta_{23} \cap W)$ will be analytic (and our proof will be finished) if $\Delta_{23} \cap W$ is analytic in $\mathcal{N}^3$. Now consider the map $F : \mathcal{N} \to \mathcal{N}^3$ defined by

$$F(a) := ((a_1, a_4, a_7, \ldots), (a_2, a_5, \ldots), (a_3, a_6, \ldots)) \qquad (a \in \mathcal{N})$$

Since F is continuous and both Δ_{23} and W are closed in $\mathcal{N}^3$, $F^{-1}(\Delta_{23} \cap W)$ is closed, hence analytic in $\mathcal{N}$. Since F is not only continuous but also surjective, $\Delta_{23} \cap W$ is analytic.

18.20. To prove Sierpiński's theorem we translate the above from $\mathcal{N}$ to $\mathbb{R}$. Let $B \subset \mathcal{N}$ be such that $\mathcal{N} \setminus B$ is analytic while B itself is not. Let f, g, Q_2 be as in Exercise 18.A. Set $C := f(\mathcal{N} \setminus B)$. Then C is analytic in $\mathbb{R}$: we show that C is not Borel. Suppose it is. Then so are $f(B)$ (which is $(0, 1] \setminus f(\mathcal{N} \setminus B)$) and $f(B) \setminus Q_2$. Then $g(f(B) \setminus Q_2)$ is analytic in $\mathcal{N}$, i.e. $B \setminus f^{-1}(Q_2)$ is analytic. Now $B \cap f^{-1}(Q_2)$, being countable, is certainly analytic. Then so is B and we have a contradiction.

COROLLARY 18.21. ***The image $f(X)$ of a Borel subset X of $\mathbb{R}$ under a Borel measurable function f need not be a Borel set.***

Exercise 18.L. If we want to obtain a Borel set $X \subset \mathbb{R}$ and a measurable $f : \mathbb{R} \to \mathbb{R}$ such that $f(X)$ is not Borel, how 'tame' can we make them? According to Theorem 18.15(ii), for X we can choose a G_δ-set and for f a function that is continuous on all of $\mathbb{R}$. Alternatively, by Exercise 18.C we can specify X to be $\mathbb{R} \setminus \mathbb{Q}$ if we allow f to be a function that is continuous only on $\mathbb{R} \setminus \mathbb{Q}$.

Show that one cannot do much better: the image of $\mathbb{R} \setminus \mathbb{Q}$ under a continuous function $f : \mathbb{R} \to \mathbb{R}$ is always Borel (and can, in fact, be obtained from a suitable interval by deleting countably many points).

Notes to Section 18

A detailed account of analytic sets (also called *Suslin sets*) and their history is given by Lusin (1972). It is shown there that all analytic sets are Lebesgue measurable and have the property of Baire. (See the notes to Section 20.)

The characterizations of analytic sets given in Exercises 18.D and 18.K may be found in W. Sierpiński, *Fund. Math.* **10** (1927), 169–71 and G. Poprougénko, *Fund. Math.* **18** (1932), 77–84, respectively. Various other descriptions exist in the literature (e.g. W. Sierpiński, *Fund. Math.* **7** (1925), 155–8, S. Mazurkiewicz & W. Sierpiński, *Fund. Math.* **6** (1924),

161–9). S. Kierst (*Fund. Math.* **27** (1936), 226–33) showed that a subset of $\mathbb{C}$ is analytic if and only if it is the set of all boundary values of an analytic function on the open unit disc.

One of the most interesting results of the theory of analytic sets is the following criterion for Borel measurability. (See Lusin 1972, pp. 114 and 163.) An uncountable subset of $\mathbb{R}$ is Borel measurable if and only if it is the union of a countable set and a set that is the range of an *injective* continuous function $\mathcal{N} \to \mathbb{R}$. It follows that every set that is homeomorphic to a Borel set is itself Borel. More generally, one can show that the image of a Borel set under an injective Borel function is Borel. Conversely, every uncountable Borel set is the image of $\mathbb{R}$ under an injective function that belongs to the first class of Baire (W. Sierpiński, *Fund. Math.* **20** (1933), 126–30).

6

INTEGRATION

19. The Lebesgue integral

Riemann integration over an interval $[a, b]$ is a way to associate with every function of a certain class (viz. $\mathscr{R}[a, b]$) a real number, the integral of the function. It turns out that there are advantages in considering larger classes of functions.

Thus, in Theorem 15.3 we have seen how a differentiable function can be reconstructed from its derivative, provided that this derivative is Riemann integrable. A more general integration theory might very well yield a stronger version of Theorem 15.3. (It does in Corollary 21.23.)

There are many ways to extend the Riemann integration theory. For instance, let F be the class of all functions f on $[a, b]$ for which there exists a $g \in \mathscr{R}[a, b]$ with $f=g$ a.e. on $[a, b]$. Then, with the aid of Exercise 15.A, for $f \in F$ we can define its 'integral' $J(f)$ by the requirement

$$J(f)=\int_a^b g(x)\,\mathrm{d}x \text{ if } g \in \mathscr{R}[a, b],\ f=g \text{ a.e. on } [a, b]$$

A particularly satisfactory extension of the Riemann integral is the one developed by Lebesgue. It encompasses a very large class of functions and it leads to powerful convergence theorems. To illustrate what is meant by 'convergence theorems', consider the following statement. *If* $f_1, f_2, \ldots$ *are Riemann integrable functions on* $[a, b]$ *such that* $0 \leqslant f_n \leqslant 1$ *for all* $n \in \mathbb{N}$ *and if* $\lim_{n\to\infty} f_n=0$ *pointwise, then* $\lim_{n\to\infty} \int_a^b f_n(x)\,\mathrm{d}x=0$. This theorem is completely elementary, but it is complicated to prove by elementary means. It will be much easier to prove the theorem with the Lebesgue theory at hand. (See Lebesgue's Theorem 19.15 for a wide generalization.)

There exists a large variety of methods to construct the Lebesgue integral. We have chosen to use as our starting point the Riemann integral for continuous functions. For simplicity of notation we first consider integration of functions defined on all of $\mathbb{R}$. Later on we study functions defined on arbitrary intervals or even more general subsets of $\mathbb{R}$.

In Sections 22 and 23 we shall develop the Perron integral, which is a further extension of the Lebesgue integral, and the Stieltjes integral, which generalizes the Riemann integral in a different way.

By $\mathscr{C}_c$ we denote the vector space of all continuous functions $f: \mathbb{R} \to \mathbb{R}$ for which the set $\{x \in \mathbb{R} : f(x) \neq 0\}$ is bounded.

If $f \in \mathscr{C}_c$, then there exist $a, b \in \mathbb{R}$ such that $f(x) = 0$ for all $x \in \mathbb{R} \setminus [a, b]$. Then the value of the Riemann integral $\int_a^b f(x)\, dx$ is independent of the choice of a and b. In this section we usually denote this number by $\int f$.

LEMMA 19.1. *The following conditions on a subset X of $\mathbb{R}$ are equivalent.*
(α) *X is a null set.*
(β) *There exist $\phi_1, \phi_2, \ldots \in \mathscr{C}_c$ with $\phi_i \geq 0$ for each i, such that $\sum_{i=1}^{\infty} \int \phi_i \leq 1$ and $\sum_{i=1}^{\infty} \phi_i(x) = \infty$ for every $x \in X$.*

Proof. First, let X be null. For every $j \in \mathbb{N}$ there exist intervals $I_{1j}, I_{2j}, \ldots$ covering X and such that $\sum_{i=1}^{\infty} L(I_{ij}) \leq 3^{-j}$. (Recall that $L(I_{ij})$ is the length of I_{ij}.) Choose nonnegative $\psi_{ij} \in \mathscr{C}_c$ such that $\psi_{ij} = 1$ on I_{ij} while $\int \psi_{ij} \leq 2L(I_{ij})$. Then

$$\sum_{i,j} \int \psi_{ij} = \sum_j \sum_i \int \psi_{ij} \leq 2 \sum_j \sum_i L(I_{ij}) \leq 2 \sum_j 3^{-j} = 1$$

If $x \in X$, then for every $j \in \mathbb{N}$ there is an i with $\psi_{ij}(x) = 1$, so $\sum_{i,j} \psi_{ij}(x) = \infty$. Now let, for example, $\phi_n := \sum_{i+j=n} \psi_{ij}$; then we have ($\beta$).

Conversely, assume the existence of $\phi_1, \phi_2, \ldots$ as described in (β). Let $\varepsilon > 0$. For $n \in \mathbb{N}$, set $\Phi_n := \phi_1 + \phi_2 + \ldots + \phi_n$. The set $U := \{x : \text{there exists an } n \text{ with } \Phi_n(x) > 1/\varepsilon\}$ is open and contains X. Let $I_1, I_2, \ldots$ be the components of U: we prove $\sum_{i=1}^{\infty} L(I_i) \leq \varepsilon$. For each i, choose a closed bounded interval $[p_i, q_i] \subset I_i$. For every i, the functions $\Phi_1 \wedge \varepsilon^{-1}$, $\Phi_2 \wedge \varepsilon^{-1}, \ldots$ form an increasing sequence, tending to ε^{-1} at each point of $[p_i, q_i]$. By Dini's theorem (Exercise 8.F)

$$(q_i - p_i)\varepsilon^{-1} = \lim_{n\to\infty} \int_{p_i}^{q_i} (\Phi_n \wedge \varepsilon^{-1})(x)\, dx \leq \lim_{n\to\infty} \int_{p_i}^{q_i} \Phi_n(x)\, dx$$

Therefore, for every $k \in \mathbb{N}$,

$$\begin{aligned}\sum_{i=1}^{k} (q_i - p_i) &\leq \varepsilon \lim_{n\to\infty} \sum_{i=1}^{k} \int_{p_i}^{q_i} \Phi_n(x)\, dx \\ &\leq \varepsilon \lim_{n\to\infty} \int \Phi_n = \varepsilon \sum_{n=1}^{\infty} \int \phi_n \leq \varepsilon\end{aligned}$$

The inequality $\Sigma_{i=1}^{k} (q_i - p_i) \leqslant \varepsilon$ holds for all k and for all choices of $[p_i, q_i] \subset I_i$. It follows easily that $\Sigma_{i=1}^{\infty} L(I_i) \leqslant \varepsilon$.

COROLLARY 19.2. *If $\phi_1, \phi_2, \ldots \in \mathscr{C}_c$ and if $\Sigma_{i=1}^{\infty} \int |\phi_i|$ is finite, then the series $\Sigma\, \phi_i$ converges almost everywhere.*

DEFINITION 19.3. A function $f: \mathbb{R} \to \mathbb{R}$ is said to be *Lebesgue integrable* if there exist $\phi_1, \phi_2, \ldots \in \mathscr{C}_c$ such that $\Sigma_{i=1}^{\infty} \int |\phi_i|$ is finite while $\Sigma_{i=1}^{\infty} \phi_i = f$ almost everywhere. Clearly, the Lebesgue integrable functions form a vector space which contains $\mathscr{C}_c$. We denote this vector space by $\mathscr{L}$. Observe that $\mathscr{L}$ contains all functions $\mathbb{R} \to \mathbb{R}$ that vanish a.e.

**Exercise* 19.A
(i) The characteristic function of a bounded open set is in $\mathscr{L}$. (Hint. Use Theorem 10.6.)
(ii) The characteristic function of a bounded closed set is in $\mathscr{L}$.

Exercise 19.B. Prove or disprove the following statement. If $f \in \mathscr{L}$, $f \geqslant 0$, then there exist $\phi_1, \phi_2, \ldots \in \mathscr{C}_c$ such that $\phi_i \geqslant 0$ for all i, $\Sigma_{i=1}^{\infty} \int \phi_i$ is finite, $\Sigma_{i=1}^{\infty} \phi_i = f$ a.e.

In the situation described in Definition 19.3 we want to define the Lebesgue integral of f as the number $\Sigma_{i=1}^{\infty} \int \phi_i$. (Note that the series $\Sigma \int \phi_i$ converges because $|\int \phi_i| \leqslant \int |\phi_i|$ for each i.) Before we can do this with a clear conscience we have to show that the number $\Sigma_{i=1}^{\infty} \int \phi_i$ depends only on f and not on the choice of the ϕ_i.

LEMMA 19.4. *Let $\phi_1, \phi_2, \ldots \in \mathscr{C}_c$ be such that $\Sigma_{i=1}^{\infty} \int |\phi_i| < \infty$ and $\Sigma_{i=1}^{\infty} \phi_i \geqslant 0$ almost everywhere. Then $\Sigma_{i=1}^{\infty} \int \phi_i \geqslant 0$.*

Proof. Let X be a null set such that $\Sigma_{i=1}^{\infty} \phi_i \geqslant 0$ everywhere on $\mathbb{R} \setminus X$. Choose $\psi_1, \psi_2, \ldots \in \mathscr{C}_c$, $\psi_i \geqslant 0$, with $\Sigma_{i=1}^{\infty} \int \psi_i < \infty$ and with $\Sigma_{i=1}^{\infty} \psi_i = \infty$ on X. (Lemma 19.1.) For convenience, take $\psi_i \geqslant 2|\phi_i|$. Then for every i $\psi_i + \phi_i \geqslant 0$ and $\psi_i \geqslant 0$. We now have

$$\sum_{i=1}^{\infty} \psi_i \leqslant \sum_{i=1}^{\infty} (\psi_i + \phi_i) \quad \text{everywhere on } \mathbb{R}.$$

(Note that $\psi_i + \phi_i \geqslant \frac{1}{2}\psi_i$, so that $\Sigma_{i=1}^{\infty} (\psi_i + \phi_i) = \infty$ on X.) The series $\Sigma \int \psi_i$ and $\Sigma \int \phi_i$ being convergent, we are done if we can prove that

$$\sum_{i=1}^{\infty} \int \psi_i \leqslant \sum_{i=1}^{\infty} \left(\int \psi_i + \int \phi_i \right)$$

i.e. that

$$\int \left(\sum_{i=1}^{n} \psi_i \right) \leqslant \sup_{k \in \mathbb{N}} \int \left(\sum_{i=1}^{k} (\psi_i + \phi_i) \right) \quad \text{for all } n \in \mathbb{N}$$

Let $n \in \mathbb{N}$. Put $f := \Sigma_{i=1}^{n} \psi_i$. For $k \in \mathbb{N}$, set $g_k := \Sigma_{i=1}^{k} (\psi_i + \phi_i)$. Then

$f, g_1, g_2, \ldots$ are nonnegative elements of $\mathscr{C}_c$, $g_1 \leqslant g_2 \leqslant \ldots$ and we have $f \leqslant \sup_{k\in\mathbb{N}} g_k$. We want to show that $\int f \leqslant \sup_{k\in\mathbb{N}} \int g_k$.

Take a closed interval $[a, b]$ outside which f vanishes identically. Now $g_1 \wedge f \leqslant g_2 \wedge f \leqslant \ldots$ and $\lim_{k\to\infty} g_k \wedge f = f$. By Dini's theorem (Exercise 8.F),

$$\int f = \int_a^b f(x)\,\mathrm{d}x = \lim_{k\to\infty} \int_a^b (g_k \wedge f)(x)\,\mathrm{d}x$$

$$\leqslant \lim_{k\to\infty} \int_a^b g_k(x)\,\mathrm{d}x \leqslant \lim_{k\to\infty} \int g_k$$

which concludes the proof of the lemma.

In particular, if $\phi_1, \phi_2, \ldots, \psi_1, \psi_2, \ldots \in \mathscr{C}_c$, if $\Sigma_{i=1}^{\infty} \int |\phi_i|$ and $\Sigma_{i=1}^{\infty} \int |\psi_i|$ are finite, and if $\Sigma_{i=1}^{\infty} \phi_i = \Sigma_{i=1}^{\infty} \psi_i$ almost everywhere, then $\Sigma_{i=1}^{\infty} \int \phi_i = \Sigma_{i=1}^{\infty} \int \psi_i$. This observation makes the following definition possible.

DEFINITION 19.5. Let $f : \mathbb{R} \to \mathbb{R}$ be Lebesgue integrable. Let $\phi_1, \phi_2, \ldots \in \mathscr{C}_c$ be such that $\Sigma_{i=1}^{\infty} \int |\phi_i| < \infty$ and $\Sigma_{i=1}^{\infty} \phi_i = f$ a.e. Then we define the *Lebesgue integral* of f to be the number $\Sigma_{i=1}^{\infty} \int \phi_i$. We indicate it by $\int f$. For $f \in \mathscr{C}_c$ this definition of the symbol $\int f$ coincides with the old one. (Choose $\phi_1 := f$ and $\phi_k := 0$ for $k > 1$.)

Theorem 19.6 follows immediately from the definitions and from 19.4.

THEOREM 19.6

(i) *$f \mapsto \int f$ is a linear function on $\mathscr{L}$.*

(ii) *If $f : \mathbb{R} \to \mathbb{R}$ and $f = 0$ a.e., then $f \in \mathscr{L}$ and $\int f = 0$.*

(iii) *If f, g are functions on $\mathbb{R}$ such that $f = g$ a.e., and if f is Lebesgue integrable, then so is g, and $\int f = \int g$.*

(iv) *If $f \in \mathscr{L}$ and $f \geqslant 0$ a.e., then $\int f \geqslant 0$. If $f, g \in \mathscr{L}$ and $f \geqslant g$ a.e., then $\int f \geqslant \int g$.*

(v) *Let $f \in \mathscr{L}$. Let $a \in \mathbb{R}$. Define $g(x) := f(x+a)$ $(x \in \mathbb{R})$. Then $g \in \mathscr{L}$ and $\int g = \int f$.*

(vi) *Let $f \in \mathscr{L}$. Let $b \in \mathbb{R}$, $b \neq 0$. Define $h(x) := f(bx)$ $(x \in \mathbb{R})$. Then $h \in \mathscr{L}$ and $\int h = |b|^{-1} \int f$.*

We have already observed that all elements of $\mathscr{C}_c$ are Lebesgue integrable, as are all functions that vanish a.e. Less trivial examples of Lebesgue integrable functions are obtained from

THEOREM 19.7. *Let $a < b$ and let $f : [a, b] \to \mathbb{R}$ be (properly) Riemann integrable. Define $g : \mathbb{R} \to \mathbb{R}$ by*

$$g(x) := \begin{cases} f(x) & x \in [a, b] \\ 0 & \text{if } x \in \mathbb{R} \setminus [a, b] \end{cases}$$

Then $g \in \mathscr{L}$ and $\int g$ is equal to the Riemann integral $\int_a^b f(x)\,dx$.

Proof. It follows from Theorem 12.1 and Exercise 10.K that $g = g^\dagger$ a.e. From Theorem 10.6 one easily obtains an increasing sequence $\phi_1, \phi_2, \ldots$ of elements of $\mathscr{C}_c$ that tends to $g^\dagger$ pointwise. Then

$$g = \phi_1 + \sum_{i=1}^{\infty} (\phi_{i+1} - \phi_i) \quad \text{a.e.}$$

As the sequence $\phi_1, \phi_2, \ldots$ is increasing, for every $n \in \mathbb{N}$ we have $\Sigma_{i=1}^n \int |\phi_{i+1} - \phi_i| = \Sigma_{i=1}^n \int (\phi_{i+1} - \phi_i) = \int \phi_{n+1} - \int \phi_1 \leqslant \int_a^b f(x)\,dx - \int \phi_1$, so that $\int |\phi_1| + \Sigma_{i=1}^\infty \int |\phi_{i+1} - \phi_i|$ is finite. Hence, $g \in \mathscr{L}$ and

$$\int g = \int \phi_1 + \sum_{i=1}^{\infty} \int (\phi_{i+1} - \phi_i) = \lim_{n \to \infty} \int \phi_n \leqslant \int_a^b f(x)\,dx$$

By applying the same reasoning to $-g$ one finds $\int g \geqslant \int_a^b f(x)\,dx$.

Thus, in a sense our Lebesgue integral generalizes the Riemann integral. The extension is nontrivial: if $f : [0, 1] \to \mathbb{R}$ is the restriction of $\xi_{\mathbb{Q}}$, then f is the classical example of a bounded function that is not Riemann integrable, but the corresponding g is Lebesgue integrable because $g = 0$ a.e. (See also Exercises 19.E and 19.R.)

COROLLARY 19.8. *Let I be an interval. If I is bounded, then ξ_I is Lebesgue integrable and $\int \xi_I = L(I)$. If I is not bounded, then ξ_I is not Lebesgue integrable.*

Exercise 19.C. The function $x \mapsto (1 + |x|)^{-1}$ $(x \in \mathbb{R})$ is not Lebesgue integrable. The function $x \mapsto e^{-|x|}$ $(x \in \mathbb{R})$ is Lebesgue integrable: its Lebesgue integral is 2. The function $g : \mathbb{R} \to \mathbb{R}$, defined by

$$g(x) := \begin{cases} x^{-1/2} & \text{if } 0 < x < 1 \\ 0 & \text{if } x \leqslant 0 \text{ or } x \geqslant 1 \end{cases}$$

is Lebesgue integrable: its Lebesgue integral is 2.

Definition 19.3 in an obvious way yields the following observation. *A function $f : \mathbb{R} \to \mathbb{R}$ is Lebesgue integrable if and only if there exist $\psi_1, \psi_2, \ldots \in \mathscr{C}_c$ such that $\Sigma_{i=1}^\infty \int |\psi_{i+1} - \psi_i| < \infty$ and $\lim_{i \to \infty} \psi_i = f$ a.e.* Consequently, if $f \in \mathscr{L}$ and if $\omega : \mathbb{R} \to \mathbb{R}$ is a function such that $\omega(0) = 0$ and $|\omega(x) - \omega(y)| \leqslant |x - y|$ for all $x, y \in \mathbb{R}$, then $\omega \circ f \in \mathscr{L}$. Taking $\omega(x) := |x|$ $(x \in \mathbb{R})$ we see that $|f| \in \mathscr{L}$:

THEOREM 19.9. *If $f \in \mathscr{L}$, then $|f| \in \mathscr{L}$ and $|\int f| \leqslant \int |f|$. If $f, g \in \mathscr{L}$, then $f \wedge g \in \mathscr{L}$ and $f \vee g \in \mathscr{L}$.*

Proof. We have just seen that $|f| \in \mathscr{L}$. As $-|f| \leqslant f \leqslant |f|$, we must have

$-\int|f|\leqslant\int f\leqslant\int|f|$, i.e. $|\int f|\leqslant\int|f|$. Finally, note that for all f and g, $2(f\wedge g)=f+g-|f-g|$ and $2(f\vee g)=f+g+|f-g|$.

**Exercise* 19.D. Let $f\in\mathscr{L}$.
(i) If $g:\mathbb{R}\to\mathbb{R}$ is continuous and $g\geqslant 0$, then $f\wedge g\in\mathscr{L}$.
(ii) If $h:\mathbb{R}\to\mathbb{R}$ is bounded and continuous, then $fh\in\mathscr{L}$.

Exercise 19.E. For each of the following two functions $f:\mathbb{R}\to\mathbb{R}$, prove that f is improperly Riemann integrable but not Lebesgue integrable. (Show that $|f|\notin\mathscr{L}$.)

(i) $f(x):=\begin{cases}0 & \text{if } x<0\\ (-1)^n n^{-1} & \text{if } n\in\mathbb{N} \text{ and } n-1\leqslant x<n\end{cases}$

(ii) $f(x):=\begin{cases}0 & \text{if } x\geqslant 1 \text{ or } x\leqslant 0\\ (-1)^n n & \text{if } n\in\mathbb{N} \text{ and } 1/(n+1)\leqslant x<1/n\end{cases}$

LEMMA 19.10. *Let $f\in\mathscr{L}$ and $\varepsilon>0$. Then there exists a $\Phi\in\mathscr{C}_c$ such that $\int|f-\Phi|\leqslant\varepsilon$. In fact, there exist $\Phi,\phi_1,\phi_2,\ldots\in\mathscr{C}_c$ with $f=\Phi+\sum_{i=1}^{\infty}\phi_i$ a.e. and $\int|f-\Phi|\leqslant\sum_{i=1}^{\infty}\int|\phi_i|\leqslant\varepsilon$.*

Proof. There exist $\psi_1,\psi_2,\ldots\in\mathscr{C}_c$ such that $f=\sum_{i=1}^{\infty}\psi_i$ a.e. and $\sum_{i=1}^{\infty}\int|\psi_i|$ is finite. There is an $n\in\mathbb{N}$ with $\sum_{i>n}\int|\psi_i|\leqslant\varepsilon$. Now set $\Phi:=\psi_1+\ldots+\psi_n$ and $\phi_i:=\psi_{n+i}(i\in\mathbb{N})$. Then $f=\Phi+\sum_{i=1}^{\infty}\phi_i$ a.e. By Corollary 19.2 there is a $g\in\mathscr{L}$ with $g=\sum_{i=1}^{\infty}|\phi_i|$ a.e. But then $|f-\Phi|\leqslant g$ a.e., so $\int|f-\Phi|\leqslant\int g=\sum_{i=1}^{\infty}\int|\phi_i|=\sum_{i>n}\int|\psi_i|\leqslant\varepsilon$.

We have extended the integral from the space $\mathscr{C}_c$ to the space $\mathscr{L}$. The following theorem shows that repeating this procedure yields nothing new.

THEOREM 19.11. *Let $f_1,f_2,\ldots\in\mathscr{L}$ be such that $\sum_{i=1}^{\infty}\int|f_n|<\infty$. Then the series $\sum f_n$ converges a.e. If $f:\mathbb{R}\to\mathbb{R}$ and $f=\sum_{i=1}^{\infty}f_n$ a.e., then $f\in\mathscr{L}$ and $\int f=\sum_{i=1}^{\infty}\int f_n$.*

Proof. Applying the preceding lemma, for every n we choose $\Phi_n,\phi_{n1},\phi_{n2},\ldots$ in $\mathscr{C}_c$ such that $f_n=\Phi_n+\sum_{i=1}^{\infty}\phi_{ni}$ a.e. and $\int|f_n-\Phi_n|\leqslant\sum_{i=1}^{\infty}\int|\phi_{ni}|\leqslant 2^{-n}$. Then $\int|\Phi_n|\leqslant\int|f_n|+\int|f_n-\Phi_n|\leqslant\int|f_n|+2^{-n}$, so that

$$\sum_{n=1}^{\infty}\int|\Phi_n|+\sum_{n,i=1}^{\infty}\int|\phi_{ni}|\leqslant\sum_{n=1}^{\infty}\int|f_n|+\sum_{n=1}^{\infty}2^{-n}+\sum_{n=1}^{\infty}2^{-n}<\infty$$

Therefore, $\sum_n|\Phi_n|+\sum_{n,i}|\phi_{ni}|$ is a.e. finite. As $|f_n|\leqslant|\Phi_n|+\sum_i|\phi_{ni}|$ a.e., it follows that $\sum_n|f_n|<\infty$ a.e., so $\sum f_n$ converges a.e.

Furthermore, if $f=\sum_n f_n$ a.e., then $f=\sum_n\Phi_n+\sum_{n,i}\phi_{ni}$ a.e., so that $f\in\mathscr{L}$ and $\int f=\sum_n\int\Phi_n+\sum_{n,i}\int\phi_{ni}=\sum_n(\int\Phi_n+\sum_i\int\phi_{ni})=\sum_n\int f_n$.

COROLLARY 19.12

(i) *If $f:\mathbb{R}\to\mathbb{R}$ and $f=0$ a.e., then $f\in\mathscr{L}$ and $\int f=0$. Conversely, if $f\in\mathscr{L}$,*

$f \geqslant 0$ and $\int f=0$, then $f=0$ a.e.
(ii) *Let $X \subset \mathbb{R}$. Then X is a null set if and only if $\xi_X \in \mathscr{L}$, $\int \xi_X=0$.*
Proof. The first part of (i) is Theorem 19.6(ii). For the second part, apply Theorem 19.11 to the sequence $f, f, f, \ldots$ Trivially, (ii) follows from (i).

**Exercise* 19.F. If $f \in \mathscr{L}$ and if I is an interval, then $f\xi_I \in \mathscr{L}$.

From Theorem 19.11 we deduce three convergence theorems that form the core of the Lebesgue theory.

THEOREM 19.13. (Monotone convergence, Levi's theorem) *Let $f_1, f_2, \ldots \in \mathscr{L}$ be such that either $f_1 \leqslant f_2 \leqslant \ldots$ a.e. or $f_1 \geqslant f_2 \geqslant \ldots$ a.e. and such that the sequence $\int f_1, \int f_2, \ldots$ is bounded. Then the sequence $f_1, f_2, \ldots$ converges a.e. If $f: \mathbb{R} \to \mathbb{R}$ and if $f=\lim_{n\to\infty} f_n$ a.e., then $f \in \mathscr{L}$ and $\int f=\lim_{n\to\infty} \int f_n$.*
Proof. Apply Theorem 19.11 to the sequence $f_1, f_2-f_1, f_3-f_2, \ldots$

THEOREM 19.14. (Fatou) *Let $f_1, f_2, \ldots \in \mathscr{L}$, $f_n \geqslant 0$ a.e. for every n. Suppose that the sequence $\int f_1, \int f_2, \ldots$ is bounded. Let $f: \mathbb{R} \to \mathbb{R}$ be such that $f=\liminf_{n\to\infty} f_n$ a.e. Then $f \in \mathscr{L}$ and $\int f \leqslant \liminf_{n\to\infty} \int f_n$.* (To see that not always $\int f=\lim_{n\to\infty} \int f_n$, even if $f=\lim_{n\to\infty} f_n$, choose $g \in \mathscr{C}_c$, $g \geqslant 0$ and set $f_n(x) := g(x+n)$ $(x \in \mathbb{R}, n \in \mathbb{N})$.)
Proof. By Theorem 19.6(ii) we may assume that $f_n \geqslant 0$ everywhere. For $n \in \mathbb{N}$, set $g_n := f_1 \wedge \ldots \wedge f_n$ and let $h := \inf_{i \in \mathbb{N}} f_i$. Then we have $g_1 \geqslant g_2 \geqslant \ldots$ and $\lim_{n\to\infty} g_n = h$. By Levi's theorem, $h \in \mathscr{L}$.
In the same way, if for every n we set $h_n := \inf_{i \geqslant n} f_i$, then $h_n \in \mathscr{L}$. Now $h_1 \leqslant h_2 \leqslant \ldots$ and $\lim_{n\to\infty} h_n = f$ a.e. Moreover, as $0 \leqslant h_n \leqslant f_n$ for all n, the sequence $\int h_1, \int h_2, \ldots$ is bounded. By another application of Levi's theorem, $f \in \mathscr{L}$ and

$$\int f = \lim_{n\to\infty} \int h_n = \lim_{n\to\infty}\left(\inf_{i \geqslant n} \int h_i\right) \leqslant \lim_{n\to\infty}\left(\inf_{i \geqslant n} \int f_i\right) = \liminf_{n\to\infty} \int f_n$$

THEOREM 19.15. (Dominated convergence, Lebesgue's theorem) *Let $g \in \mathscr{L}$, $g \geqslant 0$ a.e. Let $f_1, f_2, \ldots \in \mathscr{L}$ be such that $|f_n| \leqslant g$ a.e. for all n. Let $f: \mathbb{R} \to \mathbb{R}$, $f=\lim_{n\to\infty} f_n$ a.e. Then $f \in \mathscr{L}$ and $\int f=\lim_{n\to\infty} \int f_n$.*
Proof. Applying Fatou's theorem to the sequences $g-f_1, g-f_2, \ldots$ and $g+f_1, g+f_2, \ldots$ one sees that $f \in \mathscr{L}$, $\int f \geqslant \limsup_{n\to\infty} \int f_n$ and $\int f \leqslant \liminf_{n\to\infty} \int f_n$.

Exercise 19.G. Reconsider Exercise 19.C in the light of the above theorems.

Exercise 19.H. Let $g: \mathbb{R} \to \mathbb{R}$ be bounded, increasing and differentiable. Set

$A := \lim_{x \to -\infty} g(x)$, $B := \lim_{x \to \infty} g(x)$. Using Fatou's theorem, show that g' is Lebesgue integrable and that $\int g' \leqslant B - A$. (Yes, they are equal, but the proof is complicated. See Section 21.)

With the help of Theorem 19.15 we can show that there are many Lebesgue integrable functions, for example the characteristic functions of the bounded Borel sets:

COROLLARY 19.16. *If $f : \mathbb{R} \to \mathbb{R}$ is a Borel function, if $g \in \mathscr{L}$ and if $|f| \leqslant g$, then $f \in \mathscr{L}$.*

Proof. Let $g \in \mathscr{L}$, $g \geqslant 0$. It follows from Exercise 19.D and Theorem 19.15 that the functions $h : \mathbb{R} \to \mathbb{R}$ with $|h| \wedge g \in \mathscr{L}$ form an L-set (see Definition 16.2). Hence, if $h : \mathbb{R} \to \mathbb{R}$ is a Borel function, then $|h| \wedge g \in \mathscr{L}$. In particular, if h is Borel and $0 \leqslant h \leqslant g$, then $h \in \mathscr{L}$.

Consequently, if f is a Borel function and $|f| \leqslant g$, then $f \vee 0 \in \mathscr{L}$ and $(-f) \vee 0 \in \mathscr{L}$, whence $f = f \vee 0 - (-f) \vee 0 \in \mathscr{L}$.

**Exercise* 19.I. Let U be an open subset of $\mathbb{R}$. Denote by $L(U)$ the sum of the lengths of the components of U. Show that the following are true. If $L(U) < \infty$, then $\xi_U \in \mathscr{L}$ and $\int \xi_U = L(U)$. If $L(U) = \infty$, then $\xi_U \notin \mathscr{L}$.
If $U_1 \subset U_2 \subset \ldots$ are open sets, then $L(\bigcup_n U_n) = \sup_n L(U_n)$.
If $U_1, U_2, \ldots$ are open sets, then $L(\bigcup_n U_n) \leqslant \Sigma_n L(U_n)$.

Exercise 19.J. Let $f \in \mathscr{L}$. Prove that $\lim_{n \to \infty} f(x+n) = 0$ for almost every x. (Apply Exercise 19.F and Theorem 19.11 to $f_n : x \mapsto f(x+n)\xi_{[0,1)}(x)$.)

Give an example *disproving* the following conjecture. If $g \in \mathscr{L}$ and $\varepsilon > 0$, then there exists an $a \in \mathbb{R}$ such that $|g(x)| \leqslant \varepsilon$ for almost every $x > a$.

We have already made the (trivial) observation that, if $f \in \mathscr{L}$ and if $g = f$ a.e., then $g \in \mathscr{L}$ and $\int g = \int f$. From the point of view of Lebesgue integration, the behaviour of a function on a null set is irrelevant. We may then as well consider integration of a function f whose domain of definition is the complement of a null set: we extend f to a function $g : \mathbb{R} \to \mathbb{R}$, and if $g \in \mathscr{L}$ we call f 'integrable' and define $\int g$ to be the 'integral' of f. These definitions do not depend on the choice of g. Formally, we have:

DEFINITION 19.17. Let $S \subset \mathbb{R}$ and $f : S \to \mathbb{R}$. We say that f is *defined almost everywhere* if $\mathbb{R} \setminus S$ is a null set. (A natural example is $f = \Sigma_{n=1}^{\infty} f_n$ where $f_1, f_2, \ldots$ are as in Theorem 19.11.) For such f define $f_S : \mathbb{R} \to \mathbb{R}$ by

$$f_S(x) := \begin{cases} f(x) & \text{if } x \in S \\ 0 & \text{if } x \in \mathbb{R} \setminus S \end{cases}$$

We call f *Lebesgue integrable* if f_S is Lebesgue integrable; by definition, the *Lebesgue integral of* f is $\int f := \int f_S$.

Exercise 19.K. Formulate and prove extensions of the convergence theorems 19.11, 19.13, 19.14, 19.15 for a.e. defined functions.

So far we have considered only integration over the whole set $\mathbb{R}$. Now let us look into integration over intervals.

DEFINITION 19.18. Let I be an interval. Let $S \subset \mathbb{R}$ and $f: S \to \mathbb{R}$. f is said to be *defined almost everywhere on* I if $I \setminus S$ is a null set. For such an f we put

$$f_{S \cap I}(x) := \begin{cases} f(x) & \text{if } x \in S \cap I \\ 0 & \text{if } x \in \mathbb{R} \setminus (S \cap I) \end{cases}$$

f is *Lebesgue integrable over* I if $f_{S \cap I} \in \mathscr{L}$; its *Lebesgue integral over* I is $\int_I f := \int f_{S \cap I}$. The functions $f: I \to \mathbb{R}$ that are Lebesgue integrable over I form a set $\mathscr{L}(I)$. Thus, $\mathscr{L}(\mathbb{R}) = \mathscr{L}$.

Exercise 19.L. Prove analogues of Theorems 19.6 and 19.9 for $\mathscr{L}(I)$.

Exercise 19.M. Prove extensions of Theorems 19.11, 19.13, 19.14 and 19.15 for functions defined a.e. on some interval.

**Exercise* 19.N. Use Exercise 19.F to prove: if f is Lebesgue integrable over an interval I, then f is Lebesgue integrable over every subinterval of I.

19.19. It follows from Theorem 19.7 that, if $a \leqslant b$ and if f is (properly) Riemann integrable over $[a, b]$, then f is also Lebesgue integrable over $[a, b]$ and its Riemann and Lebesgue integrals over $[a, b]$ are equal. Henceforth we shall use the notation $\int_a^b f(x)\,dx$ (or $\int_a^b f$) also to indicate the *Lebesgue* integral of f over $[a, b]$.

Observe that, if f is Lebesgue integrable over one of the intervals (a, b), $(a, b]$, $[a, b)$, $[a, b]$, then it is Lebesgue integrable over all of them, and the Lebesgue integrals are the same.

Without explicit definitions the reader will understand what is meant by expressions like $\int_{-\infty}^{\infty} f(x)\,dx$, $\int_{-\infty}^{\infty} f(t)\,dt$, $\int_3^{\infty} f$.

**Exercise* 19.O. Let $-\infty \leqslant a \leqslant b \leqslant c \leqslant \infty$. Discuss the formula $\int_a^c f = \int_a^b f + \int_b^c f$.

**Exercise* 19.P. Let $f: [a, b] \to \mathbb{R}$ be a differentiable function whose derivative is bounded. Use dominated convergence to show that f' is Lebesgue integrable over $[a, b]$ and that $\int_a^b f'(x)\,dx = f(b) - f(a)$. ($f$ is an indefinite integral of f'. See Section 15.)

Thus, the function L' of Example 13.2 is Lebesgue integrable over every bounded interval.

Exercise 19.Q. Let $a \leqslant b$. Let $\phi: [a, b] \to \mathbb{R}$ have a continuous derivative. Assume that $\phi'(x) \geqslant 0$ for all $x \in [a, b]$. Then

$$\int_{\phi(a)}^{\phi(b)} f(x)\,dx = \int_a^b f(\phi(x))\phi'(x)\,dx$$

for every $f \in \mathscr{L}[\phi(a), \phi(b)]$.

Exercise 19.R. There exists a function $f : [0, 1] \to \mathbb{R}$ which is bounded and Lebesgue integrable without being equal a.e. to a Riemann integrable function. In fact, let X be a closed nowhere dense subset of $[0, 1]$ that is not a null set (Theorem 5.5). Let $f : [0, 1] \to \mathbb{R}$ be the characteristic function of X. Deduce from Corollary 19.16 that f is Lebesgue integrable over $[0, 1]$ and from Corollary 19.12 that $\int_0^1 f > 0$. Now suppose $g : [0, 1] \to \mathbb{R}$ is Riemann integrable and $f = g$ a.e. on $[0, 1]$. Show that $\{x \in [0, 1] : g(x) = 0\}$ is a dense subset of $[0, 1]$. Deduce that $\int_0^1 g = 0$. This contradicts the assumption $f = g$ a.e.

Exercise 19.S. For $x \in [0, 1]$, let $0 . \alpha_1(x)\alpha_2(x) \ldots$ be its standard development to the base 2. We want to prove that for 'most' numbers x the sequence $\alpha_1(x), \alpha_2(x), \ldots$ contains 'as many 0s as 1s'.

To this end, we call a number x *regular* if

$$\lim_{n\to\infty} \frac{1}{n} \# \{i : 1 \leqslant i \leqslant n, \alpha_i(x) = 1\} = \tfrac{1}{2}$$

where, for a finite set S, we denote the number of elements of S by $\# S$. Then we claim: *Almost every element of* $[0, 1]$ *is regular.*

For the proof we use a slightly different notation. Define the functions $f_1, f_2, \ldots$ on $[0, 1]$ by

$$f_i(x) := \begin{cases} 1 & \text{if } \alpha_i(x) = 1 \\ -1 & \text{if } \alpha_i(x) = 0 \end{cases}$$

and set $F_n := (1/n)(f_1 + \ldots + f_n)$. We are done if $\lim_{n\to\infty} F_n = 0$ a.e. One can prove this formula along the following lines.

(i) Show that $\int_0^1 f_i f_j = 0$ if $i \neq j$.
(ii) Deduce from this that $\int_0^1 F_n{}^2 = 1/n$ for all n.
(iii) Use Theorem 19.11 to show that $\lim_{i\to\infty} F_{i^2} = 0$ a.e. on $[0, 1]$.
(iv) Show that, if $n < p \leqslant m$, then $|F_p| \leqslant (m - n)/n + |F_n|$.
(v) Conclude that $\lim_{n\to\infty} F_n = 0$ a.e. on $[0, 1]$.

Without serious problems, similar theorems can be proved for bases other than 2. Thus, if $0.\beta_1(x)\beta_2(x) \ldots$ is the standard development of x to the base 10, then for almost every $x \in [0, 1]$ one has

$$\lim_{n\to\infty} \frac{1}{n} \# \{i : 1 \leqslant i \leqslant n, \beta_i(x) = 7\} = \tfrac{1}{10}$$

Exercise 19.T. For $f : \mathbb{R} \to \mathbb{R}$ and $s \in \mathbb{R}$ define $f_s : \mathbb{R} \to \mathbb{R}$ by

$$f_s(x) := f(x - s) \qquad (x \in \mathbb{R})$$

(i) If $f \in \mathscr{L}$ and $s \in \mathbb{R}$, then $f_s \in \mathscr{L}$ and $\int f_s = \int f$.
(ii) For every $f \in \mathscr{L}$,

$$\lim_{s \to 0} \int |f - f_s| = 0$$

(Prove this first for $f \in \mathscr{C}_c$, using, for instance, dominated convergence or uniform continuity; then apply Lemma 19.10.)

(iii) Prove the *Lemma of Riemann and Lebesgue*: if $f \in \mathscr{L}$, then

$$\lim_{|\lambda| \to \infty} \int_{-\infty}^{\infty} f(x) \sin \lambda x \, dx = 0, \qquad \lim_{|\lambda| \to \infty} \int_{-\infty}^{\infty} f(x) \cos \lambda x \, dx = 0$$

(The integrals exist according to Exercise 19.D(ii). Show that, for $\lambda \neq 0$, $\int_{-\infty}^{\infty} f(x) \sin \lambda x \, dx = -\int_{-\infty}^{\infty} f_{\pi/\lambda}(x) \sin \lambda x \, dx$, so that $\int_{-\infty}^{\infty} f(x) \sin \lambda x \, dx = \frac{1}{2} \int_{-\infty}^{\infty} \{f(x) - f_{\pi/\lambda}(x)\} \sin \lambda x \, dx$.)

Exercise 19.U. A function $f : \mathbb{R}^2 \to \mathbb{R}$ is said to be *separately continuous* if for every $c \in \mathbb{R}$ the functions $x \mapsto f(x, c)$ and $y \mapsto f(c, y)$ are continuous. Prove that every separately continuous function on $\mathbb{R}^2$ is of the first class of Baire. (Show that we may restrict ourselves to a bounded $f : \mathbb{R}^2 \to \mathbb{R}$. Define $g : \mathbb{R}^2 \to \mathbb{R}$ by $g(x, y) := \int_0^y f(x, t) dt$ $(x, y \in \mathbb{R})$. Use dominated convergence to prove that g is continuous and observe that f is a partial derivative of g.)

Exercise 19.V (The indefinite lower Riemann integral) Let $f : [a, b] \to \mathbb{R}$ be bounded. If f is Riemann integrable, then one easily sees (e.g. Exercise 12.O or Theorem 15.6) that the indefinite integral Jf of f is increasing if and only if $f \geq 0$ a.e. If f is not Riemann integrable we can still talk about its 'indefinite lower Riemann integral', the function $J_- f : x \mapsto \underline{\int}_a^x f(t) dt$ $(a \leq x \leq b)$. Even if $f \geq 0$ a.e., this $J_- f$ may fail to be increasing. (Consider $f = -\xi_{\mathbb{Q}}$.) In this exercise we investigate $J_- f$ for any bounded function f. The Lebesgue integral will be a useful tool.

Let $f : [a, b] \to \mathbb{R}$ be bounded. We use the symbols $f^{\uparrow}$ and $f^{\downarrow}$ as in Definition 10.4 and ω as in Exercise 10.K. Observe that $f^{\uparrow}$, $f^{\downarrow}$, ω are Lebesgue integrable over $[a, b]$ (Corollary 19.16).

For $x \in [a, b]$, $J_+ f(x)$ and $J_- f(x)$ will denote the upper and lower Riemann integrals of f over $[a, x]$, respectively.

(i) Show that

$$J_+ f(b) = \int_a^b f^{\downarrow}, \quad J_- f(b) = \int_a^b f^{\uparrow}, \quad J_+ f(b) - J_- f(b) = \int_a^b \omega$$

(For the first formula, note that there exist continuous functions $g_1, g_2, \ldots$ with $g_1 \geq g_2 \geq \ldots$ and $\lim_{n \to \infty} g_n = f^{\downarrow}$. Apply the monotone convergence theorem.)

(ii) Give a new proof of Theorem 12.1.

(iii) Show that $J_- f$ is increasing if and and only if for every $\varepsilon > 0$ the closure of $\{x : f(x) \leq -\varepsilon\}$ is a null set.

(iv) Set $f^-(x) := \max\{-f(x), 0\}$ $(x \in [a, b])$. Show that $J_- f$ is increasing if and only if f^- is Riemann integrable over $[a, b]$ with integral 0.

Notes to Section 19

There are so many current ways to introduce the Lebesgue integral that there is no purpose in giving references here. The one we have chosen leads without detours to the basic definitions and theorems, starting from the notions of 'null set' and 'Riemann integral' while relying heavily on Dini's theorem (proof of Lemma 19.4).

There also exist various generalizations. These are of two types. One can make the theory more abstract and consider 'integration' of functions defined on other sets than $\mathbb{R}$ or having other range spaces than $\mathbb{R}$. Here one moves into functional analysis (abstract measure spaces, vector valued measures, noncommutative integration). Closer to home are theories in which the Lebesgue integral is extended to a linear functional on a space of functions $\mathbb{R} \to \mathbb{R}$ that properly contains $\mathscr{L}(\mathbb{R})$. (An example is the Perron integral which we shall consider in Section 22.)

A reasonable question in this context is whether the Lebesgue integral can be extended to a linear functional on the space of *all* functions $\mathbb{R} \to \mathbb{R}$. The answer is yes and there even exist extensions that are invariant under translations. (Theorem 19.6(v) describes the translation invariance of the Lebesgue integral.)

20. Lebesgue measurability

In Corollary 19.16 we have seen that every Borel function on $\mathbb{R}$ that is not 'too large' is Lebesgue integrable. We are now going to investigate the relations between measurability and integrability. First we prove that every integrable function is equal to a Borel function a.e.

LEMMA 20.1. *Let* $f \in \mathscr{L}$, $f \geqslant 0$.
(i) *There exist closed sets* $C_1 \subset C_2 \subset \ldots$ *such that* $\mathbb{R} \setminus \bigcup_n C_n$ *is a null set and for every n the restriction of f to* C_n *is continuous.*
(ii) *There exist upper semicontinuous functions* $0 \leqslant f_1 \leqslant f_2 \leqslant \ldots \leqslant f$ *on* $\mathbb{R}$ *such that* $\lim_{n\to\infty} f_n = f$ *a.e.*

Proof. It suffices to prove (i), as (ii) follows by taking $f_n := f\xi_{C_n}$. By Lemma 19.10 we can choose $\phi_1, \phi_2, \ldots \in \mathscr{C}_c$ such that $\int |f-\phi_n| < 4^{-n}$ for every n. Then according to Theorem 19.11, $\lim_{n\to\infty} \phi_n = f$ everywhere outside some null set X. Choose open sets $U_1 \supset U_2 \supset \ldots$ with $U_n \supset X$ and $L(U_n) < 2^{-n}$ for all n. Take

$$C_n := \bigcap_{k \geqslant n} \{x : |\phi_{k+1}(x) - \phi_k(x)| \leqslant 2^{-k}\} \setminus U_n \qquad (n \in \mathbb{N})$$

Clearly, each C_n is closed and $C_1 \subset C_2 \subset \ldots$ Furthermore, for each n the sequence $\phi_n, \phi_{n+1}, \phi_{n+2}, \ldots$ converges to f uniformly on C_n, so the restriction of f to C_n is continuous. It remains to prove that the complement of $\bigcup_n C_n$ is a null set.

For $k \in \mathbb{N}$ let $W_k := \{x : |\phi_{k+1}(x) - \phi_k(x)| > 2^{-k}\}$. W_k is open and $\xi_{W_K} \leqslant 2^k |\phi_{k+1} - \phi_k|$, so (by Corollary 19.16) we have $\xi_{W_k} \in \mathscr{L}$ and $\int \xi_{W_K} \leqslant 2^k \int |\phi_{k+1} - \phi_k| \leqslant 2^k (\int |f - \phi_{k+1}| + \int |f - \phi_k|) \leqslant 2^{-k+1}$. Now for $n \in \mathbb{N}$ the open set $\mathbb{R} \setminus C_n$ is contained in $U_n \cup W_n \cup W_{n+1} \cup W_{n+2} \cup \ldots$: by Exercise 19.I, $L(\mathbb{R} \setminus C_n) \leqslant L(U_n) + \Sigma_{k \geqslant n} L(W_k) \leqslant 2^{-n} + 4 \cdot 2^{-n}$. It follows that $\bigcap_n (\mathbb{R} \setminus C_n)$, which is $\mathbb{R} \setminus \bigcup_n C_n$, is a null set.

If $f \in \mathscr{L}$, $f \geqslant 0$ and if $f_1, f_2, \ldots$ are as in Theorem 20.1(ii), then the sequence $f_1, f_2, \ldots$ converges pointwise to some function g. We have $g \in \mathscr{B}^2$ and $f = g$ almost everywhere. It follows that every Lebesgue integrable function is equal to an element of $\mathscr{B}^2$ a.e.

DEFINITION 20.2. A function $f : \mathbb{R} \to \mathbb{R}$ is called *Lebesgue measurable* if there exists a Borel function $g : \mathbb{R} \to \mathbb{R}$ with $f = g$ a.e. By $\mathscr{M}$ we denote the set of all Lebesgue measurable functions on $\mathbb{R}$.

Exercise 20.A
(i) $\mathscr{M}$ is an L-set (Definition 16.2) containing $\mathscr{B}$.
(ii) The inclusion $\mathscr{B} \subset \mathscr{M}$ is strict. (Hint. Use Corollary 17.12 to show the existence of a subset V of $\mathbb{D}$ that is not Borel. Consider ξ_V.)

THEOREM 20.3
(i) *Every Lebesgue integrable function is Lebesgue measurable.*
(ii) *If* $f \in \mathscr{M}$ *and* $g \in \mathbb{B}$, *then* $g \circ f \in \mathscr{M}$.
(iii) *If* $f \in \mathscr{M}$ *is bounded and* $g \in \mathscr{L}$, *then* $fg \in \mathscr{L}$.
Proof. (i) we have just proved. (ii) follows from Corollary 19.16, (iii) from (ii).

Theorem 20.4 follows almost trivially from the definition:

THEOREM 20.4
(i) *If* $f \in \mathscr{M}$ *and* $g \in \mathscr{M}$, *then* $f+g \in \mathscr{M}$, $fg \in \mathscr{M}$, $f \vee g \in \mathscr{M}$, $f \wedge g \in \mathscr{M}$, $|f| \in \mathscr{M}$ *and* $\alpha f \in \mathscr{M}$ *for every* $\alpha \in \mathbb{R}$.
(ii) *If* $f \in \mathscr{M}$ *and* $g \in \mathscr{B}$, *then* $g \circ f \in \mathscr{M}$.
(iii) *If* $f \in \mathscr{M}$ *and* $g = f$ *a.e., then* $g \in \mathscr{M}$.
(iv) *If* $f_1, f_2, \ldots \in \mathscr{M}$ *and* $f = \lim_{n \to \infty} f_n$ *a.e., then* $f \in \mathscr{M}$.
(v) *If* $f_1, f_2, \ldots \in \mathscr{M}$ *and* $\sup_{n \in \mathbb{N}} f_n$ *is everywhere finite, then* $\sup_{n \in \mathbb{N}} f_n \in \mathscr{M}$.

**Exercise* 20.B. Let $f : \mathbb{R} \to \mathbb{R}$. Show that f is Lebesgue measurable if and only if the function $x \mapsto e^{-|x|} \arctan f(x)$ $(x \in \mathbb{R})$ is Lebesgue integrable.

THEOREM 20.5. *For $f:\mathbb{R}\to\mathbb{R}$ the following conditions are equivalent.*
(α) *f is Lebesgue measurable.*
(β) *There exist $\phi_1, \phi_2, \ldots \in \mathscr{C}_c$ with $\lim_{n\to\infty}\phi_n=f$ a.e.*
(γ) *There exist closed subsets $C_1\subset C_2\subset \ldots$ of $\mathbb{R}$ such that the restriction of f to each C_n is continuous and $\lim_{n\to\infty} L(\mathbb{R}\setminus C_n)=0$.*

The equivalence (α)$\Leftrightarrow$(γ) is known as *Lusin's Theorem.*

Proof. (α)$\Rightarrow$(β). Define $g(x):=e^{-|x|}\arctan f(x)(x\in\mathbb{R})$. Then by Exercise 20.B, $g\in\mathscr{L}$. Take $\psi_1,\psi_2,\ldots\in\mathscr{C}_c$ with $\int|g-\psi_n|\leqslant 2^{-n}$ for every n (Lemma 19.10). One can choose these ψ_n in such a way that $|\psi_n(x)|<\frac{1}{2}\pi e^{-|x|}$ for all x. By Theorem 19.11, $\lim_{n\to\infty}\psi_n=g$ a.e. With $\phi_n(x):=\tan(e^{|x|}\psi_n(x))$ $(x\in\mathbb{R}, n\in\mathbb{N})$ we have $\phi_1,\phi_2,\ldots\in\mathscr{C}_c$ and $\lim_{n\to\infty}\phi_n=f$ a.e.
(β)$\Rightarrow$(α). Apply Theorem 20.4(iv).
(α)$\Rightarrow$(γ). We may assume $f\geqslant 0$. Let g be as in the first part of this proof. Lemma 20.1 and its proof yield closed sets $C_1\subset C_2\subset\ldots$ such that for each n $L(\mathbb{R}\setminus C_n)\leqslant 5\cdot 2^{-n}$ while the restriction of g to C_n is continuous. Clearly the restrictions of f to the C_n are continuous.
(γ)$\Rightarrow$(α). For every n, $f\xi_{C_n}$ is a Borel function (Exercise 16.O). Now apply Theorem 20.4(iv).

Exercise 20.C. Condition (β) of Theorem 20.5 might lead to the thought that every Lebesgue measurable function is a.e. equal to a function of the first class. The following is a counterexample.

Let $L:\mathbb{R}\to\mathbb{R}$ be as in Example 13.2. Recall that L is differentiable, on no interval monotone, and that $|L(x)-L(y)|\leqslant|x-y|$ $(x, y\in\mathbb{R})$.
(i) By Exercise 19.P, if $a, b\in\mathbb{R}$ and $a<b$, then $L(b)-L(a)=\int_a^b L'$. Let $X:=\{x\in\mathbb{R}: L'(x)>0\}$. Show that for every interval I neither $I\cap X$ nor $I\setminus X$ is a null set.
(ii) Let $h:=\xi_X$. Prove that $h\in\mathscr{B}^2$. *A fortiori*, h is Lebesgue measurable. Suppose there exists a $g\in\mathscr{B}^1$ for which $g=h$ a.e. Use (i) to prove that g takes the values 0 and 1 on every interval. Now show that this is impossible.

Exercise 20.D. A function $f:\mathbb{R}\to\mathbb{R}$ is Lebesgue measurable if and only if there exists a $g\in\mathscr{B}^2$ such that $f=g$ a.e.

DEFINITION 20.6. If $f\in\mathscr{M}$, $f\geqslant 0$ and $f\notin\mathscr{L}$, we set $\int f:=\infty$.

**Exercise* 20.E
(i) If $f, g\in\mathscr{M}$ and $0\leqslant f\leqslant g$, then $\int f\leqslant\int g$.
(ii) If $f, g\in\mathscr{M}$, $f\geqslant 0$ and $g\geqslant 0$, then $\int(f+g)=\int f+\int g$. If $f\in\mathscr{M}$, $f\geqslant 0$, $\lambda\in\mathbb{R}$ and $\lambda>0$, then $\int(\lambda f)=\lambda\int f$.
(iii) If $g, f_1, f_2,\ldots\in\mathscr{M}$, $0\leqslant f_1\leqslant f_2\leqslant\ldots$ and $\lim_{n\to\infty} f_n=g$, then $\int g=\sup_{n\in\mathbb{N}}\int f_n$.
(iv) If $g, f_1, f_2,\ldots\in\mathscr{M}$, $f_n\geqslant 0$ for all n and $\Sigma_{n=1}^{\infty} f_n=g$, then $\int g=\Sigma_{n=1}^{\infty}\int f_n$.

DEFINITION 20.7. A subset of $\mathbb{R}$ is said to be *Lebesgue measurable* if its characteristic function is Lebesgue measurable. For a Lebesgue measurable set E we define $\lambda(E) \in [0, \infty) \cup \{\infty\}$ by $\lambda(E) := \int \xi_E$. It is called the *Lebesgue measure* of E.

All Borel sets and all null sets are Lebesgue measurable. The collection of all Lebesgue measurable sets is closed for complementation and for formation of countable unions and intersections.

**Exercise* 20.F. For $f: \mathbb{R} \to \mathbb{R}$ the following conditions are equivalent. (Compare Theorem 16.7.)
(α) $f \in \mathcal{M}$.
(β) For every Borel set $E \subset \mathbb{R}$, $f^{-1}(E)$ is Lebesgue measurable.
(γ) For every $\alpha \in \mathbb{R}$, $\{x \in \mathbb{R} : f(x) > \alpha\}$ is Lebesgue measurable.
(δ) For every $\alpha \in \mathbb{Q}$, $\{x \in \mathbb{R} : f(x) > \alpha\}$ is Lebesgue measurable.
(See also Exercise 20.J.)

THEOREM 20.8. *Let E, F, E_1, E_2, ... be Lebesgue measurable sets.*
(i) *If $a \leqslant b$, then $\lambda((a, b)) = \lambda([a, b]) = b - a$. $\lambda(\emptyset) = 0$.*
(ii) *If $E \subset F$, then $\lambda(E) \leqslant \lambda(F)$. (In particular, if $\lambda(F)$ is finite, then so is $\lambda(E)$.)*
(iii) *If E is an open set, then $\lambda(E) = L(E)$ (See Exercise 19.I).*
(iv) *$\lambda(E) = 0$ if and only if E is a null set.*
(v) *If $E \subset \bigcup_n E_n$, then $\lambda(E) \leqslant \Sigma_{n=1}^{\infty} \lambda(E_n)$.*
(vi) *If E_1, E_2, ... are pairwise disjoint, then $\lambda(\bigcup_n E_n) = \Sigma_{n=1}^{\infty} \lambda(E_n)$.*
(vii) *If $E_1 \subset E_2 \subset \ldots$, then $\lambda(\bigcup_n E_n) = \sup \lambda(E_n)$.*
(viii) *If $E_1 \supset E_2 \supset \ldots$ and $\lambda(E_1) < \infty$, then $\lambda(\bigcap_n E_n) = \lim_{n\to\infty} \lambda(E_n)$.*
Proof. Most of the proof is obvious. (v) and (vii) can be obtained by applying (vi) to the sequence E_1, $E_2 \setminus E_1$, $E_3 \setminus (E_1 \cup E_2)$, ... (vi) follows from Lebesgue's theorem for the case when $\lambda(\bigcup_n E_n)$ is finite, from Levi's theorem if $\Sigma_{n=1}^{\infty} \lambda(E_n)$ is finite. (viii) can be deduced from (vi) by considering the sequence $\bigcap_n E_n$, $E_1 \setminus E_2$, $E_2 \setminus E_3$, ... (The example $E_n = (n, \infty)$ shows the relevance of the finiteness condition.)

**Exercise* 20.G. Let E be a Lebesgue measurable set, let $a \in \mathbb{R}$. Then the sets $E + a := \{x + a : x \in E\}$ and $aE := \{ax : x \in E\}$ are Lebesgue measurable and $\lambda(E + a) = \lambda(E)$, $\lambda(aE) = |a|\lambda(E)$. (Here $0 \cdot \infty := 0$.)

THEOREM 20.9. *For $E \subset \mathbb{R}$ the following statements are equivalent.*
(α) *E is Lebesgue measurable.*
(β) *For every $\varepsilon > 0$ there exist a closed set A and an open set W such that $A \subset E \subset W$ and $\lambda(W \setminus A) < \varepsilon$.*
(γ) *E is the union of an F_σ and a null set.*
(δ) *There exists a G_δ-set B, containing E and such that $B \setminus E$ is null.*

Proof. $(\alpha)\Rightarrow(\beta)$. By the implication $(\alpha)\Rightarrow(\gamma)$ of Theorem 20.5 there exists a closed set C such that $\lambda(\mathbb{R}\setminus C)<\varepsilon$ and such that the restriction of ξ_E to C is continuous. Then $A:=\{x\in C:\xi_E(x)=1\}$ is a relatively closed subset of C, hence is closed in $\mathbb{R}$. The set $\{x\in C:\xi_E(x)\neq 0\}$ (which, of course, is A) is the intersection of C with an open subset W_1 of $\mathbb{R}$. Set $W:=W_1\cup(\mathbb{R}\setminus C)$. Clearly, W is open, $A\subset E\subset W$ and $\lambda(W\setminus A)\leqslant\lambda(\mathbb{R}\setminus C)<\varepsilon$.

$(\beta)\Rightarrow(\gamma)$. Choose closed sets $A_1, A_2, \ldots$ and open sets $W_1, W_2, \ldots$ such that, for every n, $A_n\subset E\subset W_n$ and $\lambda(W_n\setminus A_n)<2^{-n}$. Now $\bigcup_n A_n$ is an F_σ, contained in E, and

$$\lambda\left(E\setminus\bigcup_n A_n\right)\leqslant\inf_{n\in\mathbb{N}}\lambda(E\setminus A_n)\leqslant\inf_{n\in\mathbb{N}}\lambda(W_n\setminus A_n)=0$$

$(\gamma)\Rightarrow(\alpha)$. All Borel sets and all null sets are Lebesgue measurable.
$(\alpha)\Leftrightarrow(\delta)$. Apply $(\alpha)\Leftrightarrow(\gamma)$ to $\mathbb{R}\setminus E$.

EXAMPLE 20.10. (Vitali's nonmeasurable set) The formula

$$x\sim y \quad\text{if}\quad x-y\in\mathbb{Q}$$

defines an equivalence relation $\sim$ on $\mathbb{R}$. Every equivalence class is a dense subset of $\mathbb{R}$. By the axiom of choice, there exists a subset V of $[0, 1]$ that has exactly one point in common with each equivalence class.

For $X\subset\mathbb{R}$ and $a\in\mathbb{R}$, define $X+a:=\{x+a:x\in X\}$.

(i) *If X is a Lebesgue measurable subset of V, then X is a null set.* Proof. Let $(r_1, r_2, \ldots)$ be an enumeration of $\mathbb{Q}\cap[0, 1]$. Then for every i we have $\lambda(X+r_i)=\lambda(X)$ (Exercise 20.G), but, by Theorem 20.8(vi),

$$\sum_{i=1}^{\infty}\lambda(X+r_i)=\lambda\left(\bigcup_{i\in\mathbb{N}}(X+r_i)\right)\leqslant\lambda([0, 2])=2$$

(ii) *The set V and the function ξ_V are not Lebesgue measurable.* Proof. If V is Lebesgue measurable, then by (i) V is null. Let $(q_1, q_2, \ldots)$ be an enumeration of $\mathbb{Q}\cap[-1, 1]$. Then each $V+q_i$ is a null set. But together the sets $V+q_i$ cover $[0, 1]$.

(iii) *Every Lebesgue measurable set that is not a null set contains a subset that is not Lebesgue measurable.* Proof. Let $A\subset\mathbb{R}$ be Lebesgue measurable. Observe that $A=\bigcup\{A\cap(V+q):q\in\mathbb{Q}\}$. Either one of the sets $A\cap(V+q)$ is nonmeasurable, or they are all null sets (see (i)). In the second case, A is null.

(iv) *V and ξ_V are not Borel measurable.* This follows from (ii).

(v) *V is not analytic.* Proof. If V is analytic, then so is $V+q$ for every $q\in\mathbb{Q}$. Then $\mathbb{R}\setminus V=\bigcup\{V+q:q\in\mathbb{Q},\ q\neq 0\}$ is analytic by Theorem 18.6. Now apply Theorem 18.10.

Exercise 20.H. Let Φ be the Cantor function (see 15.9). Define the function Ψ on $[0, 1]$ by $\Psi(x) := \frac{1}{2}(x + \Phi(x))$. Ψ is a strictly increasing continuous bijection $[0, 1] \to [0, 1]$ with a continuous inverse. Show that $\Psi(\mathbb{D})$ is a closed nowhere dense set without isolated points and of measure $\frac{1}{2}$. (If I_{jk} is one of the intervals used in the definition of $\mathbb{D}$, then $\Psi(I_{jk})$ is an interval of length $\frac{1}{2}L(I_{jk})$.)

We use this result in:

Exercise 20.I. We make $f, g \in \mathcal{M}$ with $f \circ g \notin \mathcal{M}$. Let Ψ be as in Exercise 20.H. Let W be a subset of $\Psi(\mathbb{D})$ that is not Lebesgue measurable (Example 20.10(iii)). Let f be the characteristic function of $\Psi^{-1}(W)$ and define $g : \mathbb{R} \to \mathbb{R}$ by

$$g(x) := \begin{cases} \Psi^{-1}(x) & \text{if } x \in [0, 1] \\ x & \text{if } x \notin [0, 1] \end{cases}$$

Then $f = 0$ a.e., g is strictly increasing and continuous (it even satisfies a Lipschitz condition), but $f \circ g \notin \mathcal{M}$.

Exercise 20.J. Let g and W be as in Exercise 20.I. Let $\mathrm{N} := g(W)$. Then N is a null set and $g^{-1}(\mathrm{N})$ is not Lebesgue measurable. (Compare Exercise 20.F.)

We have already defined integration over intervals. We are now going to extend this concept and introduce integration over more general subsets of $\mathbb{R}$, viz. Lebesgue measurable sets.

Let f be a function defined on some subset of $\mathbb{R}$ and let X be a Lebesgue measurable set, contained in the domain of definition of f. We define the function f_X on $\mathbb{R}$ by

$$f_X(x) := \begin{cases} f(x) & \text{if } x \in X \\ 0 & \text{if } x \in \mathbb{R} \setminus X \end{cases}$$

DEFINITION 20.11. We call f *Lebesgue measurable over* X if f_X is Lebesgue measurable. f is said to be *Lebesgue integrable over* X if f_X is Lebesgue integrable. In the latter case, the *Lebesgue integral of* f *over* X is, by definition,

$$\int_X f := \int_X f(x)\,dx := \int f_X$$

If f is Lebesgue measurable but not integrable over X and if $f \geqslant 0$ on X, we set $\int_X f := \infty$. The Lebesgue measurable (integrable) functions $X \to \mathbb{R}$ form a vector space $\mathcal{M}(X)$ ($\mathcal{L}(X)$, respectively).

Of course, Definition 20.11 can be extended for functions that are defined almost everywhere on X. We shall not waste our time carrying out the details.

The condition that X is Lebesgue measurable is a technicality and is not important for the validity of these definitions. Without it, however, we would have sets over which the constant function 1 is not measurable.

**Exercise* 20.K. Let f be a function defined on a subset D of $\mathbb{R}$.
(i) If X, Y are Lebesgue measurable subsets of D with $Y \subset X$ and such that f is measurable (integrable) over X, then f is measurable (integrable) over Y.
(ii) If X, Y are Lebesgue measurable subsets of D with $X \cap Y = \emptyset$ and such that f is integrable over X and over Y, then f is integrable over $X \cup Y$ and $\int_{X \cup Y} f = \int_X f + \int_Y f$.
(iii) Let $X_1, X_2, \ldots$ be pairwise disjoint Lebesgue measurable subsets of D and let $X = \bigcup_n X_n$. If f is integrable over X, then it is integrable over each X_n and $\int_X f = \sum_{n=1}^{\infty} \int_{X_n} f$. Conversely, if f is integrable over each X_n and if the series $\sum \int_{X_n} |f|$ converges, then f is integrable over X (and $\int_X f = \sum_{n=1}^{\infty} \int_{X_n} f$).
(iv) Let $X_1, X_2, \ldots$ and X be as in (iii). If f is Lebesgue measurable over each X_n, then f is Lebesgue measurable over X. If, in addition, $f \geqslant 0$ on X, then $\int_X f = \sum_{n=1}^{\infty} \int_{X_n} f$.

The properties mentioned in this exercise and other ones, such as the inequality $|\int_X f| \leqslant \int_X |f|$ will be used in the sequel without further reference.

Exercise 20.L. Formulate and prove generalizations of 19.11–19.15 for integrals over a Lebesgue measurable set.

We shall encounter measurable functions defined on arbitrary measurable sets in the next section. For the present, we prove only one theorem which shows that they occur naturally. (Compare Theorem 7.11.)

THEOREM 20.12. *Let $f : [a, b] \to \mathbb{R}$. The set $D := \{x \in [a, b] : f$ is differentiable at $x\}$ is Borel measurable, hence is Lebesgue measurable. f' is a Lebesgue measurable function on D.*

Proof. Let $C := \{x \in [a, b] : f$ is continuous at $x\}$. By theorem 7.5, C is a G_δ and therefore a Borel set. Observe that it contains D. Now

$$(*) \quad \{x \in C : D^+f(x) > 0\} = \bigcup_{\substack{\varepsilon \in \mathbb{Q} \\ \varepsilon > 0}} \bigcap_{\substack{\delta \in \mathbb{Q} \\ \delta > 0}} \left\{ x \in C : \text{there is a } y \in [a, b] \text{ with } 0 < |x - y| < \delta \text{ and } \frac{f(y) - f(x)}{y - x} > \varepsilon \right\}$$

For every $\varepsilon > 0$ and $\delta > 0$ the set $\{x \in C :$ there is a $y \in [a, b]$ such that $0 < |x - y| < \delta$ and $(f(y) - f(x))/(y - x) > \varepsilon\}$ is a relatively open subset of C, hence the intersection of C with an open set, hence a Borel set. It follows from (*) that $\{x \in C : D^+f(x) > 0\}$ is Borel. By applying this result to the functions $x \mapsto f(x) - \alpha x$ and $x \mapsto \alpha x - f(x)$ (which have the same points of continuity as f) one sees that for all $\alpha \in \mathbb{R}$ the sets $\{x \in C : D^+f(x) > \alpha\}$ and $\{x \in C : D^-f(x) < \alpha\}$ are Borel sets. As

$$C \setminus D = \bigcup_{\alpha \in \mathbb{Q}} \{x \in C : D^+f(x) > \alpha \text{ and } D^-f(x) < \alpha\} \cup \bigcap_{\alpha \in \mathbb{Q}} \{x \in C : D^+f(x) > \alpha\} \cup \bigcap_{\alpha \in \mathbb{Q}} \{x \in C : D^-f(x) < \alpha\}$$

we find that D is Borel. Furthermore, for all $\alpha \in \mathbb{R}$, $\{x \in D : f'(x) > \alpha\} = D \cap \{x \in C : D^+f(x) > \alpha\}$, so $\{x \in D : f'(x) > \alpha\}$ is Borel. If we define

$$g(x) := \begin{cases} f'(x) & \text{if } x \in D \\ 0 & \text{if } x \in \mathbb{R} \setminus D \end{cases}$$

then by Theorem 16.7 g is a Borel function. Hence, f' is Lebesgue measurable on D.

Exercise 20.M. Let $X \subset \mathbb{R}$ be Lebesgue measurable. Let $f: X \to \mathbb{R}$ be continuous. Then f is Lebesgue measurable.

The following exercises may serve to illustrate the usefulness of measurability as a tool.

Exercise 20.N. Let $\alpha_1, \alpha_2, \ldots$ be real numbers with $\lim_{n\to\infty} \alpha_n = \infty$. Then $A := \{x \in \mathbb{R} : \lim_{n\to\infty} \sin \alpha_n x \text{ exists}\}$ is a null set.

(Hint. Define $f : \mathbb{R} \to \mathbb{R}$ by $f(x) := \lim_{n\to\infty} \xi_A(x) \sin \alpha_n x$ $(x \in \mathbb{R})$. Let B be a Lebesgue measurable set with $\lambda(B) < \infty$. Show by applying the Lemma of Riemann and Lebesgue (Exercise 19.T.) that $\int_B f = 0$. Prove now that $f = 0$ a.e. To show that A is a null set, deduce from the Lemma of Riemann and Lebesgue that $\int_B f^2 = \frac{1}{2}\lambda(A \cap B)$ for every Lebesgue measurable set B of finite measure.)

**Exercise* 20.O. (Indefinite integrals) Let I be an interval, $f \in \mathscr{L}(I)$. A function $F : I \to \mathbb{R}$ is called an *indefinite integral* of f if

$$F(y) - F(x) = \int_x^y f(t)\,\mathrm{d}t \qquad (x, y \in I, x < y)$$

(i) Show that f has an indefinite integral, that any two indefinite integrals of f differ by a constant, and that all indefinite integrals of f are continuous. (Prove that $\lim_{n\to\infty} F(y_n) = F(y)$ if $\lim_{n\to\infty} y_n = y$.)
(ii) Let F be an indefinite integral of f. If $f(x) \geq 0$ for almost every $x \in I$, then F is increasing. Less obviously, if $f(x) > 0$ for almost every $x \in I$, then F is strictly increasing.
(iii) Let F be an indefinite integral of f. If F is increasing, then $f \geq 0$ a.e. on I. (To prove this, first assume $I = \mathbb{R}$. Show that $\int_E f \geq 0$ for every open set E, for every G_δ-set E, and, by Theorem 20.9, for every Lebesgue measurable set E. Now take $E := \{x : f(x) < 0\}$ and show that E must be null.)
(iv) Prove now: if $g \in \mathscr{L}(I)$, then *f and g have the same indefinite integrals if and only if $f = g$ almost everywhere on I.* (Compare Exercise 15.A.)

**Exercise* 20.P (Sequel to the previous exercise). We shall also need indefinite integrals of bounded measurable functions that are not integrable. More generally, let I be an interval. We call a function $f: I \to \mathbb{R}$ *locally integrable* if f is integrable over every bounded subinterval of I.
(i) Every locally integrable function on I is Lebesgue measurable. Every function integrable over I is locally integrable. Every bounded Lebesgue measurable function

on I is locally integrable. Every continuous function on I is locally integrable. Not every measurable function on I is locally integrable. More than that: there exists a positive Lebesgue measurable function on $\mathbb{R}$ that is integrable over no interval. (Hint for the proof of the last statement. Let $(r_1, r_2, \ldots)$ be an enumeration of $\mathbb{Q}$. Define $g \in \mathscr{L}$ by $g(x) := x^{-1/2}$ if $0 < |x| \leqslant 1$, $g(x) := 0$ if $x = 0$ or $|x| > 1$. Then $\Sigma_{n=1}^{\infty} 2^{-n} g(x - r_n)$ is finite for almost every $x \in \mathbb{R}$. The a.e. defined function $x \mapsto (\Sigma_{n=1}^{\infty} 2^{-n} g(x - r_n))^2$ is integrable over no interval.)
(ii) Extend the previous exercise to locally integrable functions.

Let f be a continuous function on $[a, b]$. We define the *counting function* $N_f : \mathbb{R} \to \{0, 1, 2, \ldots\} \cup \{\infty\}$; for $y \in \mathbb{R}$, $N_f(y)$ is the number of elements of the set $\{x \in [a, b] : f(x) = y\}$ (see Exercises 9.N, 16.F). Theorem 20.13 is an extension of the result obtained in Exercise 3.T.

THEOREM 20.13. (Banach) *Let $f : [a, b] \to \mathbb{R}$ be continuous. There exist Borel functions $\phi_1, \phi_2, \ldots : \mathbb{R} \to [0, \infty)$ such that $N_f = \sup_{n \in \mathbb{N}} \phi_n$. (Thus, in some generalized sense, N_f is a Borel function. In particular, if $N_f < \infty$ a.e., then N_f may be regarded as an a.e. defined Lebesgue measurable function.) If f is of bounded variation, then N_f is finite a.e., N_f is Lebesgue integrable and*

$$(*) \qquad \int_{\mathbb{R}} N_f = \operatorname*{Var}_{[a,b]} f$$

Conversely, if $N_f < \infty$ a.e. and if N_f is Lebesgue integrable, then f is of bounded variation (and (*) holds).
Proof. Without restriction, let $[a, b] = [0, 1]$. For $n \in \mathbb{N}$, put

$$I_1^n := [0, 2^{-n}]$$
$$I_i^n := ((i-1)2^{-n}, i2^{-n}] \quad \text{if} \quad i = 2, 3, \ldots, 2^n$$
$$\phi_n := \sum_{i=1}^{2^n} \xi_{f(I_i^n)}$$

For all $n \in \mathbb{N}$ and $y \in \mathbb{R}$, $\phi_n(y)$ is the number of indices i for which I_i^n intersects $f^{-1}(\{y\})$. It follows that $\phi_1 \leqslant \phi_2 \leqslant \phi_3 \leqslant \ldots$ and $N_f = \sup_{n \in \mathbb{N}} \phi_n$. Furthermore, for all n and i, I_i^n is connected: then so is $f(I_i^n)$, since f is continuous. In particular, $f(I_i^n)$ is Borel. Therefore, ϕ_n is a Borel function.
For every n,

$$\int \phi_n = \sum_{i=1}^{2^n} \lambda(f(I_i^n)) = \sum_{i=1}^{2^n} \lambda(f(\overline{I_i^n})) \leqslant \operatorname*{Var}_{[0,1]} f$$

Consequently, by Levi's theorem, if f is of bounded variation, then N_f

is finite a.e., N_f is integrable and $\int N_f = \lim_{n\to\infty} \int \phi_n \leqslant \mathrm{Var}_{[0,1]} f$.

Conversely, assume that N_f is finite a.e. and integrable. For $x \in [0, 1]$ we consider the restriction f_x of f to $[0, x]$ and the corresponding counting function N_{f_x}. Obviously, $N_{f_x} \leqslant N_f$, so we can define a function S on $[0, 1]$ by setting $S(x) := \int N_{f_x}$ $(0 \leqslant x \leqslant 1)$. This S is increasing and if $0 \leqslant x \leqslant y \leqslant 1$, then $S(y) - S(x) \geqslant |f(y) - f(x)|$. It follows easily that $\int N_f = S(1) - S(0) \geqslant \mathrm{Var}_{[0,1]} f$.

This proves the theorem.

Exercise 20.Q. Consider again Exercise 3.U.

Exercise 20.R. As we have seen, if $f : [a, b] \to \mathbb{R}$ is continuous and of bounded variation, then N_f is finite a.e. Give an example of a continuous function f on $[0, 1]$ that is not of bounded variation although N_f is everywhere finite.

The last three exercises of this section deal with the counting function and have nothing to do with integration. We want to construct *a continuous function $f : [0, 1] \to [0, 1]$ such that $N_f(y) = \infty$ for every $y \in [0, 1]$*. In Exercise 20.S we see that we have already met such a function. In Exercises 20.T and 20.U we present another one which may be somewhat easier to visualize. The latter function is differentiable a.e. (Compare Corollary 21.7.)

Exercise 20.S. If f is the restriction to $[0, 1]$ of the function ψ_2 of Exercise 16.T(ii), then $N_f(y) = \infty$ for every $y \in [0, 1]$. Actually, for every $y \in [0, 1]$, the set $f^{-1}(\{y\})$ is uncountable.

Exercise 20.T. (Preparation for 20.U) Let $c_1, c_2, \ldots$ be positive numbers such that $\lim_{k\to\infty} c_k = 0$. Let I_{jk} $(k \in \mathbb{N};\ j = 1, 2, \ldots, 2^{k-1})$ be the components of $[0, 1] \setminus \mathbb{D}$, as in 4.2. Suppose that for each I_{jk} we have a function $g_{jk} : \overline{I_{jk}} \to \mathbb{R}$ with the properties

(i) g_{jk} is continuous,

(ii) if $I_{jk} = (\alpha, \beta)$, then $g_{jk}(\alpha) = g_{jk}(\beta) = 0$,

(iii) for all $x \in \overline{I_{jk}}$, $|g_{jk}(x)| \leqslant c_k$.

Define $g : [0, 1] \to \mathbb{R}$ by

$$g(x) := \begin{cases} g_{jk}(x) & \text{if} \quad x \in I_{jk} \\ 0 & \text{if} \quad x \in \mathbb{D} \end{cases}$$

Then g is continuous.

Exercise 20.U. We construct a continuous $f : [0, 1] \to [0, 1]$ such that $N_f(y) = \infty$ for every $y \in [0, 1]$. Fig. 16 may suggest the graph of such an f. (In the graph of the function that we are going to define there are sine waves instead of the broken lines that occur in the sketch, but the idea is the same.)

Let I_{jk} be as in the previous exercise. For each k and j define

$$g_{jk}(x) := 2^{-k} \sin 2\pi(x - \alpha)(\beta - \alpha)^{-1} \qquad (x \in \overline{I_{jk}})$$

where α and β are the end points of I_{jk}.

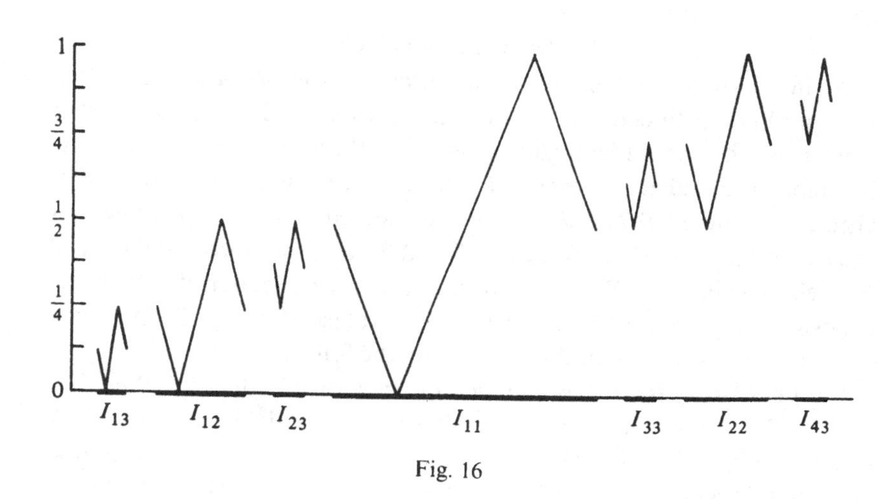

Fig. 16

Show that (i), (ii) and (iii) of Exercise 20.T hold. Let g be as above: then g is continuous.

Set $f := g + \Phi$ where Φ is the Cantor function (see 15.9). Now f is a continuous map $[0, 1] \to [0, 1]$. Show that for every k one has $[0, 1] = f(\bigcup_{j \leq 2^{k-1}} I_{jk})$. Conclude that $N_f(y) = \infty$ for all $y \in [0, 1]$.

Notes to Section 20

A subset X of $\mathbb{R}$ is said to have the *property of Baire* if there exists an open set $U \subset \mathbb{R}$ such that $X \setminus U$ and $U \setminus X$ are meagre. In Exercise 16.M we considered the collection $\mathscr{W}$ of all sets with this property. It follows from the results obtained there that a set X has the property of Baire if and only if there exists a Borel set B such that $X \setminus B$ and $B \setminus X$ are meagre. There is an interesting analogy with the theorem (easily obtainable from our definitions) which says that a set $X \subset \mathbb{R}$ is Lebesgue measurable if and only if there exists a Borel set B such that $X \setminus B$ and $B \setminus X$ are null. This analogy is one of the basic ideas in Oxtoby (1971). See also N. Lusin, *Fund. Math.* **9** (1927), 116–18.

In the notes to Section 19 we mentioned the problem of extending the Lebesgue integral. Closely related is the problem of extending the Lebesgue measure. It turns out to be possible to extend the Lebesgue measure to a finitely additive translation invariant function $\mathscr{P}(\mathbb{R}) \to [0, \infty) \cup \{\infty\}$ where $\mathscr{P}(\mathbb{R})$ is the set of all subsets of $\mathbb{R}$. However, from Example 20.10 it follows easily that such an extension can never be σ-additive.

21. Absolute continuity

At the beginning of Section 15 we discussed the operations D (differentiation) and J (indefinite Riemann integration) and the question of in what sense they could be regarded as each other's inverses. We saw that it might be useful to generalize D and J to operations that were provisionally denoted $\tilde{D}$ and $\tilde{J}$, and we considered possible candidates for D such as D_r^+ and D^+. In Sections 19 and 20 we developed the theory of the Lebesgue integral. We now return to the subject matter of Section 15, choosing for $\tilde{J}$ the indefinite Lebesgue integral (see Exercise 20.O). Some of the problems studied in this section are the following.

1. If $f \in \mathscr{L}[a, b]$ and if F is an indefinite integral of f, does it follow that $F' = f$? The trivial example $f = \xi_{\{a\}}$ shows that we do not always have $F'(x) = f(x)$ for *all* $x \in [a, b]$. But, as f is a difference of two nonnegative integrable functions, F is of bounded variation: then *a priori* we know that $F'(x)$ exists for *almost* every $x \in [a, b]$. Do we have $F'(x) = f(x)$ for almost every x? (A better result can hardly be expected, since functions that are equal a.e. have the same indefinite integrals.) In Theorem 21.18 we shall show that such a conclusion is indeed valid. Thus, in this context it seems natural to choose for our $\tilde{D}$ 'differentiation almost everywhere'. (Strictly speaking, this is not an operation, but that will not worry us.)

2. If $f : [a, b] \to \mathbb{R}$ is differentiable, is f an indefinite integral of f'? We have just observed that on $[a, b]$ indefinite integrals are of bounded variation. Besides, derivatives may fail to be Lebesgue integrable (Exercise 21.M). Thus, we have to impose restrictions on f or f', for example $f \in \mathscr{BV}$ or $f' \in \mathscr{L}$. In Corollaries 21.22 and 21.23 we shall see that either condition guarantees a positive answer.

3. Lebesgue integration works for functions defined almost everywhere. If $f : [a, b] \to \mathbb{R}$ is differentiable a.e. and if f' is Lebesgue integrable, does f have to be an indefinite integral of f'? To exclude trivial counterexamples such as $f = \xi_{\{a\}}$, assume f to be continuous. The example of the Cantor function (15.9) shows that further restrictions are needed. A workable condition on f turns out to be the so-called 'absolute continuity' (Definition 21.13). We shall see that the conditions 'f is absolutely continuous', 'f is an indefinite integral' and 'f is differentiable a.e., f' is integrable and f is an indefinite integral of f'' are equivalent.

One of the tools we need is the 'outer measure'. It is used only on the following few pages to simplify some notations and plays no role of importance in our theory.

DEFINITION 21.1. Let $X \subset \mathbb{R}$. We define $\lambda^*(X)$, the *outer measure* of X, as follows. If X is contained in a set that has a finite Lebesgue measure, then $\lambda^*(X) := \inf\{\lambda(Y) : Y$ is a Lebesgue measurable subset of $\mathbb{R}$ that contains $X\}$. Otherwise, $\lambda^*(X) := \infty$.

We have $\lambda^*(X) = \lambda(X)$ for every Lebesgue measurable set X. Furthermore, if $X \subset \mathbb{R}$ and $\lambda^*(X) = 0$, then X is a null set.

LEMMA 21.2. *Let $X_1, X_2, \ldots \subset \mathbb{R}$ and $X \subset \bigcup_n X_n$. Then $\lambda^*(X) \leqslant \Sigma_{n=1}^{\infty} \lambda^*(X_n)$.*
Proof. The inequality is trivial if $\Sigma_{n=1}^{\infty} \lambda^*(X_n) = \infty$. Otherwise, let $\varepsilon > 0$. For each n, choose a Lebesgue measurable set $Y_n \supset X_n$ for which $\lambda(Y_n) \leqslant \lambda^*(X_n) + \varepsilon 2^{-n}$. Then $X \subset \bigcup_n Y_n$, so $\lambda^*(X) \leqslant \Sigma_n \lambda(Y_n) \leqslant \Sigma_n \lambda^*(X_n) + \varepsilon$.

**Exercise* 21.A
(i) If $X, Y \subset \mathbb{R}$, then $\lambda^*(X \cup Y) \leqslant \lambda^*(X) + \lambda^*(Y)$.
(ii) There exist sets X, Y such that $X \cap Y = \varnothing$, $\lambda^*(X \cup Y) < \lambda^*(X) + \lambda^*(Y)$. (Let V be the set of Vitali, constructed in Example 20.10. Let $X = V$, $Y = [0, 1] \setminus V$.)
(iii) If $X, Y \subset \mathbb{R}$, $X \cap Y = \varnothing$ and if Y is Lebesgue measurable, then $\lambda^*(X \cup Y) = \lambda^*(X) + \lambda(Y)$.

LEMMA 21.3. *Let X be a Lebesgue measurable subset of $\mathbb{R}$. Let $M \in \mathbb{R}$ and let $f : X \to \mathbb{R}$ have the following property. For every $x \in X$ there exists an $\varepsilon > 0$ such that $|f(y) - f(x)| \leqslant M(y - x)$ for all $y \in X \cap (x, x + \varepsilon)$. Then $\lambda^*(f(X)) \leqslant M\lambda(X)$.*
Proof. For $n \in \mathbb{N}$, set

$$S_n := \{x \in X : |f(y) - f(x)| \leqslant M(y - x) \quad \text{for all } y \in X \cap (x, x + 1/n)\}$$

It is not hard to prove that $\overline{S_n} \cap X = S_n$. (Note that every element of $\overline{S_n} \cap X$ is the limit of a monotone sequence $x_1, x_2, \ldots$ of elements of S_n, and that $f(x) = \lim_{i \to \infty} f(x_i)$ if this sequence is decreasing.) It follows that S_n is Lebesgue measurable. Clearly, $S_1 \subset S_2 \subset \ldots$ and $\bigcup_n S_n = X$. Define $X_1 := S_1$, $X_n := S_n \setminus S_{n-1}$ $(n = 2, 3, \ldots)$. Because $\lambda^*(f(X)) = \lambda^*(\bigcup_n f(X_n)) \leqslant \Sigma_{n=1}^{\infty} \lambda^*(f(X_n))$ and $\lambda(X) = \Sigma_{n=1}^{\infty} \lambda(X_n)$, it suffices to prove that $\lambda^*(f(X_n)) \leqslant M\lambda(X_n)$ for each n. In other words, we may assume that $|f(y) - f(x)| \leqslant M(y - x)$ if $x \in X$ and $y \in X \cap (x, x + 1/n)$. Furthermore, as we can write X as a union of countably many pairwise disjoint Lebesgue measurable sets that each have diameter $< 1/n$, we may assume that the diameter of X itself is $< 1/n$. Then $|f(y) - f(x)| \leqslant M|y - x|$ for all $x, y \in X$.

Let $\varepsilon > 0$. Choose an open set U containing X and such that $\lambda(U) \leqslant \lambda(X) + \varepsilon$. Let $U_1, U_2, \ldots$ be the components of U. If $k \in \mathbb{N}$ and $x, y \in X \cap U_k$, then $|f(y) - f(x)| \leqslant M|y - x| \leqslant M\lambda(U_k)$. Hence, for every k $f(X \cap U_k)$ is contained in a closed interval whose length is at most $M\lambda(U_k)$.

Consequently, $\lambda^*(f(X \cap U_k)) \leq M\lambda(U_k)$ for each k. It follows that $\lambda^*(f(X)) \leq \Sigma_{k=1}^{\infty} \lambda^*(f(X \cap U_k)) \leq M\lambda(U) \leq M\lambda(X) + M\varepsilon$.

This can be done for every $\varepsilon > 0$, so $\lambda^*(f(X)) \leq M\lambda(X)$.

COROLLARY 21.4. *Let $X \subset [a, b]$ be Lebesgue measurable. Let $f: [a, b] \to \mathbb{R}$ be differentiable at every point of X and let $M \in \mathbb{R}$ be such that $|f'(x)| \leq M$ for all $x \in X$. Then $\lambda^*(f(X)) \leq M\lambda(X)$.*
Proof. By the above, $\lambda^*(f(X)) \leq (M+\delta)\lambda(X)$ for all $\delta > 0$.

Exercise 21.B. Let $X \subset [a, b]$ be Lebesgue measurable. Let $M \in \mathbb{R}$ be such that $|D_r^+ f(x)| \leq M$, $|D_r^- f(x)| \leq M$ for all $x \in X$. Then $\lambda^*(f(X)) \leq M\lambda(X)$.

COROLLARY 21.5. (Lusin). *Let $f : [a, b] \to \mathbb{R}$. Let $Z := \{x \in [a, b] : f$ is* differentiable at x and $f'(x) = 0\}$. *Then $f(Z)$ is a null set.* (Compare Theorem 7.2, 7.3 and the remark following Exercise 7.D. Note that Z is Lebesgue measurable because of Theorem 20.12.)

COROLLARY 21.6. *If $f: [a, b] \to \mathbb{R}$ is differentiable, then N_f is finite almost everywhere. (For N_f, see Definition 20.11.)*
Proof. Let Z be as above; let $A := \{y \in \mathbb{R} : N_f(y) = \infty\}$. If $y \in A$, then $f^{-1}(\{y\})$, being an infinite subset of $[a, b]$, has an accumulation point, x, say. Then $f'(x) = 0$, so $x \in Z$. It follows that $A \subset f(Z)$.

LEMMA 21.7. *If $X \subset [a, b]$ is a null set and if $f: [a, b] \to \mathbb{R}$ is differentiable at every point of X, then $f(X)$ is a null set.*
Proof. For $n \in \mathbb{N}$, let $X_n := \{x \in X : |f'(x)| < n\}$. By Corollary 21.4, $\lambda^*(f(X_n)) \leq n\lambda(X_n) = 0$ for every n, so each $f(X_n)$ is a null set. Then so is their union, which is $f(X)$.

This observation leads to:

DEFINITION 21.8. (Lusin). A function $f : [a, b] \to \mathbb{R}$ is said to have the *property* (N) if for every null set $X \subset [a, b]$ its image set $f(X)$ is a null set.

The function Φ of Cantor maps $\mathbb{D}$ onto $[0, 1]$ and therefore does not have the property (N). On the other hand, we have just seen:

THEOREM 21.9. *Every differentiable function on $[a, b]$ has the property (N).*

Exercise 21.C. Show that a function defined on $[a, b]$ that satisfies a Lipschitz condition (see Corollary 4.13) has the property (N).

Exercise 21.D. If f has the property (N), does f^2 necessarily have it? If f:

$[a, b] \to [0, \infty)$ has the property (N), does $\sqrt{f}$ necessarily have it?

Exercise 21.E. Prove the following extension of Theorem 21.9. If f is a continuous function on $[a, b]$ and if $D_r^+ f$ and $D_l^- f$ are finite at every point of $[a, b)$, then f has the property (N).

The property (N) plays a very special role in the theory of Lebesgue measurability:

Exercise 21.F (Rademacher) Let $f: [a, b] \to \mathbb{R}$ be continuous. Then f has the property (N) if and only if for every Lebesgue measurable set $X \subset [a, b]$, $f(X)$ is Lebesgue measurable. (For the 'only if', observe that every Lebesgue measurable set is a union of countably many compact sets and a null set (Theorem 20.9). For the 'if', recall that every Lebesgue measurable set that is not null contains a subset that is not Lebesgue measurable (Example 20.10(iii))).

Exercise 21.G. (Two functions having the property (N) whose sum does not)
(i) Every continuous function $f: \mathbb{D} \to \mathbb{R}$ for which $f(\mathbb{D})$ is a null set can be extended to a continuous function on $[0, 1]$ that has the property (N). (Make it differentiable outside $\mathbb{D}$.)
(ii) We can define continuous functions $\alpha_1, \alpha_2, \ldots : \mathbb{D} \to \{0, 1\}$ by

$$x = 2 \sum_{i=1}^{\infty} \alpha_i(x) 3^{-i} \qquad (x \in \mathbb{D})$$

(See Exercise 15.N(ii).) There exist continuous functions f_1 and f_2 on $[0, 1]$, both having the property (N), and such that

$$\left.\begin{aligned} f_1(x) &= \sum_{i=1}^{\infty} \alpha_{2i}(x) 3^{-i} \\ f_2(x) &= \sum_{i=1}^{\infty} \alpha_{2i+1}(x) 3^{-i} \end{aligned}\right\} \quad (x \in \mathbb{D})$$

Apply Exercise 4.M to show that $f_1 + f_2$ does *not* have the property (N).

The property (N) has also to do with the subject of Section 15. For a moment, let us return to 15.8. Let $f: [a, b] \to \mathbb{R}$ be continuous, let $A \subset [a, b]$ and suppose that $f'(x) > 0$ for all $x \in [a, b] \setminus A$. The proof of Theorem 15.7 shows that that f will be strictly increasing, provided that $f(A)$ does not contain any interval. In 15.9 we saw that $f(A)$ may very well contain an interval, even if A is a null set. But we see that this cannot happen if f has the property (N) (and A is null). Thus, we have proved part of the following theorem.

THEOREM 21.10. *Let $f: [a, b] \to \mathbb{R}$ be continuous. Suppose that $f'(x) > 0$ (or $f'(x) \geq 0$, $f'(x) = 0$) for almost every $x \in [a, b]$ and that f has the property (N). Then f is strictly increasing (or increasing, constant).*

Proof. We have already proved the first part: if $f'>0$ a.e. on $[a, b]$, then f is strictly increasing. Now assume that $f'\geqslant 0$ a.e. on $[a, b]$. Trying to follow the lines of the proof of Theorem 15.6, for $\varepsilon>0$ we consider the function $f_\varepsilon : x \mapsto f(x)+\varepsilon x$ $(a\leqslant x\leqslant b)$. Of course $f'_\varepsilon>0$ a.e. on $[a, b]$ and we would be done if every f_ε were strictly increasing. Unfortunately, there is no guarantee that f_ε has the property (N)! (See the preceding exercise.)

We have to be more careful. Assume that f is not increasing. Without restriction, let $f(a)>f(b)$. Let $A:=\{x \in [a, b] : f$ is not differentiable at x, or $f'(x)<0\}$ and $B:=\{x \in [a, b] : f'(x)=0\}$. Then $f(A)$ is null since f has the property (N), while $f(B)$ is null because of Corollary 21.5. Thus we can choose a number t with $f(b)<t<f(a)$ and $t \notin f(A)\cup f(B)$. From here on, the proof of Theorem 15.7 works without a hitch: with $c:=\max\{x : f(x)=t\}$ we have $c\neq b$ and $f(x)\leqslant f(c)$ for $c\leqslant x\leqslant b$. As $c \notin A$, f is differentiable at c, so $f'(c)\leqslant 0$. Again because $c \notin A$ we see that $f'(c)\geqslant 0$. Then $f'(c)=0$. But $c \notin B$; a contradiction.

The proof of the last part of the theorem is obvious.

COROLLARY 21.11 *Let $f : [a, b]\to\mathbb{R}$ be differentiable and let $f'(x)>0$ (or $f'(x)\geqslant 0$, $f'(x)=0$) for almost every $x \in [a, b]$. Then f is strictly increasing (or increasing, constant).*

We return to the main theme of this section.

THEOREM 21.12 *Let $X \subset [a, b]$ be Lebesgue measurable. Let $f : [a, b]\to\mathbb{R}$ be differentiable at every point of X. Then $f(X)$ is Lebesgue measurable and $\lambda(f(X))\leqslant \int_X |f'|$.*

Proof. By Theorem 20.9, X is the union of a null set Y and countably many compact sets $Y_1, Y_2, \ldots$ According to Lemma 21.7, $f(Y)$ is a null set, while each $f(Y_n)$ is compact because the restriction of f to X is continuous. Hence, $f(X)$ is Lebesgue measurable.

Now let $\varepsilon>0$ and for $n \in \mathbb{N}$ let $X_n:=\{x \in X : (n-1)\varepsilon\leqslant |f'(x)|<n\varepsilon\}$. Then X_n is Lebesgue measurable (Theorem 20.12) and by Corollary 21.4 we have

$$\lambda(f(X_n))\leqslant n\varepsilon\lambda(X_n)\leqslant \int_{X_n} (|f'(x)|+\varepsilon)\,\mathrm{d}x = \int_{X_n} |f'| + \varepsilon\lambda(X_n)$$

Consequently, $\lambda(f(X))\leqslant \overline{\Sigma}_{n=1}^{\infty} \lambda(f(X_n))\leqslant \int_X |f'| + \varepsilon\lambda(X)$. The theorem follows from the fact that this inequality holds for all $\varepsilon>0$.

Exercise 21.H. Show that in Theorem 21.12 we may have strict inequality.

DEFINITION 21.13. Let $f:[a, b]\to\mathbb{R}$. We call f *absolutely continuous* if for every $\varepsilon>0$ there exists a $\delta>0$ such that, if $n\in\mathbb{N}$, $a\leqslant a_1\leqslant b_1\leqslant a_2\leqslant b_2\leqslant\ldots b_n\leqslant b$ and $\Sigma_{i=1}^n (b_i-a_i)<\delta$, then $\Sigma_{i=1}^n |f(b_i)-f(a_i)|<\varepsilon$.

21.14. *The absolutely continuous functions on* $[a, b]$ *form a vector space. Every absolutely continuous function is continuous. Not every continuous function is absolutely continuous (see Exercise* 21.I). *Every function that satisfies a Lipschitz condition (Corollary* 4.13) *is absolutely continuous.*

Exercise 21.I. The Cantor function is not absolutely continuous.

Exercise 21.J
(i) If $f:[a, b]\to[A, B]$ is absolutely continuous and if $\phi:[A, B]\to\mathbb{R}$ satisfies a Lipschitz condition, then $\phi\circ f$ is absolutely continuous (see, however, Exercise 21.L). If f and g are absolutely continuous, then so are $|f|$, $f\vee g$ and $f\wedge g$.
(ii) The product of two absolutely continuous functions is absolutely continuous.

LEMMA 21.15. *Every absolutely continuous function has the property (N).*
Proof. Let $f:[a, b]\to\mathbb{R}$ be absolutely continuous. Let $X\subset[a, b]$ be a null set and let $\varepsilon>0$: we construct a Lebesgue measurable set that contains $f(X)$ and whose measure is at most ε.

There exists a $\delta>0$ such that, if $a\leqslant a_1\leqslant b_1\leqslant\ldots\leqslant b_n\leqslant b$ and $\Sigma_{i=1}^n (b_i-a_i)\leqslant\delta$, then $\Sigma_{i=1}^n |f(b_i)-f(a_i)|\leqslant\varepsilon$. Choose a relatively open subset W of $[a, b]$ with $W\supset X$ and $\lambda(W)\leqslant\delta$ and write W as a union of pairwise disjoint intervals $W_1, W_2,\ldots$ For each $i\in\mathbb{N}$ we can choose $s_i, t_i\in\overline{W_i}$ such that $\overline{f(W_i)}=[f(s_i), f(t_i)]$. Observing that $\Sigma_{i=1}^n |t_i-s_i|\leqslant\delta$ for every n, we see that $\Sigma_{i=1}^n |f(t_i)-f(s_i)|\leqslant\varepsilon$. Thus we obtain $\varepsilon\geqslant\Sigma_{i=1}^\infty |f(t_i)-f(s_i)|=\Sigma_{i=1}^\infty \lambda(\overline{f(W_i)})$. Now $\bigcup_i \overline{f(W_i)}$ is Lebesgue measurable and $\lambda(\bigcup_i \overline{f(W_i)})\leqslant\Sigma_{i=1}^\infty \lambda(\overline{f(W_i)})\leqslant\varepsilon$. On the other hand, we have $\bigcup_i \overline{f(W_i)}\supset f(\bigcup_i W_i)=f(W)\supset f(X)$.

LEMMA 21.16. *Every absolutely continuous function is of bounded variation.*
Proof. Let $f:[a, b]\to\mathbb{R}$ be absolutely continuous. Choose $\delta>0$ such that, if $a\leqslant a_1\leqslant b_1\leqslant\ldots\leqslant b_n\leqslant b$ and $\Sigma_{i=1}^n (b_i-a_i)<\delta$, then $\Sigma_{i=1}^n |f(b_i)-f(a_i)|\leqslant 1$.

Then in particular, if $[p, q]$ is a subinterval of $[a, b]$ of length $<\delta$, then the total variation of f over $[p, q]$ is at most 1. Now take $c_0, c_1,\ldots, c_N$ such that $a=c_0<c_1<\ldots<c_N=b$ and $c_j-c_{j-1}<\delta$ for all j: then

$$\operatorname{Var} f=\sum_{j=1}^N \operatorname*{Var}_{[c_{j-1}, c_j]} f=\sum_{j=1}^N 1<\infty.$$

LEMMA 21.17. *Every indefinite integral is absolutely continuous.*
Proof. Let $f:[a, b]\to\mathbb{R}$ be an indefinite integral of $g\in\mathscr{L}[a, b]$ and let $\varepsilon>0$. It follows from Lemma 19.10 that there exists a continuous function ϕ on $[a, b]$ such that $\int_a^b|g-\phi|<\frac{1}{2}\varepsilon$. Let $M:=\sup_{a\leqslant x\leqslant b}|\phi(x)|$ and let $\delta>0$, $2M\delta\leqslant\varepsilon$.

If $n\in\mathbb{N}$, $a\leqslant a_1\leqslant b_1\leqslant\ldots\leqslant b_n\leqslant b$ and $\Sigma_{i=1}^n(b_i-a_i)<\delta$, then, setting $U:=(a_1, b_1)\cup\ldots\cup(a_n, b_n)$ we have $\lambda(U)<\delta$, whence

$$\sum_{i=1}^n|f(b_i)-f(a_i)|=\sum_{i=1}^n\left|\int_{a_i}^{b_i}g\right|\leqslant\sum_{i=1}^n\int_{a_i}^{b_i}|g|=\int_U|g|$$
$$\leqslant\int_U|\phi|+\int_U|g-\phi|\leqslant\int_U M+\int_a^b|g-\phi|\leqslant M\delta+\frac{\varepsilon}{2}\leqslant\varepsilon.$$

**Exercise* 21.K. If $f\in\mathscr{BV}[a, b]$, then the almost everywhere defined function f' is Lebesgue integrable and $\int_a^b|f'|\leqslant\operatorname{Var}f$. In particular, if f is increasing, then $\int_a^b f'\leqslant f(b)-f(a)$. (Hint. If f is increasing, adapt Exercise 19.H to show that $\int_a^b f'\leqslant f(b)-f(a)$. For the general case, consider the indefinite variation T of f (Exercise 3.B) and note that $|f'|\leqslant T'$ a.e.)

THEOREM 21.18. *Let* $f:[a, b]\to\mathbb{R}$ *be an indefinite integral of* $g\in\mathscr{L}[a, b]$. *Then* $f'=g$ *a.e. on* $[a, b]$.
Proof. We may assume $g\geqslant0$. Then f is increasing, hence differentiable outside some null set $X\subset[a, b]$. By Exercise 21.K, f' is Lebesgue integrable over $[a, b]$ and $\int_a^b f'\leqslant f(b)-f(a)$. Taking $p, q\in[a, b]$ with $p<q$, by the same token we have $\int_p^q f'\leqslant f(q)-f(p)$. On the other hand, from Theorem 21.12 and the fact that f has the property (N) (Lemmas 21.17 and 21.15) we infer that

$$\int_p^q f'\geqslant\int_{[p,q]\setminus X}f'\geqslant\lambda(f([p, q]\setminus X))=\lambda(f([p, q]\setminus X))+\lambda(f(X))$$
$$\geqslant\lambda(f([p, q]))=f(q)-f(p)$$

Hence, $\int_p^q f'=f(q)-f(p)$ for $a\leqslant p<q\leqslant b$. Thus, f is an indefinite integral of f'. But f is also an indefinite integral of g. Therefore, $g=f'$ a.e. on $[a, b]$ (Exercise 20.O(i)).

LEMMA 21.19. *Let* $f:[a, b]\to\mathbb{R}$ *be continuous and of bounded variation and let* f *have the property* (*N*). *Then* f *is absolutely continuous.*
Proof. As f is of bounded variation, it is differentiable a.e. on $[a, b]$ and f' is Lebesgue integrable (Exercise 21.K). The function $x\mapsto\int_a^x|f'|$ is absolutely continuous (Lemma 21.17) and, by the definition of absolute continuity, we are done if

(*) $$|f(s)-f(t)| \leqslant \int_s^t |f'|$$

for $a \leqslant s < t \leqslant b$.

Choose a null set $X \subset [a, b]$ such that f is differentiable at every point of $[a, b] \setminus X$. Let $a \leqslant s < t \leqslant b$. By the continuity of f and by Lemma 21.2 we have

$$|f(s)-f(t)| \leqslant \lambda(f([s, t])) \leqslant \lambda^*(f([s, t] \setminus X)) + \lambda(f(X))$$

so that (*) follows from Theorem 21.12 and from the fact that f has the property (N).

LEMMA 21.20. *If $f : [a, b] \to \mathbb{R}$ is absolutely continuous, then f is an indefinite integral of f'.*

Proof. From Lemma 21.16 we know that f is of bounded variation, so f is differentiable a.e. on $[a, b]$ and f' is Lebesgue integrable over $[a, b]$ (Exercise 21.K). Let h be an indefinite integral of f'. Then h and f are absolutely continuous (Lemma 21.17), so $h-f$ is absolutely continuous and has the property (N) (Lemma 21.15). Furthermore, $h' = f'$ a.e. on $[a, b]$ (Theorem 21.18), i.e. $(h-f)' = 0$ a.e. on $[a, b]$. By Theorem 21.10, $h-f$ is constant and f is an indefinite integral of f'.

The rewards for the troubles that we went to consist of the following.

THEOREM 21.21. (Vitali–Banach) *Let $f : [a, b] \to \mathbb{R}$.*
(i) *The following conditions are equivalent.*
(α) *f is absolutely continuous.*
(β) *f is continuous and of bounded variation and f has the property (N).*
(γ) *f is an indefinite integral of a Lebesgue integrable function.*
(ii) *If f satisfies these conditions, then it is differentiable a.e. on $[a, b]$, f' is Lebesgue integrable and f is an indefinite integral of f'.*

Proof. (i) (α)$\Rightarrow$(β) is Lemmas 21.15 and 21.16. (β)$\Rightarrow$(α) is Lemma 21.19. (α)$\Rightarrow$(γ) is Lemma 21.20 and (γ)$\Rightarrow$(α) is Lemma 21.17.
(ii) follows from (i) and from Lemma 21.20.

Exercise 21.L. In Exercise 3.F we obtained a differentiable function $g : [0, 1] \to [0, 1]$ that is not of bounded variation while g^2 is. Deduce the existence of absolutely continuous functions $h, j : [0, 1] \to [0, 1]$ for which $h \circ j$ is not absolutely continuous and not even of bounded variation.

COROLLARY 21.22. *Every differentiable function of bounded variation is absolutely continuous and therefore is an indefinite integral of its derivative.* (Compare Exercise 21.M.)

Proof. Differentiable functions have the property (N) (Theorem 21.9).

COROLLARY 21.23. *If $f:[a, b]\to\mathbb{R}$ is differentiable and if f' is Lebesgue integrable, then f is an indefinite integral of f'.*
Proof. By Theorem 21.9, f has the property (N). If $a\leqslant p\leqslant q\leqslant b$, then (according to Theorem 21.12) $|f(p)-f(q)|\leqslant\int_p^q|f'|$. It follows easily that the total variation of f is at most equal to $\int_a^b|f'|$. Now apply Theorem 21.21(ii).

**Exercise* 21.M. Define $f:[0, 1]\to\mathbb{R}$ by

$$\begin{cases} f(x):=x^2\sin x^{-2} & \text{if } 0<x\leqslant 1 \\ f(0):=0 \end{cases}$$

(i) f is differentiable but not of bounded variation (see Exercise 3.F).
(ii) f' is not Lebesgue integrable.

COROLLARY 21.24. *If $g:[a, b]\to\mathbb{R}$ is Lebesgue integrable and has an antiderivative, f, say, then f is an indefinite integral of g.*
Proof. This is merely a restatement of Corollary 21.23.

COROLLARY 21.25. *Let $f:[a, b]\to\mathbb{R}$ be increasing.*
(i) *The (a.e. defined) function f' is Lebesgue integrable and $\int_a^b f'\leqslant f(b)-f(a)$.*
(ii) *$\int_a^b f'=f(b)-f(a)$ if and only if f is continuous and has the property (N).*
Proof. For (i), see Exercise 21.K. The 'if' part of (ii) is a special case of (ii) of Theorem 21.21. Now assume $\int_a^b f'=f(b)-f(a)$. Let g be an indefinite integral of f'. By (i), if $a\leqslant s<t\leqslant b$, then $g(t)-g(s)=\int_s^t f'\leqslant f(t)-f(s)$, so $f-g$ is increasing. But $g(b)-g(a)=\int_a^b f'=f(b)-f(a)$, so $f(a)-g(a)=f(b)-g(b)$. Then $f-g$ is actually constant. Apply Theorem 21.21(i) and Lemma 21.15.

COROLLARY 21.26. *If $g:[a, b]\to\mathbb{R}$ has an antiderivative and if $g\geqslant 0$ almost everywhere, then g is Lebesgue integrable and its indefinite integrals are its antiderivatives.*
Proof. Let f be an antiderivative of g. By Corollary 21.11, f is increasing. Now apply Corollaries 21.25 and 21.23.

COROLLARY 21.27. *Let $f:[a, b]\to\mathbb{R}$ be absolutely continuous. Then*

$$\operatorname{Var} f=\int_a^b|f'|$$

The indefinite variation, T, of f is an indefinite integral of $|f'|$. In particular, T is absolutely continuous and $T'=|f'|$ a.e.

(See also Exercise 4.J.)

Proof. If $a=a_0<a_1<\ldots<a_n=b$, then

$$\sum_{i=1}^{n}|f(a_i)-f(a_{i-1})|=\sum_{i=1}^{n}\left|\int_{a_{i-1}}^{a_i}f'\right|\leqslant\sum_{i=1}^{n}\int_{a_{i-1}}^{a_i}|f'|=\int_a^b|f'|$$

so Var $f\leqslant\int_a^b f'$. The converse inequality is contained in Exercise 21.K.

It follows that, if $a\leqslant s<t\leqslant b$, then $T(t)-T(s)=\mathrm{Var}_{[s,t]}f=\int_s^t|f'|$, so T is an indefinite integral of $|f'|$. The rest is obvious.

Exercises 21.N–21.V are further applications of the theorem of Vitali and Banach.

Exercise 21.N. Let $f:[a,b]\to\mathbb{R}$ be continuous. In Exercise 15.H we saw that f is increasing (strictly increasing) as soon as $D^-f(x)\geqslant 0$ $(D^-f(x)>0)$ for every $x\in[a,b]$. We have not yet answered the natural question: if $D^-f=0$, must f be constant? We are now in a position to do so.

(i) Let H, K be differentiable functions on $[a,b]$ such that $H'\leqslant D^-f\leqslant K'$. Then

(*) $$H(y)-H(x)\leqslant f(y)-f(x)\leqslant K(y)-K(x)\qquad(a\leqslant x\leqslant y\leqslant b)$$

We give an outline of a proof, leaving the details to the reader. Without loss of generality, assume $K'=0$. Then H is decreasing. As $D^-(f-H)=D^-f-H'\geqslant 0$, $f-H$ is increasing (Exercise 15.F). This proves half of (*). (We need this half to prove the rest, i.e. that f is decreasing.) The function $g:x\mapsto(f-2H)(x)+x$ is a strictly increasing map of $[a,b]$ onto a closed interval $[A,B]$. Let ϕ be its inverse map and set $f_1:=f\circ\phi$. It suffices to prove that f_1 is decreasing. Now ϕ is strictly increasing and

$$\phi(v)-\phi(u)\leqslant v-u\qquad(A\leqslant u\leqslant v\leqslant B)$$

From the fact that $D^-f\leqslant 0$ it follows easily that $D^-f_1=D^-(f\circ\phi)\leqslant 0$.

Now note that for all $x\in[a,b]$ we have $(g-f)(x)=-2H(x)+x$ and $(g+f)(x)=2(f-H)(x)+x$, so that both $g-f$ and $g+f$ are increasing. Therefore, if $a\leqslant x\leqslant y\leqslant b$, then $|f(y)-f(x)|\leqslant g(y)-g(x)$. Consequently, f_1 satisfies a Lipschitz condition. In particular, f_1 is absolutely continuous, hence is an indefinite integral of f_1'. But $f_1'(u)=D^-f_1(u)$ for every u for which $f_1'(u)$ exists. Thus, $f_1'\leqslant 0$ a.e. and f_1 is decreasing.

(ii) In particular, if $D^-f=0$, then f is constant.

(iii) In Exercise 15.H we have already observed that, even if $D^-f(x)\leqslant 0$ for every x, f may still not be decreasing. Paradoxically, in (i) we have proved: if there exists a differentiable function H such that $H'(x)\leqslant D^-f(x)\leqslant 0$ $(a\leqslant x\leqslant b)$, then f is decreasing.

Exercise 21.O. (Generalization of 14.I) Let $g:[0,1]\to\mathbb{R}$ be absolutely continuous. Then $fg\in\mathscr{D}'[0,1]$ for every $f\in\mathscr{D}'[0,1]$. In fact, if $f\in\mathscr{D}'[0,1]$ and if F is an antiderivative of f, then

$$h:x\mapsto F(x)g(x)-\int_0^x F(t)g'(t)\,\mathrm{d}t$$

is an antiderivative of fg. Hint. If $0\leqslant x<y\leqslant 1$, then

$$\frac{h(y)-h(x)}{y-x}=\frac{F(y)-F(x)}{y-x}\,g(y)-\int_x^y \frac{F(t)-F(x)}{t-x}\cdot\frac{t-x}{y-x}\cdot g'(t)\,dt.$$

See also Exercise 14.F(ii).

Exercise 21.P. (The arc length, again) For a continuous function $f: [a, b]\to\mathbb{R}$ we denote by $L(f)$ the length of its graph (see Section 3). We already know (Definition 3.1 and Exercise 3.E):

(i) $L(f)$ is finite if and only if f is of bounded variation.

(ii) If f has a continuous derivative, then $L(f)=\int_a^b \sqrt{(1+(f')^2)}$.

We now prove:

(iii) If f is of bounded variation, then $L(f)\geqslant\int_a^b \sqrt{(1+(f')^2)}$.

(iv) If f is absolutely continuous, then $L(f)=\int_a^b \sqrt{(1+(f')^2)}$.

To prove (iii): For $x\in[a, b]$ let $l(x)$ be the length of the graph of f, restricted to $[a, x]$. Show that l is increasing and that

$$l(y)-l(x)\geqslant\sqrt{[(y-x)^2+(f(y)-f(x))^2]} \quad \text{if} \quad a\leqslant x\leqslant y\leqslant b$$

Deduce that $l'\geqslant\sqrt{(1+(f')^2)}$ a.e. and apply Exercise 21.K.

To prove (iv): Let $\varepsilon>0$. Choose a partition P of $[a, b]$ with $L_P(f)\geqslant L(f)-\varepsilon$. (For $L_P(f)$, see Section 3.) Choose a continuous function ϕ on $[a, b]$ such that $\int_a^b|f'-\phi|$ is sufficiently small (Lemma 19.10) and show that $|L_P(f)-L_P(\Phi)|$ is very small, where Φ is an indefinite integral of ϕ. Prove that $|\int_a^b \sqrt{(1+(f')^2)}-\int_a^b \sqrt{(1+\phi^2)}|\leqslant\int_a^b|f'-\phi|$ and observe that $L_P(\Phi)\leqslant L(\Phi)=\int_a^b \sqrt{(1+\phi^2)}$.

Exercise 21.Q. Compute the length of the graph of the Cantor function (see 15.9).

Exercise 21.R. Let $g, f_1, f_2,\ldots$ be continuous functions on $[a, b]$ such that $g=\lim_{n\to\infty} f_n$. Show that $L(g)\leqslant\liminf_{n\to\infty} L(f_n)$.

Exercise 21.S. For $f:\mathbb{R}\to\mathbb{R}$ and $s\in\mathbb{R}$ define $f_s:\mathbb{R}\to\mathbb{R}$ by

$$f_s(x):=f(x-s) \qquad (x\in\mathbb{R})$$

(i) *Let f be a Lebesgue measurable function such that for all $s\in\mathbb{R}$ one has $f=f_s$ a.e. Then there exists a number c such that $f=c$ a.e.* (For the proof, one may assume f to be bounded. Let F be the indefinite integral of f with $F(0)=0$. Set $c:=F(1)$. From the given property of f it follows that $F(b)-F(a)=F(s+b)-F(s+a)(a, b, s\in\mathbb{R};$ $a<b)$. Deduce that $F(s+t)=F(s)+F(t)$ for all $s, t\in\mathbb{R}$. Prove that $F(s)=sF(1)$ for all $s\in\mathbb{Q}$ and, by continuity, for all $s\in\mathbb{R}$. It follows that F is an indefinite integral of the function $x\mapsto c$. Apply Exercise 20.O(i).)

(ii) (A strengthening of (i)) *Let $f:\mathbb{R}\to\mathbb{R}$ be Lebesgue measurable and let $S:=\{s\in\mathbb{R}:$ $f=f_s$ a.e.$\}$ be dense in $\mathbb{R}$. Then there is a $c\in\mathbb{R}$ with $f=c$ a.e.* (Proceeding as above one obtains $F(b)-F(a)=F(s+b)-F(s+a)$ $(a, b\in\mathbb{R};\, a<b)$ for $s\in S$. Then, since F is continuous, the same identity holds for all $c\in\mathbb{R}$.)

Exercise 21.T. *Let $g:\mathbb{R}\to\mathbb{R}$ be Lebesgue measurable and such that*

$$g(x+y)=g(x)+g(y) \quad \textit{for all} \quad x, y\in\mathbb{R}$$

Then there exists a $c\in\mathbb{R}$ such that $g(x)=cx$ for all $x\in\mathbb{R}$. (Take $c:=g(1)$. First show that $g(x)=cx$ for all $x \in \mathbb{N}$, all $x \in \mathbb{Z}$ and all $x \in \mathbb{Q}$. Set $f(x) := g(x)-cx$ and apply the preceding exercise.)

Exercise 21.U. *Let G be a Lebesgue measurable subset of* $\mathbb{R}$ *that is a group under addition. Then either* $G=\mathbb{R}$ *or G is a null set.* (If $G \cap (0, \infty)$ has a smallest element, a, say, then $G=\{na : n \in \mathbb{Z}\}$, so G is countable. Otherwise, G is dense. Apply Exercise 21.S to show that either G or $\mathbb{R} \setminus G$ is a null set. But if $\mathbb{R} \setminus G$ is null, then for every $x \in \mathbb{R}$ the sets G and $x+G$ have nonempty intersection, so $x \in G-G=G$.)

Exercise 21.V. We call $f: \mathbb{R} \to \mathbb{R}$ *doubly periodic* if there exist $s, t \in \mathbb{R}$ with $f=f_s=f_t$ (see Exercise 21.S) and there does not exist a $p \in \mathbb{R}$ such that both s and t are multiples of p. Prove: if f is doubly periodic and Lebesgue measurable, then there exists a $c \in \mathbb{R}$ with $f=c$ a.e.

After these exercises we go back to the general theory. Theorem 21.29 is essentially nothing but a reformulation of Theorem 21.18 for indefinite integrals of characteristic functions.

DEFINITION 21.28. Let X be a Lebesgue measurable subset of $\mathbb{R}$. A real number c is called a *density point* of X if

$$\lim_{\delta \downarrow 0} \frac{\lambda(X \cap (c-\delta, c+\delta))}{\lambda((c-\delta, c+\delta))} = 1$$

If f is an indefinite integral of ξ_X, then for all $c \in \mathbb{R}$ and $\delta > 0$,

$$\frac{\lambda(X \cap (c-\delta, c+\delta))}{\lambda((c-\delta, c+\delta))} = \frac{f(c+\delta)-f(c-\delta)}{2\delta}$$
$$= \frac{1}{2}\left(\frac{f(c+\delta)-f(c)}{\delta} + \frac{f(c)-f(c-\delta)}{\delta}\right)$$

where $(f(c+\delta)-f(c))/\delta \leq 1$ and $(f(c)-f(c-\delta))/\delta \leq 1$. Therefore, c is a density point of X if and only if

$$\lim_{\delta \downarrow 0} \frac{f(c+\delta)-f(c)}{\delta} = 1 \quad \text{and} \quad \lim_{\delta \downarrow 0} \frac{f(c)-f(c-\delta)}{\delta} = 1,$$

i.e. $f'(c)=1$.

However, it follows from Theorem 21.18 that $f'=\xi_X$ a.e. Thus, we have the so-called density theorem:

THEOREM 21.29. (Lebesgue) *Let X be a Lebesgue measurable subset of* $\mathbb{R}$. *Let* X_d *be the set of all density points of X.*
(i) $\xi_X = \xi_{X_d}$ *a.e. Almost every element of X is a density point of X. Almost all density points of X are elements of X.*
(ii) *X is a null set if and only if it has no density points.*

(iii) *X and X_d have the same density points.*
(The proof of Theorem 21.29 relies on Theorem 21.18 and thereby on Theorem 4.10 on differentiability a.e. of monotone functions. In Appendix D we present a direct proof of Theorem 21.29 that does not involve 4.10 and its complicated proof.)

From the density theorem we see that there does not exist a measurable set X with

$$0<\lambda(X\cap J)/\lambda(J)<\tfrac{4}{5} \text{ for every bounded interval } J$$

On the other hand, it is possible to have an X for which

$$0<\lambda(X\cap J)/\lambda(J)<1 \quad \text{for every bounded interval } J$$

(Let $L:\mathbb{R}\to\mathbb{R}$ be a differentiable function, monotone on no interval (Example 13.2). Set $X:=\{x\in\mathbb{R}: L'(x)>0\}$. Then X is Lebesgue measurable. If $\lambda(X\cap J)=\lambda(J)$ for some bounded interval J we would have $L'>0$ a.e. on J. By Corollary 21.11 L would be increasing on J, a contradiction. If $\lambda(X\cap J)=0$ for some interval J, similar reasoning shows that L is decreasing on J.)

Exercises 21.W and 21.X are applications of the density theorem.

Exercise 21.W. For $X, Y\subset\mathbb{R}$, define $X+Y:=\{x+y: x\in X, y\in Y\}$.
(i) *If X, Y are Lebesgue measurable and if neither is a null set, then $X+Y$ contains an interval.* More explicitly, for every $a\in X_d+Y_d$ there exists an $\varepsilon>0$ such that $(a-\varepsilon, a+\varepsilon)\subset X+Y$. (Let $a=x+y$ with $x\in X_d$, $y\in Y_d$, Choose $\varepsilon>0$ with $\lambda(X\cap(x-2\varepsilon, x+2\varepsilon))>3\varepsilon$, $\lambda(Y\cap(y-\varepsilon, y+\varepsilon))>\varepsilon$, and prove for every $t\in(-\varepsilon,\varepsilon)$ that $(X-x-t)\cap(y-Y)\neq\emptyset$ by showing that both $X-x-t$ and $y-Y$ have intersections with $(-\varepsilon,\varepsilon)$ whose Lebesgue measures are $>\varepsilon$.)
(ii) Although $\mathbb{D}$ is a null set, $\mathbb{D}+\mathbb{D}=[0, 2]$ (Exercise 4.M).

Exercise 21.X. Let $f:\mathbb{R}\to\mathbb{R}$ be such that $f(\frac{1}{2}(x+y))\leqslant\frac{1}{2}(f(x)+f(y))$ for all $x, y\in\mathbb{R}$. It follows from Exercise 2.I that f is convex provided f is continuous. We now prove that the continuity condition may be weakened to Lebesgue measurability.

Assume that f is Lebesgue measurable: we prove f to be continuous. It suffices to show that f is continuous at 0.
(i) Let $\alpha\in\mathbb{R}$, $a>0$ and $f\leqslant\alpha$ on $[-a, a]$. Let $\beta:=\max(\alpha, f(-3a), f(3a))$. Then $f\leqslant\frac{1}{2}(\alpha+\beta)$ on $[-2a, 2a]$.
(ii) If f is bounded above on some interval, then f is bounded above on every bounded interval.
(iii) There exists an $\alpha>0$ such that the set $X:=\{x\in\mathbb{R}: f(x)\leqslant\alpha\}$ has positive measure. By Exercise 21.W the set $\{\frac{1}{2}(x+y): x, y\in X\}$ contains an interval I. Then f is bounded above on I.
(iv) By (ii) and (iii) there exists an $\alpha\in\mathbb{R}$ such that $f\leqslant f(0)+\alpha$ on $[-1, 1]$. It follows that for every $n\in\mathbb{N}$

$$f(x)-f(0)\leqslant 2^{-n}\alpha \qquad (|x|\leqslant 2^{-n})$$

Further, $f(x)-f(0)\geqslant f(0)-f(-x)$ for every x. Now prove the continuity of f at 0.
(v) Obtain another solution for Exercise 21.T.

The density theorem (21.29) can be used to prove a theorem due to Denjoy and Young which essentially says that, if $f: I\to\mathbb{R}$, then at almost every point of I either D^+f and D^-f both behave very badly or they both behave very well. The proof is somewhat laborious, but the theorem has interesting consequences.

LEMMA 21.30. *Let $f:[a,b]\to\mathbb{R}$. Define*

$$E:=\{x\in[a,b]:f(y)\geqslant f(x) \quad \text{for all } y\geqslant x;$$
$$f(y)\leqslant f(x) \quad \text{for all } y\leqslant x\}$$

Then for almost all $x\in E$ one has $D_l^-f(x)\geqslant D_r^+f(x)$ and $D_r^+f(x)\neq\infty$.
Proof. Without restriction, assume that E contains more than one point. Then $A:=\inf E$ differs from $B:=\sup E$. There is an increasing function ϕ on (A,B) which on $E\cap(A,B)$ coincides with f (e.g. $\phi(x):\sup\{f(y):y\leqslant x\}$ $(A<x<B)$). By the theorem of Lebesgue (4.10) ϕ is differentiable a.e. on (A,B), hence differentiable at almost every point of E. Let

$$E_0:=\{x\in E: x \text{ is a density point of } E \text{ and } \phi \text{ is differentiable at } x\}$$

Then $E_0\subset E$ and $E\setminus E_0$ is a null set. Observe that $\phi=f$ on E_0 and that E and E_0 have the same density points. We show that

$$D_l^-f\geqslant\phi'\geqslant D_r^+f \quad \text{on} \quad E_0$$

(From this, the lemma follows.)
Take $c\in E_0$ and $\varepsilon\in(0,1)$. It suffices to find a $\delta>0$ such that
(i) if $c<x<c+\frac{1}{2}\delta$, then $f(x)-f(c)\leqslant(\phi'(c)+\varepsilon)(1+\varepsilon)(x-c)$
(ii) if $c-\frac{1}{2}\delta<y<c$, then $f(c)-f(y)\leqslant(\phi'(c)-\varepsilon)(1-\varepsilon)(c-y)$

c is a density point of E, hence a density point of E_0. By the definition of E_0 there exists a $\delta>0$ with the following properties (1) and (2).
(1) If $0<\beta<\delta$, then the Lebesgue measure of $E_0\cap(c-\beta,c+\beta)$ is strictly larger than $2\beta-(\varepsilon/(1+\varepsilon))\beta$. (In particular, if $0<\beta<\delta$, then every subinterval of $(c-\beta,c+\beta)$ of length $(\varepsilon/(1+\varepsilon))\beta$ must intersect E_0.)
(2) If $|x-c|<\delta$ and $x\neq c$, then $(\phi(x)-\phi(c))(x-c)^{-1}\leqslant\phi'(c)+\varepsilon$.

For such a δ we prove (i) and (ii).

Take $x\in(c,c+\frac{1}{2}\delta)$. Choose $\beta:=(1+\varepsilon)(x-c)$: then $0<\beta<(1+\varepsilon)\frac{1}{2}\delta<\delta$. Now $(x,c+\beta)$ is a subinterval of $(c-\beta,c+\beta)$ whose length is $c+\beta-x$ $=\varepsilon(x-c)=(\varepsilon/(1+\varepsilon))\beta$. By (1), $E_0\cap(x,c+\beta)$ contains an element, s, say. As $s\in E_0\subset E$, by the definition of E we have $f(x)\leqslant f(s)$. Consequently, because $\phi = f$ on E_0 and ϕ is increasing,

$$\frac{f(x)-f(c)}{x-c} \leqslant \frac{f(s)-f(c)}{x-c} = \frac{\phi(s)-\phi(c)}{x-c} = \frac{\phi(s)-\phi(c)}{s-c}\,\frac{s-c}{x-c}$$

$$\leqslant \frac{\phi(s)-\phi(c)}{s-c}\,\frac{\beta}{x-c} \leqslant (\phi'(c)+\varepsilon)(1+\varepsilon)$$

which proves (i).

Similarly, let $y \in (c-\frac{1}{2}\delta,\ c)$. This time we take $\beta := (1+\varepsilon)(c-y)$ and from (1) we obtain a point t of $E_0 \cap (y,\ y+\varepsilon c-\varepsilon y)$. Now $f(y) \leqslant f(t)$ and

$$\frac{f(c)-f(y)}{c-y} \geqslant \frac{f(c)-f(t)}{c-y} = \frac{\phi(c)-\phi(t)}{c-y} = \frac{\phi(c)-\phi(t)}{c-t}\,\frac{c-t}{c-y}$$

$$\geqslant \frac{\phi(c)-\phi(t)}{c-t}\,\frac{c-(y+\varepsilon c-\varepsilon y)}{c-y} = \frac{\phi(c)-\phi(t)}{c-t}(1-\varepsilon) \geqslant (\phi'(c)-\varepsilon)(1-\varepsilon)$$

Thus, (ii) holds also.

THEOREM 21.31. (Denjoy–Young) *Let $f : [a,\ b] \to \mathbb{R}$. Then for almost all $x \in [a,\ b]$ one has*

either $D^+f(x) = \infty$ and $D^-f(x) = -\infty$

or f is differentiable at x.

Proof. (Part 1) Let $S := \{x \in [a,\ b] : D^-f(x) \neq -\infty\}$. We first show that for almost every $x \in S$,

(*) $$D_l^-f(x) \geqslant D_r^+f(x) \quad \text{and} \quad D_r^+f(x) \neq \infty.$$

For $m \in \mathbb{N}$ and $q, r \in [a,\ b] \cap \mathbb{Q}$ with $q<r$, define

$$E_{m,q,r} := \left\{x \in [q,\ r] : \text{if } y \in [q,\ r] \text{ and } y \neq x, \text{ then } \frac{f(y)-f(x)}{y-x} \geqslant -m\right\}$$

Applying the previous lemma to the function $x \mapsto f(x) - mx$ ($x \in [q,\ r]$) we see that (*) holds for almost every $x \in E_{m;q,r}$. But the union of these (countably many) sets $E_{m;q,r}$ contains S except possibly for the end points a and b. The desired result follows.

(Part 2) We have proved that for almost every $x \in [a, b]$ one of the following statements is true.

(1) $D^-f(x) = -\infty$.

(2) $D_l^-f(x) \geqslant D_r^+f(x)$ and $D_r^+f(x) \neq \infty$.

In a similar way, replacing f by the function $t \mapsto -f(-t)$ $(-b \leqslant t \leqslant -a)$, for almost every $x \in [a,\ b]$ one obtains one of the alternatives (1) and (2′):

(1) $D^-f(x) = -\infty$.

(2′) $D_r^-f(x) \geqslant D_l^+f(x)$ and $D_l^+f(x) \neq \infty$.

Hence, for almost every x either (1) holds or (2) and (2′) hold simultaneously. Observing that $D_r^- f \leqslant D_r^+ f$ and $D_l^- f \leqslant D_l^+ f$ we see that for almost every x either $D^- f(x) = -\infty$ or f is differentiable at x. Similarly (consider the function $-f$), for almost all x either $D^+ f(x) = \infty$ or f is differentiable at x. This proves the theorem.

The following corollary generalizes Lebesgue's theorem (4.10).

COROLLARY 21.32. *Let f be a function on an interval I and let D be the set of those points of I where f is increasing. Then f is differentiable at almost every point of D.*

COROLLARY 21.33. (Marchaud) *Let $f: [a, b] \to \mathbb{R}$ be continuous and let N_f be everywhere finite. Then f is differentiable a.e. (N_f is the counting function of f; see Theorem 20.11.)*

Proof. We prove that at every point of (a, b) either f is increasing or f is decreasing or f attains a strict local extremum. Then, by the preceding corollary and by Theorem 7.2, we are done.

Let $c \in (a, b)$. The set $\{x : f(x) = f(c)\}$ is finite, so there exists an $\varepsilon > 0$ such that $(c - \varepsilon, c + \varepsilon) \subset [a, b]$ while on $(c, c + \varepsilon) \cup (c - \varepsilon, c)$ f does not take the value $f(c)$. As f has the Darboux property this implies that

$$\begin{array}{llll} \text{either} & f > f(c) & \text{on} & (c, c+\varepsilon) \\ \text{or} & f < f(c) & \text{on} & (c, c+\varepsilon) \end{array}$$

and

$$\begin{array}{llll} \text{either} & f > f(c) & \text{on} & (c-\varepsilon, c) \\ \text{or} & f < f(c) & \text{on} & (c-\varepsilon, c) \end{array}$$

Thus, we get the following four possible situations. Either f is increasing at c, or f is decreasing at c, or f attains a strict local minimum at c, or f attains a strict local maximum at c.

Two more consequences of the theorem of Denjoy and Young:

COROLLARY 21.34. *Let f be a function on an interval I such that each of its Dini derivatives is finite a.e. Then f is differentiable a.e.*

COROLLARY 21.35. *Let f be a function on an interval such that $f^\nabla(x)$ exists for almost every $x \in I$. Then f is differentiable almost everywhere, i.e. f^∇ is finite a.e.*

(For f^∇, see Exercise 15.P.)

Notes to Section 21

All subjects covered in this section are treated extensively in Saks (no date) where the reader will also find a wealth of information on the history of the theory. The class of absolutely continuous functions has been thoroughly investigated by N. Bary (*Math. Ann.* **103** (1930), 185–248 and 598–653).

The density theorem (21.29) has led to the following cluster of ideas. Let f be a Lebesgue measurable function defined a.e. on a set S of positive measure and let a be a density point of S. We say that a real number L is the *approximate limit* of f at a,

$$L = \operatorname*{limap}_{x \to a} f(x)$$

if for every $\varepsilon > 0$ a is a density point of the set $\{x \in S : |f(x) - L| < \varepsilon\}$. This definition evokes quite naturally the notions of *approximate continuity* and the *approximate* (or *asymptotic*) *derivative*. Information on these subjects and further references can be found in Saks (no date) and Bruckner (1978); see also V. Jarník (*Fund. Math.* **22** (1934), 4–16), J. S. Lipiński (*Coll. Math.* **5** (1957), 172–5 and *Coll. Math.* **10** (1963), 103–9), K. Krzyzewski (*Coll. Math.* **10** (1963), 281–5), T. Swiątkowski (*Fund. Math.* **59** (1966), 189–201).

Some of the results: Denjoy proved that every Lebesgue measurable function on $\mathbb{R}$ is approximately continuous almost everywhere. Lipiński showed that a function $f : \mathbb{R} \to \mathbb{R}$ is approximately continuous if and only if for all $a, b \in \mathbb{R}$ with $a < b$ the following function is a derivative:

$$x \mapsto \begin{cases} a & \text{if } f(x) < a \\ f(x) & \text{if } a \leqslant f(x) \leqslant b \\ b & \text{if } f(x) > b \end{cases}$$

If $f : \mathbb{R} \to \mathbb{R}$ is Lebesgue measurable, then for almost every $x \in \mathbb{R}$ one has either that f is approximately differentiable at x or that $D_r^+ f(x) = D_l^+ f(x) = \infty$, $D_r^- f(x) = D_l^- f(x) = -\infty$.

Mrs G. C. Young has shown that for every $f : \mathbb{R} \to \mathbb{R}$ the set $\{x \in \mathbb{R} : D_l^- f(x) < D_r^+ f(x)\}$ is countable. (It follows that, if $D_l f(x)$ and $D_r f(x)$ both exist for all $x \in \mathbb{R}$, then f is differentiable everywhere outside a countable set.) The proof, which is based upon an ingenious idea of W. Sierpiński, is reproduced by A. Rajchman & S. Saks in *Fund. Math.* **4** (1923), 204–13, but we invite the reader to try and find a proof by himself before looking it up.

In this context we should also mention the work of C. J. Neugebauer (*Acta Sci. Math.* **23** (1962), 79–81), who showed that for a continuous $f:\mathbb{R}\to\mathbb{R}$ the sets $\{x : D_r^+f(x)\neq D_l^+f(x)\}$ and $\{x : D_r^-f(x)\neq D_l^-f(x)\}$ are meagre.

22. The Perron integral

Let $f:[a, b]\to\mathbb{R}$ be differentiable. We have seen (Theorem 15.3) that, if f' is Riemann integrable, then

$$(*) \qquad \int_a^b f'(x)\,\mathrm{d}x=f(b)-f(a)$$

but we have also seen (Example 14.3) that there is no *a priori* reason for f' to *be* Riemann integrable. In an attempt to generalize (*) we have extended the Riemann integral to the Lebesgue integral. It turned out (Corollary 21.23) that now (*) already holds if f' is Lebesgue integrable, but even this wider theory does not catch all differentiable functions (see Exercise 21.M).

The Perron integral is a further extension of the Lebesgue integral in which (*) is valid for every differentiable $f:[a, b]\to\mathbb{R}$. It has the advantage over the Lebesgue integral of being easier to define. However, to prove the natural generalizations of the convergence theorems of Levi and Lebesgue (Theorems 19.13 and 19.15) we have to fall back upon Lebesgue theory. In fact, it will turn out that a positive Perron integrable function is Lebesgue integrable (Corollary 22.8) and that the inequality $|f|\leqslant g$ where f and g are Perron integrable implies Lebesgue integrability of f (Exercise 22.E).

In the sequel we shall need the operations D^+ and D^- that were introduced in Definition 15.4. The following exercise yields some useful formulas.

**Exercise* 22.A

(i) If $f:[a, b]\to\mathbb{R}$, then $D^+(-f)=-D^-f$.

(ii) Let $f, g:[a, b]\to\mathbb{R}$ and let $x\in[a, b]$. Then

$$D^-(f+g)(x)\geqslant D^-f(x)+D^-g(x)$$

if the right hand member is defined.

If g is differentiable at x, then

$$D^-(f+g)(x)=D^-f(x)+g'(x)$$

(iii) Let $f:[a, b]\to\mathbb{R}$ be Lebesgue integrable and upper semicontinuous. If v is an indefinite integral of f, then $D^+v\leqslant f$.

If $u : [a, b] \to \mathbb{R}$ and if $x, y \in [a, b]$, we define

$$\Big/_x^y u := u(y) - u(x)$$

Let $f : [a, b] \to \mathbb{R}$. A continuous function u on $[a, b]$ is called an *upper function* of f if $D^- u \geqslant f$. A continuous function v on $[a, b]$ is a *lower function* of f if $D^+ v \leqslant f$.

**Exercise* 22.B. If u and v are upper and lower functions of f, respectively, then $D^-(u - v) \geqslant 0$. Hence, in this case $u - v$ is increasing. In particular, $/_a^b u \geqslant /_a^b v$.

This exercise makes the following definition reasonable.

DEFINITION 22.1. f is said to be *Perron integrable over* $[a, b]$ if for every $\varepsilon > 0$ there exist an upper function u and a lower function v for which $/_a^b u \leqslant /_a^b v + \varepsilon$. For such a Perron integrable f,

$$\sup\left\{\Big/_a^b v : v \text{ is a lower function of } f\right\}$$

$$= \inf\left\{\Big/_a^b u : u \text{ is an upper function of } f\right\}$$

The number $\sup\{/_a^b v : v \text{ is a lower function of } f\}$ is then called the *Perron integral of f over* $[a, b]$. We denote it by $\mathscr{P}\int_a^b f$.

Theorems 22.2–22.5 are left to the reader as exercises.

THEOREM 22.2. *If $f : [a, b] \to \mathbb{R}$ has an antiderivative F, then f is Perron integrable and $\mathscr{P}\int_a^b f = /_a^b F$.*

THEOREM 22.3. *The Perron integrable functions on $[a, b]$ form a vector space on which $f \mapsto \mathscr{P}\int_a^b f$ is a linear function.*

THEOREM 22.4. *Let $a < b < c$ and let $f : [a, c] \to \mathbb{R}$.*
(i) *f is Perron integrable over $[a, c]$ if and only if f is Perron integrable over $[a, b]$ and over $[b, c]$.*
(ii) *If f is Perron integrable over $[a, c]$, then we have $\mathscr{P}\int_a^c f = \mathscr{P}\int_a^b f + \mathscr{P}\int_b^c f$.*

THEOREM 22.5. *If f is Perron integrable over $[a, b]$ and if $f \geqslant 0$, then $\mathscr{P}\int_a^b f \geqslant 0$.*

The Perron integral is an extension of the Lebesgue integral:

THEOREM 22.6. *Let $f:[a,b]\to\mathbb{R}$. f is Lebesgue integrable if and only if for every $\varepsilon>0$ f has an absolutely continuous upper function u and an absolutely continuous lower function v with $/_a^b u\leqslant /_a^b v+\varepsilon$. In particular, every Lebesgue integrable function is Perron integrable. If $f:[a,b]\to\mathbb{R}$ is Lebesgue integrable, then its Perron integral is equal to its Lebesgue integral.*

Proof. (*a*) Suppose that for every $n\in\mathbb{N}$ there exist absolutely continuous upper and lower functions, u_n and v_n, respectively, that satisfy $/_a^b u_n\leqslant /_a^b v_n+1/n$. For each n we have, a.e. on $[a,b]$,

$$u_n'=D^-u_n\geqslant f\geqslant D^+v_n=v_n'$$

where u_n' and v_n' are Lebesgue integrable and $\int_a^b(u_n'-v_n')=/_a^b u_n-/_a^b v_n\leqslant 1/n$ (see Theorem 21.21). Set $g:=\inf_{n\in\mathbb{N}}u_n'$, $h:=\sup_{n\in\mathbb{N}}v_n'$. Then g and h are Lebesgue measurable and $u_n'\geqslant g\geqslant f\geqslant h\geqslant v_n'$ a.e. It follows from Theorem 20.3(ii) that g and h are Lebesgue integrable, and that for all n, $\int_a^b(g-h)\leqslant\int_a^b(u_n'-v_n')\leqslant 1/n$. As $g-h\geqslant 0$ a.e. we see that $g=h$ a.e. (by Corollary 19.12(i)) and therefore $f=g=h$ a.e. Consequently, f is Lebesgue integrable.

(*b*) In the above language, for every n we have $/_a^b u_n=\int_a^b u_n'\geqslant\int_a^b f\geqslant\int_a^b v_n'=/_a^b v_n$. Therefore, $\mathscr{P}\int_a^b f=\int_a^b f$.

(*c*) Now let us assume that f is Lebesgue integrable and let $\varepsilon>0$: we construct absolutely continuous upper and lower functions u and v with $/_a^b u\leqslant /_a^b v+2\varepsilon$. (This will finish the proof of the theorem.) Without restriction, assume $f\geqslant 0$. (Compare your proof of Theorem 22.3.)

By Lemma 20.1(ii) there exists an upper semicontinuous function h on $[a,b]$ with $0\leqslant h\leqslant f$ and $\int_a^b(f-h)<\varepsilon$. If v is an indefinite integral of h, then v is absolutely continuous and $D^+v\leqslant h\leqslant f$ (see Exercise 22.A(iii)).

The construction of u requires more care. (Why?) First, choose Lebesgue integrable functions $f_1, f_2,\dots$ on $[a,b]$ with $0\leqslant f_n\leqslant 1$ for each n and $\sum_{n=1}^\infty f_n=f$ (e.g. $f_n:=f\wedge n-f\wedge(n-1)$). By applying Lemma 20.1(ii) to $1-f_n$ we see that for every n there is a lower semicontinuous function g_n on $[a,b]$ with $f_n\leqslant g_n\leqslant 1$ and $\int_a^b g_n\leqslant\int_a^b f_n+\varepsilon 2^{-n}$. Then $\sum_{n=1}^\infty\int_a^b g_n\leqslant\sum_{n=1}^\infty\int_a^b f_n+\sum_{n=1}^\infty\varepsilon 2^{-n}=\int_a^b f+\varepsilon<\infty$, so $g:=\sum_{n=1}^\infty g_n$ is an a.e. defined Lebesgue integrable function. Let u and u_n ($n\in\mathbb{N}$) be indefinite integrals of g and g_n, respectively. Then u is absolutely continuous. For each n, $u-(u_1+\dots+u_n)$ is increasing, so $D^-u\geqslant D^-(u_1+\dots+u_n)$. But, since $g_1+\dots+g_n$ is lower semicontinuous, we have $D^-(u_1+\dots+u_n)\geqslant g_1+\dots+g_n$ (Exercise 22.A(iii)). Hence, $D^-u\geqslant\sum_{n=1}^\infty g_n\geqslant\sum_{n=1}^\infty f_n=f$.

Finally, $/_a^b u=\int_a^b g=\sum_{n=1}^\infty\int_a^b g_n\leqslant\int_a^b f+\varepsilon\leqslant /_a^b v+2\varepsilon$.

We see that we could have given the definition of the Lebesgue integral completely analogous to the one of the Perron integral, merely by restricting the class of the upper and lower functions. A further restriction yields the Riemann integral:

**Exercise* 22.C. For any class $\mathscr{F}$ of continuous functions on $[a, b]$ one can define '$\mathscr{F}$-integrability' by calling a function $f:[a, b]\to\mathbb{R}$ *$\mathscr{F}$-integrable* if for every $\varepsilon>0$ f has upper and lower functions $u, v \in \mathscr{F}$ such that $\int_a^b u \leqslant \int_a^b v+\varepsilon$. The class of all continuous functions on $[a, b]$ gives us the Perron integral, the class of all absolutely continuous functions defines the Lebesgue integral.

Prove now that each of the following classes of functions leads to the Riemann integral: the class of all piecewise linear continuous functions; the class of all functions that have continuous derivatives.

THEOREM 22.7. *Let f be Perron integrable over $[a, b]$. Define a function F on $[a, b]$ by $F(x):=\mathscr{P}\int_a^x f(a\leqslant x\leqslant b)$. Then F is continuous and $F'=f$ a.e. In particular, f is Lebesgue measurable.*

Proof. Let u and v be upper and lower functions of f, respectively. For $a\leqslant x\leqslant y\leqslant b$ we have $\mathscr{P}\int_x^y f=F(y)-F(x)$ and $\int_x^y v\leqslant\mathscr{P}\int_x^y f\leqslant\int_x^y u$, i.e. $v(y)-v(x)\leqslant F(y)-F(x)\leqslant u(y)-u(x)$. The continuity of F follows.

Now let $\varepsilon>0$, $\delta>0$. Choose upper and lower functions, u and v, of f such that $\int_a^b u\leqslant\int_a^b v+\delta\varepsilon$. Set $H:=u-F$. For $a\leqslant x\leqslant y\leqslant b$ we see that $H(y)-H(x)=\int_x^y u-\mathscr{P}\int_x^y f\geqslant 0$. Therefore, H is increasing and differentiable a.e. Let $E:=\{x\in[a, b]: H'(x)\geqslant\varepsilon\}$. According to Exercise 21.K,

$$\lambda(E)\cdot\varepsilon\leqslant\int_a^b H'\leqslant\int_a^b H=\int_a^b u-\mathscr{P}\int_a^b f\leqslant\int_a^b u-\int_a^b v\leqslant\delta\varepsilon$$

whence $\lambda(E)\leqslant\delta$. On the other hand, for every $x\in[a, b]\setminus E$ for which $H'(x)$ exists, by Exercise 22.A(ii) one obtains

$$D^-F(x)=D^-(u-H)(x)=D^-u(x)-H'(x)\geqslant f(x)-\varepsilon$$

Consequently, $\{x: D^-F(x)<f(x)-\varepsilon\}\subset E\cup\{x: H \text{ is not differentiable at } x\}$. Thus, the outer measure of $\{x: D^-F(x)<f(x)-\varepsilon\}$ is at most δ.

For given $\varepsilon>0$ this is true for all $\delta>0$; hence, the set $\{x: D^-F(x)<f(x)-\varepsilon\}$ is null. This, in turn, is true for all $\varepsilon>0$, so $D^-F\geqslant f$ a.e.

Similarly, $D^+F\leqslant f$ a.e. But $D^+F\geqslant D^-F$. We see that $F'=f$ a.e. By Theorem 20.12, f must be Lebesgue measurable.

Exercise 22.D. Show that F need not be absolutely continuous.

COROLLARY 22.8. *If $f:[a, b]\to\mathbb{R}$ is Perron integrable and $\geqslant 0$, then f is also Lebesgue integrable.*

Proof. By Theorem 22.5, F is increasing. Now apply Exercise 21.K.

**Exercise* 22.E. If $f:[a, b]\to\mathbb{R}$ is Perron integrable and if $|f|\leqslant g$ for some Perron integrable $g:[a, b]\to\mathbb{R}$, then f is Lebesgue integrable.

**Exercise* 22.F. Prove analogues of the convergence theorems of Levi, Fatou and Lebesgue (19.13, 19.14, 19.15) for the Perron integral.

Exercise 22.G. Find a Perron integrable function on $[a, b]$ whose absolute value is not Perron integrable.

Improper Perron integrals are meaningless:

THEOREM 22.9. *Let $f:[a, b]\to\mathbb{R}$. Suppose that, for every $\beta\in[a, b)$, f is Perron integrable over $[a, \beta]$ and that $\lim_{\beta\uparrow b}\mathscr{P}\int_a^\beta f$ exists. Then f is Perron integrable over $[a, b]$ and $\mathscr{P}\int_a^b f=\lim_{\beta\uparrow b}\mathscr{P}\int_a^\beta f$.*

Proof. Let $L:=\lim_{\beta\uparrow b}\mathscr{P}\int_a^\beta f$. Let $\varepsilon>0$. We make an upper function $\tilde{u}$ of f on $[a, b]$ for which $\big/_a^b\,\tilde{u}\leqslant L+3\varepsilon$. (In the same way one can find a lower function $\tilde{v}$ such that $\big/_a^b\,\tilde{v}\geqslant L-3\varepsilon$. The theorem then follows.)

Choose a strictly increasing sequence $a_0, a_1, \ldots$ with $a_0=a$ and $b=\lim_{n\to\infty} a_n$. It is not difficult to construct a continuous function u on $[a, b)$ such that for every n

the restriction of u to $[a_{n-1}, a_n]$ is an upper function of f on $[a_{n-1}, a_n]$

$$\Big/_{a_{n-1}}^{a_n} u\leqslant\mathscr{P}\int_{a_{n-1}}^{a_n} f+\varepsilon 2^{-n}$$

We proceed to prove that $\lim_{x\uparrow b} u(x)$ exists.

For $a_{n-1}\leqslant x\leqslant a_n$,

$$\Big/_{a_{n-1}}^{x} u=\Big/_{a_{n-1}}^{a_n} u-\Big/_{x}^{a_n} u\leqslant\mathscr{P}\int_{a_{n-1}}^{a_n} f+\varepsilon 2^{-n}-\mathscr{P}\int_{x}^{a_n} f$$

$$=\mathscr{P}\int_{a_{n-1}}^{x} f+\varepsilon 2^{-n}$$

It follows that

$$(*)\qquad \Big/_{a_p}^{x} u\leqslant\mathscr{P}\int_{a_p}^{x} f+\varepsilon 2^{-p}\qquad (p=0, 1, 2, \ldots;\ a_p\leqslant x<b)$$

Further, if $a_p\leqslant y<b$, then $\big/_{a_p}^{y} u\geqslant\mathscr{P}\int_{a_p}^{y} f$ because $D^- u\geqslant f$ on $[a_p, y]$. We infer that, if $x, y\in[a_p, b)$, then

$$u(x)-u(y)\leqslant\mathscr{P}\int_{a_p}^{x} f-\mathscr{P}\int_{a_p}^{y} f+\varepsilon 2^{-p}=\mathscr{P}\int_{a}^{x} f-\mathscr{P}\int_{a}^{y} f+\varepsilon 2^{-p}$$

whence, by symmetry,

$$|u(x)-u(y)|\leqslant\left|\mathscr{P}\int_{a}^{x} f-\mathscr{P}\int_{a}^{y} f\right|+\varepsilon 2^{-p}$$

Consequently, $\lim_{x\uparrow b} u(x)$ exists. Thus, u can be extended to a continuous function on $[a, b]$. We denote this extension again by u. It follows from (*) that $/_a^b u \leqslant L+\varepsilon$. Since $D^- u \geqslant f$ on $[a, b)$, we would be done if only $D^- u(b) \geqslant f(b)$. Unfortunately, $D^- u(b)$ may very well be $-\infty$.

Choose $c \in (a, b)$ such that $|u(x)-u(y)| \leqslant \varepsilon$ for all $x, y \in [c, b]$. Define $\phi : [a, b] \to \mathbb{R}$ by

$$\phi(x) := \begin{cases} \sup\{|u(s)-u(b)| : s \geqslant x\} & \text{if} \quad x \in [c, b] \\ \phi(c) & \text{if} \quad x \in [a, c) \end{cases}$$

This ϕ is continuous and decreasing; $\phi(b)=0$ and $\phi(a) \leqslant \varepsilon$. Define $\tilde{u} : [a, b] \to \mathbb{R}$ as follows:

$$\tilde{u}(x) := u(x) - \phi(x) - \varepsilon \sqrt{\left(\frac{b-x}{b-a}\right)} \qquad (a \leqslant x \leqslant b)$$

This $\tilde{u}$ is going to solve our problem. First, note that $\tilde{u}$ is continuous. Also, $u-\tilde{u}$, being the sum of two decreasing functions, is decreasing, so $D^- u \leqslant D^- \tilde{u}$. Therefore, $D^- \tilde{u} \geqslant f$ on $[a, b)$. Furthermore, if $c \leqslant x \leqslant b$, then $\phi(x) \geqslant u(x)-u(b)$, so

$$\tilde{u}(b) - \tilde{u}(x) = u(b) - u(x) + \phi(x) + \varepsilon \sqrt{\left(\frac{b-x}{b-a}\right)} \geqslant \varepsilon \sqrt{\left(\frac{b-x}{b-a}\right)}$$

Apparently, $D^- \tilde{u}(b) = \infty > f(b)$.

We see that $\tilde{u}$ is an upper function of f. Finally, observe that $/_a^b \tilde{u} = /_a^b u - /_a^b \phi + \varepsilon \leqslant (L+\varepsilon)+\varepsilon+\varepsilon = L+3\varepsilon$.

Notes to Section 22

Our *Perron integral* is, by definition, the same as the $\mathscr{P}_0$-integral introduced in Saks (no date). By Saks' results VIII.3.9 and 3.11, *a posteriori* it turns out to be identical to the objects called the *Perron integral* (or *$\mathscr{P}$-integral*) and the *restricted Denjoy integral* (or *$\mathscr{D}_*$-integral*) defined in Saks' book. It also coincides with the *Riemann-complete* integral devised by R. Henstock (*Can. J. Math.* **20** (1968), 79–87. See H. W. Pu, *Coll. Math.* **28** (1973), 105–10).

For further theory we again refer to Saks.

23. The Stieltjes integral

The integral, $\int_0^1 f(x)\,dx$, of a continuous function f on $[0, 1]$ may be regarded as the 'average value' of f. (For large n the integral is approximately equal to the arithmetic average of the numbers $f(i/n)$ where $i=1, 2, \ldots,$

n). In practical cases it is sometimes important to consider a different 'average' of f which is more influenced by the behaviour of f on one part of $[0, 1]$ than on another. For instance, the operation $f \mapsto 3 \int_0^1 f(x)x^2 \, dx$ is also an averaging of the values of f, but this time the points of $[0, 1]$ close to 1 carry more weight than those near 0. (The factor 3 is a normalization constant, designed to give a constant function $x \mapsto c$ the average value c.)

Often such a biased average (for functions on $[0, 1]$) can be described by a weight function $w : [0, 1] \to [0, \infty)$; the averaging operation determined by the weight function w is

$$(1) \qquad f \mapsto \int_0^1 f(x)w(x) \, dx \bigg/ \int_0^1 w(x) \, dx \qquad (f \in \mathscr{C}[0, 1])$$

However, occasionally one wants to concentrate all the weight in a few points. The formula

$$f \mapsto \tfrac{1}{2} f(\tfrac{1}{4}) + \tfrac{1}{2} f(\tfrac{3}{4}) \qquad (f \in \mathscr{C}[0, 1])$$

defines a way of averaging that takes into account only the values of f at $\frac{1}{4}$ and at $\frac{3}{4}$. In this case, the intervals $[0, \frac{1}{4})$, $(\frac{1}{4}, \frac{3}{4})$ and $(\frac{3}{4}, 1]$ are comparable to the null sets of the Lebesgue integration. This procedure is easily generalized to

$$(2) \qquad f \mapsto \sum_{n=1}^{\infty} p_n f(a_n) \qquad (f \in \mathscr{C}[0, 1])$$

where $a_1, a_2, \ldots \in [0, 1]$, $p_n \geqslant 0$ and $\Sigma_{n=1}^{\infty} p_n = 1$.

In this section we build a theory in which (1) and (2) occur as special cases. (A third special case is considered in Exercises 23.J and 23.K.) We shall do so by generalizing the definition of the Riemann integral. It is possible to push the generality farther by extending the Lebesgue theory in a similar way, but this will not interest us. We use Lebesgue integrals only in one or two exercises.

Let $a < b$. By a *Riemann sequence on* $[a, b]$ we mean a (finite) sequence $(x_0, \xi_1, x_1, \xi_2, x_2, \ldots, \xi_m, x_m)$ of elements of $[a, b]$ such that $a = x_0 \leqslant \xi_1 \leqslant x_1 \leqslant \ldots \leqslant \xi_m \leqslant x_m = b$. If $V := (x_0, \xi_1, x_1, \ldots, \xi_m, x_m)$ is such a Riemann sequence, the *mesh* of V is the number $\max_i(x_i - x_{i-1})$. It is well known (and we shall see it again in Theorem 23.5) that the following two assertions about a function $f : [a, b] \to \mathbb{R}$ and a real number I are equivalent.

(α) f is Riemann integrable and $\int_a^b f(x) \, dx = I$.

(β) For every $\varepsilon > 0$ there exists a $\delta > 0$ such that $|\Sigma_{i=1}^m f(\xi_i)(x_i - x_{i-1}) - I| \leqslant \varepsilon$ for every Riemann sequence $(x_0, \xi_1, x_1, \ldots, \xi_m, x_m)$ on $[a, b]$ whose mesh is $\leqslant \delta$.

This observation leads to Definitions 23.1 and 23.2.

DEFINITION 23.1. Let $a<b$. For every Riemann sequence V on $[a, b]$ let S_V be a real number. If $I \in \mathbb{R}$, then by

$$\lim_V S_V = I$$

we mean that for every $\varepsilon>0$ there exists a $\delta>0$ such that $|S_V - I| \leqslant \varepsilon$ for all Riemann sequences V with mesh $\leqslant \delta$. (Note that, if $\lim_V S_V = I$ and $\lim_V S_V = J$, then $I = J$.)

**Exercise* 23.A. Let $a<b$. For every Riemann sequence V on $[a, b]$ let $S_V \in \mathbb{R}$. Show that the following statements are equivalent.
(α) $\lim_V S_V = I$ for some $I \in \mathbb{R}$.
(β) For every $\varepsilon>0$ there exists a $\delta>0$ such that $|S_V - S_W| \leqslant \varepsilon$ for all Riemann sequences V, W with mesh $\leqslant \delta$.

For two functions, f and g, on $[a, b]$ and for a Riemann sequence $V = (x_0, \xi_1, x_1, \ldots, \xi_m, x_m)$ on $[a, b]$ we set

$$S_V(f\,;g) := \sum_{i=1}^{m} f(\xi_i)(g(x_i) - g(x_{i-1}))$$

DEFINITION 23.2. Let $a<b$. Let f and g be functions on $[a, b]$. f is called *Stieltjes integrable over* $[a, b]$ *with respect to* g if there exists a number I such that

$$\lim_V S_V(f; g) = I$$

If this is so, then the number I for which this formula is valid is called the *Stieltjes integral of* f *over* $[a, b]$ *with respect to* g. It is denoted by $\int_a^b f\,dg$.

**Exercise* 23.B. Let $g : [a, b] \to \mathbb{R}$. Prove the following. The functions $f : [a, b] \to \mathbb{R}$ that are Stieltjes integrable with respect to g form a vector space on which $f \mapsto \int_a^b f\,dg$ is a linear function. If g is increasing, if $f \geqslant 0$ and if the integral $\int_a^b f\,dg$ exists, then $\int_a^b f\,dg \geqslant 0$. If g is constant, then $\int_a^b f\,dg = 0$ for every function f on $[a, b]$.

**Exercise* 23.C. Let $a<b$. Let $c \in [a, b]$ and let $f : [a, b] \to \mathbb{R}$.
(i) $\int_a^b f\,d\xi_{\{c\}}$ exists if and only if f is continuous at c. Show that in that case $\int_a^b f\,d\xi_{\{c\}} = 0$ if $c \neq a, b$; $\int_a^b f\,d\xi_{\{c\}} = -f(a)$ if $c = a$; $\int_a^b f\,d\xi_{\{c\}} = f(b)$ if $c = b$.
(ii) Let $c \neq a$. $\int_a^b f\,d\xi_{[c,b]}$ exists if and only if f is continuous at c. Show that in that case $\int_a^b f\,d\xi_{[c,b]} = f(c)$.
(iii) Let $c \neq b$. $\int_a^b f\,d\xi_{(c,b]}$ exists if and only if f is continuous at c. Show that in that case $\int_a^b f\,d\xi_{(c,b]} = f(c)$.

Exercise 23.D. $\int_0^1 f\,d\xi_{\mathbb{Q}}$ exists if and only if f is constant.

THEOREM 23.3. *Let f and ϕ be Riemann integrable functions on $[a, b]$. Let $g : [a, b] \to \mathbb{R}$ be an indefinite integral of ϕ. Then the Stieltjes integral $\int_a^b f \, dg$ exists and we have*

$$\int_a^b f \, dg = \int_a^b f(x)\phi(x) \, dx$$

Proof. Let $M := \sup\{|\phi(x)| : a \leqslant x \leqslant b\}$.

Take $\varepsilon > 0$. There exists a $\delta > 0$ with the following property.

(*) If $a = x_0 \leqslant x_1 \leqslant \ldots \leqslant x_n = b$ and if $x_i \leqslant x_{i-1} + \delta$ for each i,

$$\text{then } \sum_{i=1}^{m} \left(\sup_{[x_{i-1}, x_i]} f - \inf_{[x_{i-1}, x_i]} f \right)(x_i - x_{i-1}) \leqslant \varepsilon$$

(see Exercise 23.E). Now let $V = (x_0, \xi_1, x_1, \ldots, \xi_n, x_n)$ be a Riemann sequence on $[a, b]$ whose mesh is $\leqslant \delta$. Then

$$\left| S_V(f; g) - \int_a^b f(x)\phi(x) \, dx \right| = \left| \sum_{i=1}^{n} \int_{x_{i-1}}^{x_i} (f(\xi_i) - f(x))\phi(x) \, dx \right|$$

$$\leqslant M \cdot \sum_{i=1}^{n} \int_{x_{i-1}}^{x_i} |f(\xi_i) - f(x)| dx \leqslant \varepsilon M$$

The theorem follows.

**Exercise* 23.E. Let $f : [a, b] \to \mathbb{R}$ be Riemann integrable. Let $\varepsilon > 0$. Show that there exists a $\delta > 0$ such that the assertion (*) used in the above proof is valid. (Hint. There exist a partition $a = a_0 < a_1 < \ldots < a_m = b$ of $[a, b]$ and numbers $h_1, \ldots, h_m, k_1, \ldots, k_m$ such that for $x \in (a_{i-1}, a_i)$ we have $k_i \leqslant f(x) \leqslant h_i$, while $\sum_{i=1}^{m} (h_i - k_i)(a_i - a_{i-1}) < \frac{1}{2}\varepsilon$. There exists a number A with $|f(x)| \leqslant A$ for all $x \in [a, b]$. Choose $\delta > 0$ such that $4mA\delta < \frac{1}{2}\varepsilon$.)

With the help of Exercise 8.C(i) we obtain from Theorem 23.3:

COROLLARY 23.4. *Let $f : [a, b] \to \mathbb{R}$ be Riemann integrable. Let $g : [a, b] \to \mathbb{R}$ be differentiable and such that g' is Riemann integrable. Then $\int_a^b f \, dg = \int_a^b f(x)g'(x) \, dx$.*

Exercise 23.F. Let $f : [a, b] \to \mathbb{R}$ be continuous and let $g : [a, b] \to \mathbb{R}$ be absolutely continuous. Then $\int_a^b f \, dg$ exists and is equal to the Lebesgue integral $\int_a^b f(x)g'(x) \, dx$.

THEOREM 23.5. *Let $\chi : [a, b] \to \mathbb{R}$ be the identity function:*

$$\chi(x) := x \qquad (a \leqslant x \leqslant b)$$

Let $f : [a, b] \to \mathbb{R}$. Then $\int_a^b f \, d\chi$ exists if and only if f is Riemann integrable. In that case,

$$\int_a^b f \, d\chi = \int_a^b f(x) \, dx$$

Proof. If f is Riemann integrable, we apply Corollary 23.4.

Now assume that $\int_a^b f\, d\chi$ exists. It suffices to show that f is Riemann integrable. Let $\varepsilon>0$. Choose $\delta>0$ such that for every Riemann sequence V on $[a, b]$ with mesh $\leqslant\delta$ we have $|S_V(f;\chi)-\int_a^b f\, d\chi|\leqslant\frac{1}{2}\varepsilon$.

Let $a=x_0<x_1<\ldots<x_n=b$ be a partition of $[a, b]$ with $x_i\leqslant x_{i-1}+\delta$ for $i=1, 2, \ldots, n$. For every choice of $\xi_i\in[x_{i-1}, x_i]$ $(i=1,\ldots,n)$ we obtain

$$\int_a^b f\, d\chi-\tfrac{1}{2}\varepsilon\leqslant\sum_{i=1}^n f(\xi_i)(x_i-x_{i-1})\leqslant\int_a^b f\, d\chi+\tfrac{1}{2}\varepsilon$$

It follows easily that f is bounded on each $[x_{i-1}, x_i]$, hence on $[a, b]$. With $s_i:=\sup_{[x_{i-1},x_i]} f$ and $t_i:=\inf_{[x_{i-1},x_i]} f$ $(i=1,\ldots,n)$ we have

$$\int_a^b f\, d\chi-\tfrac{1}{2}\varepsilon\leqslant\sum_{i=1}^n t_i(x_i-x_{i-1})\leqslant\sum_{i=1}^n s_i(x_i-x_{i-1})\leqslant\int_a^b f\, d\chi+\tfrac{1}{2}\varepsilon$$

Thus, for any $\varepsilon>0$ we have found an upper and a lower estimate of f (viz. $\Sigma_{i=1}^n s_i(x_i-x_{i-1})$ and $\Sigma_{i=1}^n t_i(x_i-x_{i-1})$) whose difference is at most ε. By Exercise 12.A, f is Riemann integrable.

THEOREM 23.6. (Integration by parts) *Let f and g be functions on $[a, b]$ such that $\int_a^b f\, dg$ exists. Then $\int_a^b g\, df$ exists and*

$$\int_a^b f\, dg+\int_a^b g\, df=f(b)g(b)-f(a)g(a)$$

Proof. Let $\varepsilon>0$. Let $\delta>0$ be such that $|S_V(f;g)-\int_a^b f\, dg|\leqslant\varepsilon$ for every Riemann sequence V whose mesh is at most 2δ. Take any Riemann sequence $W=(x_0, \xi_1, x_1,\ldots, \xi_m, x_m)$ whose mesh is $\leqslant\delta$. Then $V:=(x_0, x_0, \xi_1, x_1,\ldots, \xi_m, x_m, x_m)$ is a Riemann sequence with mesh $\leqslant 2\delta$. By substitution one obtains

$$S_W(g;f)+S_V(f;g)=f(b)g(b)-f(a)g(a)$$

whence

$$\left|S_W(g;f)-\left(\int_a^b f\, dg-f(b)g(b)+f(a)g(a)\right)\right|\leqslant\varepsilon$$

The theorem follows.

Exercise 23.G. Let f, g be functions on $[a, b]$ such that $\int_a^b f\, dg$ exists. Let $a<c<b$. Then $\int_a^c f\, dg$ and $\int_c^b f\, dg$ exist and their sum is $\int_a^b f\, dg$. (Hint. Consider two Riemann sequences that coincide on $[c, b]$.)

One might expect that the existence of $\int_a^c f\, dg$ and $\int_c^b f\, dg$ (where $a<c<b$) would entail the existence of $\int_a^b f\, dg$. This is not the case. In fact, if we define $f:=\xi_{(0,1]}$ and $g:=\xi_{[0,1]}$, then trivially $\int_0^1 f\, dg=0$ (as g

is constant on $[0, 1]$) while $\int_{-1}^{0} f\, dg=0$ (since $f=0$ on $[-1,0]$), but $\int_{-1}^{1} f dg$ does not exist (Exercise 23.C(ii)). We have, however, the following.

THEOREM 23.7. *Let f, g be functions on $[a, b]$. Let $a<c<b$ and assume that $\int_a^c f\, dg$ and $\int_c^b f\, dg$ exist. If f is continuous at c and g is bounded, then $\int_a^b f dg$ exists and is equal to $\int_a^c f dg+\int_c^b f dg$.*

Proof. Let $\varepsilon>0$. Take $\delta_1>0$ such that

$$\left|S_V(f; g)-\int_a^c f\, dg\right|\leqslant\frac{\varepsilon}{3} \quad \text{and} \quad \left|S_W(f; g)-\int_c^b f\, dg\right|\leqslant\frac{\varepsilon}{3}$$

for all Riemann sequences V, W on $[a, c]$ and $[c, b]$, respectively, with mesh $\leqslant\delta_1$.

Take $\delta_2>0$ such that, if $\xi\in[a, b]$ and $|\xi-c|\leqslant\delta_2$, then $|f(\xi)-f(c)|\cdot 2M \leqslant\frac{1}{3}\varepsilon$, where $M:=\sup_{[a,b]}|g|$. Let $U=(x_0, \xi_1, x_1, \ldots, \xi_n, x_n)$ be a Riemann sequence on $[a, b]$ and let the mesh of U be less than $\min(\delta_1, \delta_2)$. There is a $j\in\{1, \ldots, n\}$ for which $c\in[x_{j-1}, x_j]$. Consider the following Riemann sequences V and W on $[a, c]$ and $[c, b]$, respectively:

$$V:=(x_0, \xi_1, \ldots, \xi_{j-1}, x_{j-1}, c, c)$$
$$W:=(c, c, x_j, \xi_{j+1}, x_{j+1}, \ldots, x_n)$$

We have $|S_V(f; g)-\int_a^c f\, dg|\leqslant\frac{1}{3}\varepsilon$ and $|S_W(f; g)-\int_c^b f\, dg|\leqslant\frac{1}{3}\varepsilon$. It is easy to see that

$$S_U(f; g)=S_V(f; g)+S_W(f; g)+(f(\xi_j)-f(c))(g(x_j)-g(x_{j-1}))$$

As $|\xi_j-c|\leqslant\delta_2$ and $|g(x_j)-g(x_{j-1})|\leqslant 2M$, it follows that

$$\left|S_U(f; g)-\left(\int_a^c f\, dg+\int_c^b f\, dg\right)\right|\leqslant\varepsilon$$

We have proved the theorem.

Exercise 23.H. Let $g:[a, b]\to\mathbb{R}$ be of bounded variation. Denote by J_g the set of all functions $f:[a, b]\to\mathbb{R}$ that are Stieltjes integrable over $[a, b]$ with respect to g.
(i) If $f\in J_g$ is bounded, then

$$\left|\int_a^b f\, dg\right|\leqslant \sup_{a\leqslant x\leqslant b}|f(x)|\cdot \operatorname*{Var}_{[a,b]} g$$

(ii) If $f_1, f_2, \ldots \in J_g$ and if $\lim_{n\to\infty} f_n=f$ uniformly, then $f\in J_g$ and $\int_a^b f\, dg =\lim_{n\to\infty}\int_a^b f_n\, dg$.
(iii) If x is a point of continuity of g, then $\xi_{[x,b]}\in J_g$, $\xi_{(x,b]}\in J_g$.
(iv) If $f:[a, b]\to\mathbb{R}$ is continuous, then $f\in J_g$. (Hint. Approximate f by suitable step functions, using (ii) and (iii).)

In the beginning of this section we mentioned two types of averaging operations on $\mathscr{C}[0, 1]$ and we claimed that both can be given in the form of Stieltjes integrals. Theorem 23.3 shows how this can be done with the averaging that uses a Riemann integrable weight function. (See Exercise 23.F for a Lebesgue integrable weight function.) Exercise 23.I deals with the averaging operations of the second type, while in Exercise 23.J we consider a third kind.

**Exercise* 23.I. Let $x_1, x_2, \ldots \in [a, b]$. Let $p_1, p_2, \ldots \in \mathbb{R}$, $\Sigma_{n=1}^{\infty} |p_n| < \infty$. Show that there exists a function g on $[a, b]$ such that

$$\sum_{n=1}^{\infty} p_n f(x_n) = \int_a^b f \, \mathrm{d}g \qquad (f \in \mathscr{C}[a, b])$$

Exercise 23.J. Let $g : [a, b] \to \mathbb{R}$ be increasing, $A = g(a)$, $B = g(b)$. Define $h : [A, B] \to [a, b]$ by

$$h(y) := \inf\{x \in [a, b] : g(x) \geqslant y\} \qquad (y \in [A, B])$$

h is increasing and $g(h(y)) = y$ for every continuity point y of h. Show that

$$(*) \qquad \int_a^b f(g(x)) \, \mathrm{d}x = \int_A^B f \, \mathrm{d}h$$

for every continuous function f on $[A, B]$. (Prove (*) first for $f = \xi_{[y, B]}$ where y is a continuity point of h. Now use the techniques of Exercise 23.H.)

Exercise 23.K. (A special case of 23.J) Let Φ be the Cantor function. Φ maps $\mathbb{D}$ onto $[0, 1]$, so we can define $g : [0, 1] \to [0, 1]$ by

$$g(y) := \inf\{x \in \mathbb{D} : \Phi(x) = y\} \qquad (y \in [0, 1])$$

q is increasing and maps $[0, 1]$ into $\mathbb{D}$. We have

$$\int_0^1 f(g(x)) \, \mathrm{d}x = \int_0^1 f \, \mathrm{d}\Phi \qquad (f \in \mathscr{C}[0, 1])$$

In particular, $\int_0^1 f \, \mathrm{d}\Phi$ depends only on the restriction of f to $\mathbb{D}$.

Exercise 23.L. Let $g : [a, b] \to [A, B]$ be Lebesgue integrable. For $x \in [A, B]$, let $h(x)$ denote the Lebesgue measure of the set $\{t \in [a, b] : g(t) \leqslant x\}$. Then h is increasing. Let χ be the function $x \mapsto x$ $(x \in [A, B])$. The Stieltjes integral $\int_A^B \chi \, \mathrm{d}h$ exists and is equal to the Lebesgue integral $\int_a^b g$. (Thus, the Lebesgue integral of a bounded function can be expressed in terms of a Stieltjes integral and the Lebesgue measure.)

Exercise 23.M. There does not exist a function g on $[0, 1]$ such that

$$f'(0) = \int_0^1 f \, \mathrm{d}g$$

for all polynomial functions f on $[0, 1]$. (Hint. Suppose that such a g exists. Deduce from Theorems 23.6 and 23.5 that g is Riemann integrable. Show that $-h(0) = \int_0^1 h(x)(g(x) - g(1)) \, \mathrm{d}x$ for all polynomial functions h on $[0, 1]$, hence for all $h \in \mathscr{C}[0, 1]$. Derive a contradiction.)

APPENDIXES

Appendix A. The real number system

Without attempting to compress an entire analysis course into an appendix we want to present a survey of the most elementary facts concerning the real numbers and establish some terminology. We shall not take pains to be systematic about it: we only mention a few points that may not be known to every reader and define terms that otherwise might be ambiguous.

A.1. The real numbers form a set $\mathbb{R}$ that is provided with an addition, a multiplication and an ordering that render $\mathbb{R}$ a totally (= linearly) ordered field. Other fundamental properties of $\mathbb{R}$ will be set out in A.4 and A.5.

In $\mathbb{R}$ we have the subsets $\mathbb{N}$, $\mathbb{Z}$ and $\mathbb{Q}$ consisting of the natural numbers (= positive integers), the integers and the rational numbers, respectively. There is no need to go into further details here. We merely remark that, if $\mathbb{R}$ is considered as a known object, one may define $\mathbb{N}$ to be the intersection of all subsets A of $\mathbb{R}$ having the property

$$1 \in A; \quad \text{if } a \in A, \text{ then } a+1 \in A$$

After this, $\mathbb{Z}$ and $\mathbb{Q}$ are easily introduced. (Alternatively, one may start with a set $\mathbb{N}$ satisfying Peano's axioms and construct $\mathbb{R}$ out of it.)

A.2. For $a, b \in \mathbb{R}$, we set, as usual,

$$\begin{aligned}
(a, b) &:= \{x \in \mathbb{R} : a < x < b\} \\
(a, b] &:= \{x \in \mathbb{R} : a < x \leqslant b\} \\
(a, \infty) &:= \{x \in \mathbb{R} : a < x\} \\
[a, \infty) &:= \{x \in \mathbb{R} : a \leqslant x\} \\
(-\infty, \infty) &:= \mathbb{R}
\end{aligned}$$

and so on.

An *interval*, by definition, is any set of one of these types that contains infinitely many points. (Thus, $[1, 0]$, $[1, 1)$ and $[1, 1]$ are legitimate sets but not intervals.)

A.3. A subset U of $\mathbb{R}$ is *open* if for every $a \in U$ there exists an $\varepsilon > 0$ such that $(a-\varepsilon, a+\varepsilon) \subset U$. An interval is open if and only if it is of one of the types (a, b), (a, ∞), $(-\infty, b)$, $(-\infty, \infty)$.

A subset C of $\mathbb{R}$ is *closed* if its complement is open. This is the case if and only if the following is true.

$$\text{If } x_1, x_2, \ldots \in C \text{ and } x = \lim_{n\to\infty} x_n, \text{ then } x \in C$$

The closed intervals are precisely the ones of the types $[a, b]$, $[a, \infty)$, $(-\infty, b]$, $(-\infty, \infty)$.

$\varnothing$ and $\mathbb{R}$ are both open and closed.

A.4. There are many totally ordered fields that are in no way isomorphic. We single out $\mathbb{R}$ by postulating some additional properties. The first is the axiom of Archimedes:

$$\mathbb{N} \text{ is not bounded}$$

This is not enough to characterize $\mathbb{R}$, but it has far-reaching consequences. To illustrate its content we make a brief detour and show how it can be used to represent real numbers by decimals.

It is an elementary property of $\mathbb{N}$ that every nonempty subset of $\mathbb{N}$ has a smallest element. It follows easily that, if S is a nonempty subset of $\mathbb{Z}$ that is bounded above by an element of $\mathbb{Z}$, then S has a largest element. The axiom of Archimedes implies that we may drop the clause 'by an element of $\mathbb{Z}$': as soon as a set has an upper bound in $\mathbb{R}$, then it also has one in $\mathbb{Z}$. For every $x \in \mathbb{R}$ we can now define the *entire part* $[x]$ of x by

$$[x] := \max \mathbb{Z} \cap (-\infty, x]$$

Once x is given, $[x]$ is completely determined by

$$[x] \in \mathbb{Z} \text{ and } x - 1 < [x] \leqslant x$$

If $x \in \mathbb{R}$ and $n \in \mathbb{N}$, then $nx - 1 < [nx] \leqslant nx$, so $x - n^{-1} < n^{-1}[nx] \leqslant x$. As $\lim_{n\to\infty} n^{-1} = 0$ (which is another way to formulate Archimedes' axiom), it follows that

$$(*) \qquad x = \lim_{n\to\infty} \frac{[nx]}{n} \qquad (x \in \mathbb{R})$$

(Thus, every real number is a limit of rational numbers.)

Now let $p \in \mathbb{N}$, $p > 1$; put $P := \{0, 1, 2, \ldots, p-1\}$. If $x \in [0, 1]$ and $\alpha_1, \alpha_2, \ldots \in P$, we call the sequence $\alpha_1, \alpha_2, \ldots$ a *development of x to the base*

p if $x=\Sigma_{n=1}^{\infty} \alpha_n p^{-n}$. For $p=2$ and $p=10$ we also use the terms *dyadic development* and *decimal development*, respectively. Further, in this context we write

$$0.\alpha_1\alpha_2\alpha_3 \ldots$$

instead of

$$\alpha_1, \alpha_2, \ldots$$

Every element of $[0, 1]$ *has a development to the base* p. Indeed, we obtain the so-called *standard development* $0.\alpha_1\alpha_2 \ldots$ of x if, for all $n \in \mathbb{N}$, we set

$$\alpha_n := \begin{cases} p-1 & \text{if } x=1 \\ [p^n x] - p[p^{n-1}x] & \text{if } x \in [0, 1) \end{cases}$$

(It is clear that then $\alpha_n \in P$. For $x \in [0, 1)$ the identity $x=\Sigma_{n=1}^{\infty} \alpha_n p^{-n}$ follows from (*) and the fact that $\alpha_1 p^{-1} + \ldots + \alpha_N p^{-N} = p^{-N}[p^N x]$ for all N.)

A.5. So far, we have used only that $\mathbb{R}$ is a totally ordered field satisfying the axiom of Archimedes. There is another cluster of properties of $\mathbb{R}$ that we need for our theory. No doubt the reader is familiar with at least some of the following propositions.

(1) Every nonempty subset of $\mathbb{R}$ that is bounded above has a least upper bound.

(2) Every increasing sequence in $\mathbb{R}$ that is bounded above has a limit.

(3) Every bounded sequence in $\mathbb{R}$ has a convergent subsequence (Bolzano–Weierstrass).

(4) Every Cauchy sequence in $\mathbb{R}$ converges (Cauchy).

(5) If Ω is a set of open subsets of $\mathbb{R}$ whose union contains $[0, 1]$ then there exist $n \in \mathbb{N}$ and $U_1, \ldots, U_n \in \Omega$ with $[0, 1] \subset U_1 \cup \ldots \cup U_n$ (Heine–Borel).

(6) If $[a_1, b_1]$, $[a_2, b_2], \ldots$ are closed intervals in $\mathbb{R}$ such that $[a_1, b_1] \supset [a_2, b_2] \supset \ldots$, then their intersection is nonempty (Cantor).

In developing a theory on $\mathbb{R}$ one may choose any one of these propositions and declare it to be an axiom (in addition to the axioms of a totally ordered field and the one of Archimedes); it is then possible to obtain the other propositions of our list as valid theorems. To us, this is immaterial: we simply accept each of the propositions (1)–(6) as being true.

A.6. Now we have the necessary tools for building the whole theory of continuity, differentiation and integration. We are not going to repeat all the definitions, as they are well known. We make an exception for the definition of continuity, partly because there are several current equivalent definitions, partly because we shall need continuity in a more general setting than is customary.

Let X be a subset of $\mathbb{R}$ and let $f: X \to \mathbb{R}$. The following three statements about X and f imply each other. Whenever they are true we say that f is *continuous*.

(α) If $x_1, x_2, \ldots \in X$, $a \in X$ and $a = \lim_{n\to\infty} x_n$, then $f(a) = \lim_{n\to\infty} f(x_n)$.
(β) If $a \in X$ and $\varepsilon > 0$, then there exists a $\delta > 0$ such that for all $x \in X$ with $|x-a| < \delta$ one has $|f(x) - f(a)| < \varepsilon$.
(γ) If $U \subset \mathbb{R}$ is open, then there exists an open $W \subset \mathbb{R}$ with $f^{-1}(U) = X \cap W$.

In the remaining part of this appendix we introduce a bit of terminology and point out some properties of subsets of $\mathbb{R}$ that are quite elementary but may not be sufficiently well known.

A.7. First, a few terms.

Let X be a subset of $\mathbb{R}$. The *closure* of X (denoted $\bar{X}$) is the smallest closed subset of $\mathbb{R}$ that contains X; it is equal to the set $\{a \in \mathbb{R} :$ there exist $x_1, x_2, \ldots \in X$ with $a = \lim_{n\to\infty} x_n\}$. The *interior* of X (denoted X°) is the largest open subset of $\mathbb{R}$ that is contained in X; one has $X^\circ = \{a \in \mathbb{R} :$ there exists a positive δ with $(a-\delta, a+\delta) \subset X\}$.

If $X \subset Y \subset \mathbb{R}$, then X is said to be *dense in* Y if $Y \subset \bar{X}$, i.e. if for every $y \in Y$ and every $\delta > 0$ there is an $x \in X$ with $|x-y| < \delta$. In particular, X is dense in $\mathbb{R}$ if and only if $\bar{X} = \mathbb{R}$; if and only if every interval intersects X. (Example: $\mathbb{Q}$ is dense in $\mathbb{R}$.)

We call a subset A of $\mathbb{R}$ *connected* if it has the following property. If $x, y \in A$ and $x \leqslant y$, then $[x, y] \subset A$.

A.8. THEOREM. *The connected subsets of* $\mathbb{R}$ *are just* $\varnothing$, *the singleton sets and the intervals.*

Proof. It is clear that all sets of the three types mentioned in the theorem are connected.

Conversely, let $A \subset \mathbb{R}$ be connected, $A \neq \varnothing$. Take $a \in A$. Define $A_+ := A \cap [a, \infty)$, $A_- := A \cap (-\infty, a]$. If A is not bounded above, there exist $x_1, x_2, \ldots \in A_+$ with $\lim_{n\to\infty} x_n = \infty$; then $A \cap [a, \infty) \supset \bigcup_{n\in\mathbb{N}} [a, x_n] = [a, \infty)$, so $A_+ = [a, \infty)$. If A is bounded above, let $q := \sup A$; then $A_+ \subset [a, q]$. If $q \in A$, then $[a, q] \subset A$, so $A_+ = [a, q]$. If $q \notin A$, then $q > a$ and there exist $x_1, x_2, \ldots \in A_+$ with $\lim_{n\to\infty} x_n = q$; in this case we obtain $A_+ = A \cap [a, \infty) \supset \bigcup_{n\in\mathbb{N}} [a, x_n] = [a, q)$, so $A_+ = [a, q)$.

We see that there are only three possibilities: either $A_+ = [a, \infty)$ or $A_+ = [a, q]$ for some $q \geqslant a$ or $A_+ = [a, q)$ for some $q > a$. It is obvious how one can analyse A_- in a similar way and complete the proof.

A.9. THEOREM. *Let $U \subset \mathbb{R}$ be open.*
(i) *U is a union of countably many intervals (a, b) with $a, b \in \mathbb{Q}$.*
(ii) *U can also be written as a union of countably many pairwise disjoint open intervals. More explicitly: the formula*

$$x \equiv y \text{ if there is an interval } I \text{ with } \{x, y\} \subset I \subset U$$

defines an equivalence relation $\equiv$ in U. The equivalence classes (the 'components' of U) are open intervals. U has only countably many components.
Proof. (i) Let Ω be the set of all intervals (a, b) with $a, b \in \mathbb{Q}$ and $(a, b) \subset U$. As $\mathbb{Q}$ is countable, so are $\mathbb{Q}^2$ and, consequently, Ω. It is easy to see that the union of all elements of Ω is exactly U.
(ii) It is elementary that $\equiv$ is an equivalence relation. For $x \in U$, let U_x denote the equivalence class that contains x. If I is an interval and $I \subset U$, then I is contained in one equivalence class. It follows that the equivalence classes are connected, hence, that they are intervals. If $a \in U$, there is an $\varepsilon > 0$ with $(a - \varepsilon, a + \varepsilon) \subset U$; then $(a - \varepsilon, a + \varepsilon) \subset U_a$. Consequently, each equivalence class is open. Finally, $U \cap \mathbb{Q}$ is countable and, as every interval contains a rational number, $x \mapsto U_x$ maps $U \cap \mathbb{Q}$ *onto* the set of all equivalence classes. Therefore, the latter set is countable.

Appendix B. Cardinalities

B.1. DEFINITION. Let X, Y be sets. We write

$$X < Y \text{ if there exists an injection } X \to Y,$$
$$X \approx Y \text{ if there exists a bijection } X \to Y.$$

There are some obvious consequences: if X, Y, Z are sets and $X < Y$, $Y < Z$, then $X < Z$; if $X \approx Y$, then $Y \approx X$; if $X \subset Y$, then $X < Y$; etc.

B.2. EXAMPLES.
(i) $(0, \infty) \approx \mathbb{R}$. ($x \mapsto \log x$ is a bijection.)
(ii) $(0, \infty) \approx (0, 1)$. ($x \mapsto e^{-x}$ is a bijection.)
(iii) $(0, \infty) \approx [0, \infty)$. (Set $f(x) := x$ if $x \notin \mathbb{N}$, $f(x) := x - 1$ if $x \in \mathbb{N}$.)
(iv) $[0, 1) \approx (0, 1)$, since $[0, 1) \approx [0, \infty) \approx (0, \infty) \approx (0, 1)$.
(v) $[0, 1] \approx (0, 1)$, since $[0, 1] = [0, \frac{1}{2}] \cup (\frac{1}{2}, 1] \approx [0, \frac{1}{2}] \cup (\frac{1}{2}, 1) = [0, 1) \approx (0, 1)$.
(vi) $\mathbb{Z} \approx \mathbb{N}$. (Define $f(n) := 2n$ if $n \in \mathbb{N}$, $f(n) := 2|n| + 1$ if $n \in \mathbb{Z} \setminus \mathbb{N}$. Then f is a bijection $\mathbb{Z} \to \mathbb{N}$.)
(vii) $\mathbb{N} \times \mathbb{N} \approx \mathbb{N}$. ($(n, m) \mapsto 2^n(2m-1)$ is a bijection $\mathbb{N}^2 \to \mathbb{N}$.)
(viii) If $X < \mathbb{N}$ and X is not a finite set, then $X \approx \mathbb{N}$ (Without restriction, let $X \subset \mathbb{N}$. Then we can define a bijection $f : \mathbb{N} \to X$ by setting $f(1) := \min X$, $f(n+1) := \min X \setminus \{f(1), \ldots, f(n)\}$ $(n \in \mathbb{N})$.)

B.3. LEMMA. *Let X, Y be sets, $X \neq \varnothing$. Then $X < Y$ if and only if there exists a surjection $Y \to X$.*
Proof. Let $a \in X$. If $f : X \to Y$ is an injection we can define a surjection $g : Y \to X$ by

$$g(y) := \begin{cases} f^{-1}(y) & \text{if } y \in f(X) \\ a & \text{if } y \in Y \setminus f(X) \end{cases}$$

Conversely, if a surjection $h : Y \to X$ is given, then by the axiom of choice for each $x \in X$ we can select an element $k(x)$ of $h^{-1}(\{x\})$: then k is an injection $X \to Y$.

B.4. THEOREM. (Cantor) *Let X be any set, $\mathcal{P}(X)$ the set of all subsets of X. Then the formula $\mathcal{P}(X) < X$ is false.*
Proof. Suppose $\mathcal{P}(X) < X$. By Lemma B.3 there exists a surjection f of X onto $\mathcal{P}(X)$. Put $A := \{x \in X : x \notin f(x)\}$. For all $x \in X$ the formulas $x \in A$ and $x \notin f(x)$ are equivalent, so $A \neq f(x)$. Then f is not surjective, a contradiction.

B.5. Trivially, if $X \approx Y$, then $X < Y$ and $Y < X$. The converse holds also:

THEOREM. (Cantor, Schroeder, Bernstein) *Let X, Y be sets for which* $X < Y$ and $Y < X$. *Then* $X \approx Y$.

Proof. Let $f : X \to Y$ and $g : Y \to X$ be injections. Consider the subsets $X_1, X_2, \ldots$ of X and $Y_1, Y_2, \ldots$ of Y defined by

$$\left.\begin{array}{ll} X_1 := X, & X_{n+1} := (g \circ f)(X_n) \\ Y_1 := Y, & Y_{n+1} := (f \circ g)(Y_n) \end{array}\right\} \quad (n \in \mathbb{N})$$

We have

$$(1) \qquad \begin{array}{l} X = X_1 \supset g(Y_1) \supset X_2 \supset g(Y_2) \supset X_3 \supset \ldots \\ Y = Y_1 \supset f(X_1) \supset Y_2 \supset f(X_2) \supset Y_3 \supset \ldots \end{array}$$

For each n, f induces a bijection of $X_n \setminus g(Y_n)$ onto $f(X_n) \setminus Y_{n+1}$ and g induces a bijection of $Y_n \setminus f(X_n)$ onto $g(Y_n) \setminus X_{n+1}$; hence, for all n,

$$(2) \qquad X_n \setminus X_{n+1} \approx Y_n \setminus Y_{n+1}$$

Furthermore, f yields a bijection of $\bigcap_{n \in \mathbb{N}} X_n$ onto $\bigcap_{n \in \mathbb{N}} f(X_n)$ and the latter set is $\bigcap_{n \in \mathbb{N}} Y_n$, so

$$(3) \qquad \bigcap_{n \in \mathbb{N}} X_n \approx \bigcap_{n \in \mathbb{N}} Y_n$$

Now combine (1), (2) and (3).

B.6. COROLLARY. *Let $\mathscr{P}(\mathbb{N})$ be the set of all subsets of $\mathbb{N}$. Then*

$$\mathscr{P}(\mathbb{N}) \approx [0, 1] \approx \mathbb{R}$$

Proof. The following formula defines an injection $\mathscr{P}(\mathbb{N}) \to [0, 1]$:

$$A \mapsto \sum_{n=1}^{\infty} \xi_A(n) 3^{-n}$$

so $\mathscr{P}(\mathbb{N}) < [0, 1]$. On the other hand,

$$A \mapsto \sum_{n=1}^{\infty} \xi_A(n) 2^{-n}$$

is a *surjection* $\mathscr{P}(\mathbb{N}) \to [0, 1]$, so $[0, 1] < \mathscr{P}(\mathbb{N})$. The corollary follows from Theorem B.5 and Examples B.2.

B.7. DEFINITION. Let X be a set. We call X *countable* if $X < \mathbb{N}$; we say that X *has cardinality* $\leqslant c$ if $X < \mathbb{R}$, and that X *has cardinality* c (also called *the cardinality of the continuum*) if $X \approx \mathbb{R}$. If X is countable and infinite, then $X \approx \mathbb{N}$ (B.2(viii)). A bijection $\mathbb{N} \to X$ is an *enumeration* of X.

B.8. EXAMPLES.

(i) If X, Y are countable sets, then $X \cup Y$ and $X \times Y$ are countable. (For $X \times Y$, note that by Example B.2(vii), $\mathbb{N}^2$ is countable.)

(ii) $\mathbb{Q}$ is countable. ($(m, n)\mapsto m/n$ is a surjection $\mathbb{Z}\times\mathbb{N}\to\mathbb{Q}$, so, by Lemma B.3 and Examples B.2, $\mathbb{Q}<\mathbb{Z}\times\mathbb{N}\approx\mathbb{N}\times\mathbb{N}\approx\mathbb{N}$.)
(iii) $\mathscr{P}(\mathbb{N})$, $[0, 1]$ and $\mathbb{R}$ are not countable (Theorem B.4, Corollary B.6).
(iv) Every interval has cardinality c. (Every interval I contains an open interval (a, b); then $(a, b)<I<\mathbb{R}$. Now apply Theorem B.5.)
(v) The set $\mathscr{C}(I)$ of all continuous functions on an interval I has cardinality c. (All constant functions are continuous, so $\mathbb{R}<\mathscr{C}(I)$. On the other hand, if for every $g\in\mathscr{C}(I)$ we set $F(g):=\{(x, y)\in\mathbb{Q}^2 : x\in I,\ y\leqslant g(x)\}$, then we have an injection F of $\mathscr{C}(I)$ into the set $\mathscr{P}(\mathbb{Q}^2)$ of all subsets of $\mathbb{Q}^2$. As $\mathbb{Q}<\mathbb{N}$, we see that $\mathbb{Q}^2<\mathbb{N}^2\approx\mathbb{N}$, and therefore $\mathscr{C}(I)<\mathscr{P}(\mathbb{Q}^2)<\mathscr{P}(\mathbb{N})\approx\mathbb{R}$. By Theorem B.5. $\mathscr{C}(I)\approx\mathbb{R}$.)
(vi) $\mathbb{R}^2\approx\mathbb{R}$. ($\mathbb{R}\approx\mathscr{P}(\mathbb{N})\approx\mathscr{P}(\mathbb{Z}\setminus\mathbb{N})$, so $\mathbb{R}\times\mathbb{R}\approx\mathscr{P}(\mathbb{N})\times\mathscr{P}(\mathbb{Z}\setminus\mathbb{N})\approx\mathscr{P}(\mathbb{Z})$ $\approx\mathscr{P}(\mathbb{N})\approx\mathbb{R}$.)
(vii) The set $\mathbb{R}^{\mathbb{R}}$ of *all* functions $\mathbb{R}\to\mathbb{R}$ does not have cardinality $\leqslant c$. (By assigning to every subset of $\mathbb{R}$ its characteristic function we obtain an injection $\mathscr{P}(\mathbb{R})\to\mathbb{R}^{\mathbb{R}}$. Apply Theorem B.4.)

B.9. THEOREM. *Let $(Y_x)_{x\in X}$ be a family of sets.*
(i) *If X and each Y_x are countable, then $\bigcup_{x\in X} Y_x$ is countable.*
(ii) *If X and each Y_x have cardinality $\leqslant c$, then $\bigcup_{x\in X} Y_x$ has cardinality $\leqslant c$.*
Proof. Assume that neither X nor any Y_x is empty. In case (i) we have surjections $f : \mathbb{N}\to X$ and $g_x : \mathbb{N}\to Y_x$ $(x\in X)$. Combining these we can make a surjection $\mathbb{N}\times\mathbb{N}\to\bigcup_{x\in X} Y_x$:

$$(n, m)\mapsto g_{f(n)}(m) \qquad (n, m\in\mathbb{N})$$

Hence, $\bigcup_{x\in X} Y_x<\mathbb{N}^2\approx\mathbb{N}$ (Example B.2(vii)).
The proof of (ii) is quite similar, using Example B.8(vi) instead of B.2(vii).

B.10. THEOREM. *Let X be a set with cardinality $\leqslant c$ and let $X^{\mathbb{N}}$ be the set of all sequences $(x_1, x_2, \ldots)$ of elements of X. Then $X^{\mathbb{N}}$ has cardinality $\leqslant c$.*
Proof. Without restriction, let $X\subset\mathbb{R}$. The following formula defines an injection $f : X^{\mathbb{N}}\to\mathscr{P}(\mathbb{N}\times\mathbb{Q})$:

$$f(x_1, x_2, \ldots):=\{(n, q)\in\mathbb{N}\times\mathbb{Q} : q<x_n\} \qquad (x_1, x_2, \ldots\in X)$$

Thus, $X^{\mathbb{N}}<\mathscr{P}(\mathbb{N}\times\mathbb{Q})\approx\mathscr{P}(\mathbb{N}\times\mathbb{N})\approx\mathscr{P}(\mathbb{N})\approx\mathbb{R}$.

Appendix C. An uncountable well-ordered set: a characterization of the functions of the first class of Baire

C.1. In Section 17 we have seen that every well-ordered subset D of $\mathbb{Q}$ leads to a hypersequence $\mathscr{A}_D$; for every $x \in D$ we made a set $\mathscr{A}_D(x)$ of functions. $\mathscr{B}_D := \bigcup_{x\in D} \mathscr{A}_D(x)$ may be viewed as the set of all functions that can be constructed 'inductively in D steps'. For each D, $\mathscr{B}_D$ is a proper subset of $\mathscr{B}$, the set of Borel measurable functions, but the union of all these sets $\mathscr{B}_D$ fills the whole of $\mathscr{B}$.

It is possible to combine all hypersequences in one 'super-hypersequence' Ω that generates $\mathscr{B}$. To do so, we must work on a higher level of abstraction than we have done so far. We assume the reader to be familiar with the notions of a totally (=linearly) ordered set and of an order isomorphism between two totally ordered sets.

C.2. Consider

$$\Omega := \{\mathscr{B}_D : D \text{ is a well-ordered subset of } \mathbb{Q}\}$$

The elements of Ω are certain subsets of $\mathscr{B}$. In Theorem 17.11 we saw that the union of all elements of Ω is $\mathscr{B}$.

C.3. It is useful to observe that every element of Ω can be written as $\mathscr{B}_D$ for some well-ordered subset D of $\mathbb{Q}$ that has a largest element. Indeed, let $D \subset \mathbb{Q}$ be well ordered and suppose that D has no largest element. Let D' and a be as in Exercise 17.E(ii). Then D' has a largest element, viz. a. By Exercise 17.F we have

$$\mathscr{B}_D = \bigcup_{x\in D} \mathscr{A}_D(x) = \bigcup_{\substack{y\in D' \\ y\neq a}} \mathscr{A}_{D'}(y)$$

The latter set is just $\mathscr{A}_{D'}(a)$, since in D' a is not the successor of any element. But $\mathscr{A}_{D'}(a) = \mathscr{B}_{D'}$, so we obtain $\mathscr{B}_D = \mathscr{B}_{D'}$.

C.4. If D_1, D_2 are well-ordered subsets of $\mathbb{Q}$ and D_1 is isomorphic to an initial interval of D_2, then by Exercise 17.F, $\mathscr{B}_{D_1} \subset \mathscr{B}_{D_2}$. Hence, (apply Theorem 17.7) Ω is totally ordered by inclusion.

C.5. Let us call a subset Ω_0 of Ω *an initial interval* of Ω if every element of Ω that is contained in an element of Ω_0 is itself an element of Ω_0.

As in the proof of Theorem 17.9, if $D \subset \mathbb{Q}$ is well-ordered and $x \in D$, we set $\bar{D}^x := \{y \in D : y \leqslant x\}$. Then, by Exercise 17.F, $\mathscr{A}_{\bar{D}^x}$ is the restriction of

$\mathscr{A}_D$ to $\bar{D}^x$, so that

(*) $$\mathscr{A}_D(x) = \mathscr{A}_{\bar{D}^x}(x) = \mathscr{B}_{D^x}$$

Consequently, if $D \subset \mathbb{Q}$ is well-ordered and $x \in D$, then $\mathscr{A}_D(x) \in \Omega$.

The crucial point is now that for every well-ordered $D \subset \mathbb{Q}$, $\{\mathscr{A}_D(x) : x \in D\}$ is an initial interval of Ω. To prove this, let $D \subset \mathbb{Q}$ be well-ordered, $x \in D$ and let D_1 be a well-ordered subset of $\mathbb{Q}$ having a largest element (C.3) and such that $\mathscr{B}_{D_1} \subset \mathscr{A}_D(x)$: we want to find a $y \in D$ with $\mathscr{B}_{D_1} = \mathscr{A}_D(y)$. We may assume that the inclusion $\mathscr{B}_{D_1} \subset \mathscr{A}_D(x)$ is proper. Then, in view of (*), $\mathscr{B}_{D_1} \not\supset \mathscr{A}_D(x) = \mathscr{B}_{D^x}$. This means that $\bar{D}^x$ is not isomorphic to any initial interval of D_1 (see C.4). Applying Theorem 17.7 and observing that D_1 has a largest element, we see that there must exist a $y \in \bar{D}^x \subset D$ such that D_1 is isomorphic to $\bar{D}^y$. Then (C.4 and (*)) $\mathscr{B}_{D_1} = \mathscr{B}_{\bar{D}^y} = \mathscr{A}_D(y)$.

From the above and from the last part of Theorem 17.10 we infer: if $D \subset \mathbb{Q}$ is well-ordered and has a largest element, then $x \mapsto \mathscr{A}_D(x)$ is an order isomorphism of D onto the initial interval of Ω whose largest element is $\mathscr{B}_D$.

(Thus, roughly speaking, our Ω is the union of all hypersequences. The scrupulous reader who wants to interpret Ω as a 'sequence' may consider the identity map $\Omega \to \Omega$.)

C.6. We can now prove that, under inclusion, Ω is well-ordered, i.e. every nonempty subset of Ω has a smallest element. Indeed, let $\Omega_0 \subset \Omega$, $\Omega_0 \neq \varnothing$. There is a well-ordered $D \subset \mathbb{Q}$ with a largest element and such that $\mathscr{B}_D \in \Omega_0$. The set $\{x \in D : \mathscr{A}_D(x) \in \Omega_0\}$ has a smallest element, a, say: then $\mathscr{A}_D(a)$ is easily seen to be the smallest element of Ω_0.

C.7. Let $\mathscr{F} \in \Omega$. We prove that $\mathscr{F}$ has a successor (or, equivalently, that $\mathscr{F}$ is not the largest element of Ω). In fact, let $\mathscr{F} = \mathscr{B}_D$ where (C.3) $D \subset \mathbb{Q}$ is well-ordered and has a largest element, a. Set $E := D \cup \{a+1\}$. Then in E, $a+1$ is the successor of a. It follows that in Ω, $\mathscr{A}_E(a+1)$ is the successor of $\mathscr{A}_E(a)$. Now $\mathscr{A}_E(a) = \mathscr{A}_D(a) = \mathscr{B}_D$ and, using the asterisk as in the definition of 'hypersequence' in Section 17, we have $\mathscr{A}_E(a+1) = \mathscr{A}_E(a)^*$. Thus: for every $\mathscr{F} \in \Omega$, $\mathscr{F}^*$ is the successor of $\mathscr{F}$.

C.8. If $D_1, D_2, \ldots$ are well-ordered subsets of $\mathbb{Q}$ and if D is as in Exercise 17.E(iii), then by Exercise 17.F we have $\mathscr{B}_{D_n} \subset \mathscr{B}_D$ for all n. Hence, for all $\mathscr{F}_1, \mathscr{F}_2, \ldots \in \Omega$ there is an $\mathscr{F} \in \Omega$ with $\mathscr{F}_n \subset \mathscr{F}$ for all n; then $\mathscr{F}^* \in \Omega$ and $\mathscr{F}^* \neq \mathscr{F}_n (n \in \mathbb{N})$. It follows that Ω is uncountable.

C.9. From here on we indicate the elements of Ω by Greek letters $\alpha, \beta, \ldots$ and the inclusion relation between them by $\leqslant$. The successor of $\alpha \in \Omega$ is α^*. The following properties of the ordered set Ω will be of importance to us.

(1) *Ω is well-ordered* (C.6).

(2) *If $\alpha_0 \in \Omega$ then $\{\alpha \in \Omega : \alpha \leqslant \alpha_0\}$ is countable* (C.5).

(3) *Ω is uncountable* (C.8).

From (1), (2) and (3) we derive:

(4) *If $\Omega_0 \subset \Omega$ is countable, then $\Omega_0 \subset \{\alpha \in \Omega : \alpha \leqslant \alpha_0\}$ for some $\alpha_0 \in \Omega$.* (By (2), $\Omega_1 := \{\beta \in \Omega :$ there is an $\alpha \in \Omega_0$ with $\beta \leqslant \alpha\}$ is countable. By (3), $\Omega_1 \neq \Omega$. Take any $\alpha_0 \in \Omega \setminus \Omega_1$.)

These are the only properties of Ω that we shall use in the next few pages: we shall no longer be interested in the fact that each element of Ω is actually a set of functions on $\mathbb{R}$.

C.10. There are other ways to make an ordered set Ω with the properties (1)–(3): see, for example Sierpiński (1956). If we had constructed Ω differently we could have simplified the contents of Section 17 considerably, as the following exercise may suggest.

**Exercise.* Let X be any set, $\mathscr{F}$ any set of functions $X \to \mathbb{R}$.

(i) There exists exactly one map that assigns to every element α of Ω a set $\mathscr{F}_\alpha$ of functions $X \to \mathbb{R}$ such that

1. if α_0 is the smallest element of Ω, then $\mathscr{F}_{\alpha_0} = \mathscr{F}$;
2. for every $\alpha \in \Omega$, $\mathscr{F}_{\alpha^*} = \mathscr{F}_\alpha{}^*$;
3. if $\alpha \in \Omega$ and if α is neither the successor of any element of Ω nor the smallest element of Ω, then $\mathscr{F}_\alpha = \bigcup_{\beta < \alpha} \mathscr{F}_\beta$.

(The difficulties here are often underestimated. The simplest procedure seems to be this: for $\alpha \in \Omega$ define a notion of 'hypersequence of length α' that ties in with the given problem; show that for each α there exists at most one hypersequence of length α; next, show that there really does exist one; finally, combine all hypersequences.)

(ii) For $\alpha \in \Omega$, let $\mathscr{F}_\alpha$ be as indicated in (i). Then $\bigcup_{\alpha \in \Omega} \mathscr{F}_\alpha$ is the smallest set of functions $X \to \mathbb{R}$ that contains $\mathscr{F}$ and is closed for limits. (To prove that $\bigcup_{\alpha \in \Omega} \mathscr{F}_\alpha$ is closed for limits, use property (4) of Ω.)

To short-circuit Section 17, we apply this exercise to $\mathscr{F} = \mathscr{C}$ to obtain sets of functions $\mathscr{C}_\alpha$ for $\alpha \in \Omega$. By (ii), the union of all these sets $\mathscr{C}_\alpha$ is $\mathscr{B}$. It follows from property (2) of Ω and from Lemma 16.9 that $\{\alpha \in \Omega : \mathscr{C}_\alpha$ has a catalogue$\}$ has no smallest element, hence must be empty. Then $\mathscr{C}_\alpha \neq \mathscr{B}$ for all α.

C.11. We shall use Ω and a construction similar to the one in C.10 to prove Baire's characterization of the functions of the first class. (The theorem has already been mentioned in Section 11.)

THEOREM. (Baire). *Let $f:\mathbb{R}\to\mathbb{R}$. Then f belongs to the first class of Baire if and only if for every nonempty closed subset C of $\mathbb{R}$ the restriction of f to C has a continuity point.*

The proof is laborious but ingenious and interesting. Half of the theorem we leave to the reader:

**Exercise.* If $f\in\mathscr{B}^1(\mathbb{R})$ and if $C\subset\mathbb{R}$ is closed, nonempty, then the restriction of f to C has a continuity point. (See the proof of Theorem 11.4.)

For a proof of the converse we need three exercises.

*C.12. *Exercise.* Suppose that for every closed subset C of $\mathbb{R}$ there is given a closed set C^* which is contained in C. Then there is exactly one way to associate with every $\alpha\in\Omega$ a closed set $C_\alpha\subset\mathbb{R}$ such that

1. if α_0 is the smallest element of Ω, then $C_{\alpha_0}=\mathbb{R}$;
2. for every $\alpha\in\Omega$, $C_{\alpha^*}=C_\alpha{}^*$;
3. if $\alpha\in\Omega$ and if α is neither the smallest element of Ω nor the successor of any element of Ω, then $C_\alpha=\bigcap_{\beta<\alpha}C_\beta$.

Moreover, if the C_α are chosen in this way, then $C_\alpha\supset C_\beta$ as soon as $\alpha\leqslant\beta$. Finally, the sets $C_\alpha\setminus C_{\alpha^*}$ $(\alpha\in\Omega)$ are pairwise disjoint and their union is $\mathbb{R}\setminus\bigcap_{\alpha\in\Omega}C_\alpha$.

The following result is more surprising.

*C.13. *Exercise.* Suppose that for each $\alpha\in\Omega$ there is given a closed set $C_\alpha\subset\mathbb{R}$ and that $C_\alpha\supset C_\beta$ as soon as $\alpha\leqslant\beta$. Then there exists an $\alpha\in\Omega$ with $C_\alpha=C_{\alpha^*}$. (Hint. Otherwise, for each α there exists an interval J_α with rational end points such that $J_\alpha\cap C_\alpha\neq\varnothing$, $J_\alpha\cap C_{\alpha^*}=\varnothing$. Show that now $J_\alpha\neq J_\beta$ for $\alpha\neq\beta$.)

The third exercise has nothing to do with Ω.

*C.14. *Exercise.* Let $f:\mathbb{R}\to\mathbb{R}$. Suppose that for all $a,b\in\mathbb{R}$ with $a<b$ there exists a subset X of $\mathbb{R}$ which is both an F_σ and a G_δ, such that $f>a$ on X and $f<b$ on $\mathbb{R}\setminus X$. Then $f\in\mathscr{B}^1$. (Hint. Without restriction, let $0\leqslant f\leqslant 1$. Take $n\in\mathbb{N}$. For $i=1,\ldots,n$ find an $X_i\subset\mathbb{R}$ that is both an F_σ and a G_δ while $f>(i-1)/n$ on X_i and $f<i/n$ on $\mathbb{R}\setminus X_i$. Set $g:=\Sigma_{i=1}^n(1/n)\xi_{X_i}$. Show that $g\in\mathscr{B}^1$ and that $g-1/n\leqslant f\leqslant g+1/n$. Now apply Theorem 11.7.)

C.15. We return to the proof of the theorem of Baire. Assume that for every nonempty closed $C\subset\mathbb{R}$, the restriction of f to C has a continuity point. Let $a,b\in\mathbb{R}$, $a<b$: we construct a set X with the properties mentioned in Exercise C.14. For every closed set $C\subset\mathbb{R}$ we choose a closed set $C^*\subset C$ as follows. If $C=\varnothing$ we have little choice: let $C^*:=\varnothing$. Otherwise, first choose a continuity point p of the restriction of f to C. If $f(p)>a$ there is an open interval I containing p and such that $f>a$ on I; otherwise, $f(p)<b$

and we have an open interval I with $p \in I$ and with $f<b$ on I. In either case we set $C^* := C \setminus I$: then C^* is a closed proper subset of C and we either have $f(x)>a$ for all $x \in C \setminus C^*$ or $f(x)<b$ for all $x \in C \setminus C^*$.

We have obtained an operation $C \mapsto C^*$ to which we can apply Exercise C.12. The result is a map assigning to each element α of Ω a closed subset C_α of $\mathbb{R}$. In the same exercise we saw that $C_\alpha \supset C_\beta$ for $\alpha \leqslant \beta$. Hence, Exercise C.13 tells us that there is a $\gamma \in \Omega$ with $C_\gamma = C_\gamma{}^* = C_{\gamma^*}$, i.e. (by our choice of $*$) $C_\gamma = \varnothing$.

Then $C_\alpha = \varnothing$ for all $\alpha \geqslant \gamma$. By the last part of Exercise C.12 the sets $C_\alpha \setminus C_\alpha{}^*$ ($\alpha \in \Omega$, $\alpha < \gamma$) are pairwise disjoint and cover $\mathbb{R}$. Moreover, for each α we either have $f>a$ on all of $C_\alpha \setminus C_\alpha{}^*$ or $f<b$ on all of $C_\alpha \setminus C_\alpha{}^*$. Let $\Omega_1 := \{\alpha \in \Omega : \alpha < \gamma;\ f>a \text{ on } C_\alpha \setminus C_\alpha{}^*\}$, $\Omega_2 := \{\alpha \in \Omega : \alpha < \gamma;\ \alpha \notin \Omega_1\}$ and let X be the union of all sets $C_\alpha \setminus C_\alpha{}^*$ with $\alpha \in \Omega_1$. Then $\mathbb{R} \setminus X$ is just the union of all sets $C_\alpha \setminus C_\alpha{}^*$ for which $\alpha \in \Omega_2$. Further, $f>a$ on X and $f<b$ on $\mathbb{R} \setminus X$. As C_α and $C_\alpha{}^*$ are closed, each of the sets $C_\alpha \setminus C_\alpha{}^*$ is an F_σ. But Ω_1 and Ω_2 are countable (property (2) of Ω!). Hence, both X and $\mathbb{R} \setminus X$ are F_σ-sets, i.e. X is an F_σ and also a G_δ. Now we apply Exercise C.14 and conclude that $f \in \mathscr{B}^1$.

Appendix D. An elementary proof of Lebesgue's density theorem

The proof we gave of the density theorem (21.29) is short but requires much preparatory work. A direct proof was recently given by L. Zajícek (*Am. Math. Monthly* **86** (1979), 297–8). In this appendix we present another proof, due to W. Sierpiński (*Fund. Math.* **4** (1923), 167–71) which is at least as simple and can also be applied to nonmeasurable sets.

We make use of the outer Lebesgue measure λ^* introduced in Definition 21.1. If $E \subset \mathbb{R}$ and $a \in \mathbb{R}$, we call a an *exterior density point* of E if

$$\lim_{r \downarrow 0} \frac{1}{2r} \lambda^*(E \cap (a-r, a+r)) = 1$$

If E is Lebesgue measurable, the exterior density points of E are just its density points. Thus, the following theorem implies the Lebesgue density theorem.

THEOREM. *Let E be a subset of $\mathbb{R}$. Then almost every element of E is an exterior density point of E.*

Proof. Without restriction, assume E to be bounded. Let $0<t<1$. It suffices to prove that the set

$$A := \left\{ a \in E : \liminf_{r \downarrow 0} \frac{1}{2r} \lambda^*(E \cap (a-r, a+r)) < 1-t \right\}$$

is null. Take $\varepsilon > 0$: we are done if we can prove that $\lambda^*(A) < \varepsilon(1+3t^{-1})$.

There exists an open set $U \subset \mathbb{R}$ with $A \subset U$ and $\lambda(U) < \lambda^*(A) + \varepsilon$. Then for every measurable subset T of U we have $\lambda^*(A) \leqslant \lambda^*(A \cap T) + \lambda^*(U \setminus T)$ $= \lambda^*(A \cap T) + \lambda(U) - \lambda(T)$, so

$$\lambda^*(A \cap T) \geqslant \lambda(T) - \varepsilon \tag{1}$$

For every $a \in A$ we can choose an open interval S for which $a \in S \subset U$ and $\lambda^*(E \cap S) < (1-t)\lambda(S)$; and we can arrange for S to have rational end points. (If necessary, make S a little shorter.) It follows that there exist open intervals $S_1, S_2, \ldots,$ covering A and contained in U, and such that

$$\lambda^*(E \cap S_n) < (1-t)\lambda(S_n) \qquad (n = 1, 2, \ldots) \tag{2}$$

For each n, let S_n be the interval $(x_n - \delta_n, x_n + \delta_n)$ and define $S'_n :=$ $(x_n - 3\delta_n, x_n + 3\delta_n)$.

As $\lambda^*(A) \leqslant \lambda(\bigcup_{n \in \mathbb{N}} S_n)$, there exists an $N \in \mathbb{N}$ with $\lambda^*(A) - \varepsilon < \lambda(\bigcup_{n \leqslant N} S_n)$. Assume $\delta_1 \geqslant \delta_2 \geqslant \ldots \geqslant \delta_N$.

Define a subset I of $\{1, \ldots, N\}$ such that for every $n \in \{1, \ldots, N\}$, $(S_i)_{i \in I \cap \{1, \ldots, n\}}$ is a maximal disjoint subfamily of $(S_i)_{i \in \{1, \ldots, n\}}$. (This can be done by successively constructing $I \cap \{1\}$, $I \cap \{1, 2\}$, $I \cap \{1, 2, 3\}, \ldots$)

For every $n \in \{1, \ldots, N\}$ there is an $i \in I \cap \{1, \ldots, n\}$ for which $S_n \cap S_i \neq \emptyset$. Then $\delta_i \geq \delta_n$, whence $S_n \subset S_i'$. Thus,

$$S_1 \cup \ldots \cup S_N \subset \bigcup_{i \in I} S_i'$$

and therefore, setting $T := \bigcup_{i \in I} S_i$,

$$\lambda^*(A) - \varepsilon \leq \lambda\left(\bigcup_{i \in I} S_i'\right) \leq \sum_{i \in I} \lambda(S_i') = 3 \sum_{i \in I} \lambda(S_i) = 3\lambda(T) \tag{3}$$

On the other hand, by (2) we have

$$\lambda^*(A \cap T) \leq \sum_{i \in I} \lambda^*(A \cap S_i) \leq \sum_{i \in I} \lambda^*(E \cap S_i) \leq \sum_{i \in I} (1-t)\lambda(S_i) = (1-t)\lambda(T) \tag{4}$$

By combining (4) with (1) (note that $T \subset U$!) we obtain $\lambda(T) \leq \varepsilon t^{-1}$; together with (3) this yields the desired inequality $\lambda^*(A) < \varepsilon + 3\varepsilon t^{-1}$.

FURTHER READING

THE CLASSICS

Baire, R. (1905) *Leçons sur les fonctions discontinues*, Gauthier-Villars, Paris.

Borel, E. (1914) *Leçons sur la théorie des fonctions*, Gauthier-Villars, Paris.

Hobson, E. W. (1907) *The Theory of Functions of a Real Variable*, Cambridge University Press.

Lusin, N. (1972) *Les ensembles analytiques*, Chelsea Publishing Company, New York.

Saks, S. (no date) *Theory of the Integral*, Hafner Publishing Company, New York.

Sierpiński, W. (1956) *Hypothèse du continu*, Chelsea Publishing Company, New York.

THE MODERNS

Bruckner, A. M. (1978) *Differentiation of Real Functions*, Lecture Notes in Mathematics 659, Springer-Verlag, New York–Heidelberg–Berlin.

Gelbaum, B. R. & Olmsted, J. M. H. (1964) *Counterexamples in Analysis*, Holden-Day, Inc., San Francisco.

Hawkins, T. (1970) *Lebesgue's Theory of Integration; its Origins and Development*, University of Wisconsin Press, Madison (Wisc.).

Hewitt, E. & Stromberg, K. (1965) *Real and Abstract Analysis*, Springer-Verlag, New York–Heidelberg–Berlin.

Oxtoby J. C. (1971) *Measure and Category*, Springer-Verlag, New York–Heidelberg–Berlin.

NOTATION

The pages indicated in this list are those on which the symbols are first defined.

$\varnothing$	the empty set
ξ_X	characteristic function of X
$g \circ f$	composition of f and g
$f \wedge g$	the function $x \mapsto \min\{f(x), g(x)\}$
$f \vee g$	the function $x \mapsto \max\{f(x), g(x)\}$
$\underline{\int}_a^b f(x)\,\mathrm{d}x$	lower Riemann integral
$f(x+)$, $f(x-)$	2, 10
$\mathscr{M}on$	the set of monotone functions 2
$\mathscr{D}$	the set of differentiable functions 2, 78
$\mathscr{C}$, $\mathscr{C}(I)$	the set of continuous functions 2, 52
$\mathscr{R}$, $\mathscr{R}[a, b]$	the set of Riemann integrable functions 2, 74
$\mathscr{D}'$, $\mathscr{D}'(I)$	the set of derivative functions 2, 85
$\mathscr{DC}$	the set of Darboux continuous functions 2, 55
f_l, f_r	10
$\Phi_1 f$, $\Phi_2 f$	15
$D_l f$, $D_r f$	one-sided derivatives of f 15
$L(f)$	length of graph of f 19
Var f, $\mathrm{Var}_{[a,b]} f$	total variation of f 20
$\mathscr{BV}$, $\mathscr{BV}[a, b]$	the set of functions of bounded variation 20
$L(I)$	length of interval I 24
$\mathbb{D}$	the Cantor set 25
$\overline{\lim}_{y \downarrow a}\, g(y)$, etc.	27
$D_r^+ f$, etc.	the Dini derivatives of f 27
F_σ	40
G_δ	40
Γ_f	graph of f 52
$\mathscr{C}^+$, $\mathscr{C}^+(I)$	the set of lower semicontinuous functions 60
$\mathscr{C}^-$, $\mathscr{C}^-(I)$	the set of upper semicontinuous functions 60
$f^\downarrow$, $f^\uparrow$	62
$\mathscr{B}^1$, $\mathscr{B}^1(X)$	the first class of Baire 66

$\mathscr{B}^n$	69
$\int_a^b f(x)\,dx$	74, 138
D	91
J	indefinite integral 91
D^+, D^-	96
f^s	the symmetric derivative of f 96
f^∇	100
$\mathscr{B}^\omega$	104
$\mathscr{B}, \mathscr{B}(I)$	the set of Borel functions 104, 107
Ω	the set of Borel sets 104
$\mathscr{A}^*$	108
$\sim$	113
$\mathscr{A}_D$	hypersequence of length D 116
$\mathscr{B}_D$	117
$\mathscr{N}$	$=\mathbb{N}^\mathbb{N}$ 118
$\mathscr{C}_c$	131
$\int$	131, 133, 143
$\mathscr{L}, \mathscr{L}(X)$	the set of Lebesgue integrable functions 132, 146
$L(U)$	137
$\int_X f, \int_X f(x)\,dx$	138, 146
$\mathscr{M}, \mathscr{M}(X)$	the set of Lebesgue measurable functions 142, 146
$\lambda(E)$	Lebesgue measure of E 144
$\lambda^*(E)$	outer Lebesgue measure of E 153
X_d	the set of density points of X 163
$/_a^b u$	$=u(b)-u(a)$ 170
$\mathscr{P}\int_a^b f$	Perron integral of f 170
$S_V(f;g)$	176
$\int_a^b f\,dg$	Stieltjes integral 176
$0.\alpha_1\alpha_2\alpha_3\ldots$	183
$\bar{X}$	closure of X 183
X°	interior of X 184
$X<Y$	186
$X\approx Y$	186

INDEX